SOLUTIONS

Introduction to
Simulation and SLAM II

A. Alan B. Pritsker
President, Pritsker & Associates, Inc.
Adjunct Professor, Purdue University

Laurie J. Rolston
Systems Consultant, Pritsker & Associates, Inc.

Peter Floss
Systems Analyst, Systems Publishing Corp.

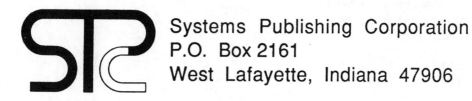

Systems Publishing Corporation
P.O. Box 2161
West Lafayette, Indiana 47906

ISBN 0-938974-01-7

Printed in the United States of America
10 9 8 7 6 5 4 3 2 1

PREFACE

This manual contains solutions to 179 of the 183 exercises contained in Introduction to Simulation and SLAM II® by A. Alan B. Pritsker. The kernel of the solutions manual is the more than 40 network models and 144 computer models which provide solutions to the exercises and all the embellishments in Chapters 5 through 16. Each computer model was run at Pritsker & Associates' Computer Center. Both statistical and trace outputs were reviewed and analyzed to insure that the models were operational. The embellishments included within the exercises were also solved and are presented in this manual. Each of the embellishments was also run. A copy of the tape that includes all computer model solutions to the exercises and most of the embellishments is distributed by Pritsker & Associates, Inc., P.O. Box2413, West Lafayette, IN 47906. In preparing this solutions manual, it became difficult to present all the material related to a solution and yet to provide a concise answer to each exercise. A middle course was taken. Concise answers were provided where possible. The description of the computer model is limited to explaining novel and difficult aspects of the model. In situations where output statistics would help clarify the solution to the exercise, they are included. It should be recognized that the solutions obtained are for a particular random number generator and will not be identical when different random number generators or computers are employed. Remember most of the solutions are modeling exercises. Learning how to model is the key feature of the text. The solution to each exercise is typically a stand alone presentation, and it was decided that figures and tables should reference the exercise number. This is accomplished by referring to any figure, table, or computer listing as a diagram and using the letter D as an abbreviation. Following the D is a reference to the exercise number presented as the chapter number in the text, a dash, and then a number representing the order of the exercise in the chapter. Within a particular exercise, diagrams are differentiated using a period followed by the number of the diagram for the exercise. Thus, D7-1.2 refers to the second diagram of the first exercise in Chapter 7. It was also decided to place references directly in a solution to an exercise since the exercises will be used on an individual basis.

The preparation of this manual was a larger project than anticipated. After reviewing the contents, we feel that the solutions manual makes a significant contribution to the understanding of simulation and the building of models using SLAM II. We found that for many of the exercises that there are alternative procedures and models that could be built. We encourage you to write and let us know of new and novel solutions to these exercises. In preparing this solutions manual, we received assistance from our co-workers at Pritsker & Associates. We gratefully acknowledge their help. In particular, we thank Jean O'Reilly, Bill Lilegdon, Cathy Stein, Carole Vasek, Jim Whitford and Dave Yancey for their contributions. We also thank Jeff Pritsker for developing and drawing SLAM II networks on the McIntosh DTP System and for helping us publish this book of solutions and Miriam Walters for typing and coordination of the preparation of the manuscript; and Jim Ketterer for page preparation and layout.

A. Alan B. Pritsker
Laurie J. Rolston
Peter Floss

West Lafayette, IN
September, 1986

SLAM II is a registered trademark of Pritsker & Associates, Inc.

Table of Contents

CHAPTER 1

Introduction to Modeling
and Simulation

1-1.   A good choice of a system to show the diversity of models is the university system. The students' and professors' views of the university system typically are different. The student as a four-to-ten year transient has a different set of problems than the instructor. The student may be concerned with education, training, grades, social life, and athletics. The instructor may view the university system as an employee and be most concerned about his paycheck and his fringe benefits. The view of the university from the president's office, again, may be quite different. Judging from the amount of time a president spends raising funds and preparing the budget, the university may be perceived as a financial institution. At the same time, the students, instructors, and president also view the university as an academic institution.

1-2.   The solution to this exercise is dependent upon the program of study of the student. The exercise should force the student to think about the courses he has taken and how they fit into an overall program. It also provides a mechanism to show how a course in simulation integrates material from other courses.

1-3.   A simulation language provides a starting point for modeling by providing a framework within which models must be built. A framework by its very nature limits flexibility but in so doing allows a modeler to focus on the concerns about how inputs are to be processed; the procedures for performing bookkeeping operations; and the form in which output reports are to be generated. By making these elements subservient to the technical aspects of the model, a simulation language allows a modeler to concentrate on the problem to be solved.

1-4.   The average number of customers in the bank is equal to the average arrival rate multiplied by the average time in the bank per customer. This can be seen by comparing the sum of the time in the bank per customer (column 6 in Table 1-2) which is 58.1, and the time integrated number in the bank obtained from Table 1-3 which is also 58.1. To get the average time in the bank per customer, 58.1 is divided by 10 to yield 5.81. To get the average number in the bank, 58.1 is divided by the total time of 40 to yield an average number in the bank of 1.4525. Thus, the conversion factor is the arrival rate or 10 customers in 40 time units which is 0.25. In this example, the relationship holds exactly as all customers depart the system at time 40. The same relationship also holds between the average number in the queue and the average time in the queue.

1-5.   Problem formulation - A simulation language provides a viewpoint to decompose a system into elements. By looking at a system and seeing entities, attributes, files, events, processes, state variables and the like, an understanding of a system and, hence a problem formulation is made clearer. In fact, simulation analysts have been said to look at a system through glasses tinted by the simulation language they know.

Model building - Models are built using the concepts included in the simulation language and it is in this area that simulation languages have their largest impact.

Data acquisition - Simulation languages normally do not provide any assistance in data acquisition. Simulation models do, however, specify the type of data that should be collected.

Model translations - Simulation languages provide efficient mechanisms for translating the model into code for computer processing. In many instances, the model translation is a direct result of the model building.

Verification - Simulation languages provide debuggers,   trace features, and output reports to verify that the translated model executes as intended. Animations showing entity movement and status changes is a good technique for visual verification.

Validation - Simulation languages provide little assistance in this area. With the advent of computer graphics, model operation can be displayed on terminals; future languages will provide outputs that resemble system outputs from which subjective validation by model builders, decision makers and analysts can be facilitated.

Strategic and tactical planning - Simulation termination conditions, random number seed and streams, and the like are typically available in simulation languages to assist in tactical planning. In the strategic planning area, simulation languages provide little assistance although some capability exists for collecting statistics over multiple runs. The advent of simulation data languages, databases and support systems will assist in this area.

Experimentation - simulation languages provide the capability of executing the model to obtain desired output values.

Analysis of results - Simulation languages provide the means to obtain plots, graphics and statistical estimates but provide only minor assistance in analyzing simulation outputs statistically.

Implementation and documentation - Simulation languages document the model by providing a standard set of symbols, subroutines, and variable definitions which allow a model   to be understood by different indivduals. Less help is provided by simulation languages in helping to implement decisions. Simulation support systems are available to aid in the process of implementing decisions.

1-6.   No solution given.

1-7.   No solution given.

1-8.   Stages 2 and 4, model building and model translation, are often combined. Stages 7 and 8 are often done iteratively although at first glance it appears that they should be separated. Experimentation with the simulation model typically leads to a desire for more experimentation or more analytic developments which in turn calls for more strategic and tactical planning. In current studies, experimentation and analysis of results are sometimes combined. In the future, less of this will be done as the outputs from a simulation will be stored in a database for future analysis. To reduce the number of stages from 10 to 7, one approach is to combine stages 2 and 4 and stages 7, 8 and 9.

1-9.   The solution to this exercise is contained in Chapter 17 , Sections 17.1 through 17.6.

CHAPTER 2

Probability and Statistics

2-1. Using the formulas on page 37, we obtain the estimate of the mean as 2.69; the estimate of the variance as 1.39; the estimate of the standard deviation as 1.18; and the estimate of the coefficient of variation as 0.439. The histogram will have one value in the first cell, seven in the second cell, three in the third cell, seven in the fourth cell, and two in the fifth cell.

2-2. Using the formulas in Table 2-1 for time persistent variables, the average is 2.067 and the standard deviation is 1.062.

2-3. The average is 15.65 and the standard deviation is 1.63. Because the samples are independent and identically distributed, an estimate of the standard deviation of the average is 1.63 divided by the square root of 20 or 0.364. Using formula 2-2, the 99 percent confidence interval on the mean is (14.61, 16.69).

2-4. Let the null hypothesis be that the average weekly cost for policy A equals the average weekly cost for policy B. The alternative hypothesis is that the average weekly cost for policy A is less than the average weekly cost for policy B. The average for policies A and B are 1.213 and 1.5, respectively. The estimates of the standard deviations for policies A and B are .162 and .181, respectively. This yields estimates for the standard deviation of the averages of .0418 and .0467 for policies A and B, respectively. Using the formulas given in Table 2-2, the degrees of freedom for the test of hypothesis of means is approximately 30. Inserting the above values to calculate the test statistic results in a value of -4.58 which is less than the t-statistic of -1.697 and, therefore, we reject the null hypothesis that the average weekly cost for policy A equals the average weekly cost for policy B.

2-5. Using the formula for I in Section 2.13.4, the number of runs required is 389 where the value of the t-statistic is 3.250; the estimate of the standard deviation is 6.07; and the half length, g, is 1.

2-6. Let N_i be Poisson distributed with mean μ_i. The probability mass function for N_i is

$$P[N_i = n] = \frac{e^{-\mu_i}\mu_i^n}{n!}, \quad n = 0, 1, 2, \ldots$$

The moment generating function of N_i is

$$M_{N_i}(s) = E[e^{sN_i}] = e^{\mu_i(e^s - 1)}$$

Let $N = N_1 + N_2$ then $M_N(s) = M_{N_1}(s) M_{N_2}(s)$ since N_1 and N_2 are given as independent. Substituting the moment generating function for the Poisson, we have

$$M_N(s) = e^{\mu_1(e^s - 1)} e^{\mu_2(e^s - 1)} = e^{(\mu_1 + \mu_2)(e^s - 1)}$$

3

We recognize that $M_N(s)$ has the form of the moment generating function for the Poisson with mean $\mu_1 + \mu_2$. Similarly, the normal distribution has a moment generating function of $e^{\mu s + \sigma^2 s}$ and the sum of independent normal random variables is normal.

2-7.   To obtain the expected value, a recursive equation can be used. Let X be the time until the thief selects the door to freedom. Then the expected value of X, E[X], can be expressed as

$$E[X] = 0.20\{E[\text{time in short tunnel}]+E[X]\}$$
$$+0.50\{E[\text{time in long tunnel}]+E[X]\} + 0.30(0)$$

since the thief starts over when he returns to the dungeon. Substituting and solving yields

$$E[X] = 0.2(3)\ 0.2E[X] + 0.5(6) + 0.5E[X],$$
$$.3E[X] = 3.6,$$
and $E[X] = 12.$

In general, this problem can be solved using a GERT representation (see Pritsker, A.A.B. and G.E. Whitehouse, "GERT: Graphical Evaluation and Review Technique Part II". Probabilistic and Industrial Engineering Applications, The Journal of Industrial Engineering, Vol. XVII, No. 6, June 1966, pp. 293-301; or Whitehouse, G.E., Systems Analysis and Design Using Network Techniques, Englewood Cliffs, N.J., Prentice-Hall, Inc., 1973). The moment generating function for X is

$$M_X(s) = \frac{3}{1-.2M_y(s) -.5M_Z(s)}$$

where Y and Z are the random variables representing the times in the short and long tunnels respectively.

2-8.   Assumptions:
 1) a heat ends when the first runner crosses the finish line, that is, the 15 minute delay between heats starts at this time; 2) there are no false starts; 3) individual racer's times are independent and identically distributed; and, 4) each race time is independent.

$$T_8 = \sum_{i=1}^{8} t_i + 7(15)$$

where T_8 is time to complete the 8th race and t_i is time for the ith race. By assumption 1,

$$t_i = \min [h_j], \quad j = 1,2,...,6$$

where h_j is uniformly distributed.

$$P[t_i \leq r] = \left[\frac{r-A}{B-A}\right]^6 \quad \text{where A=8 and B=12.}$$

The distribution function for T_8 is

$$P[T_8 \le t] = P[\sum_{i=1}^{8} t_i + 105 \le t]$$

$$= P[\sum_{i=1}^{8} t_i \le t - 105].$$

For the uniform case, a convolution integral needs to be evaluated to obtain the desired expression. To obtain the moments of T_8, the use of moment generating functions is appropriate. For the exponential case, h_j is exponential with mean 10 and t_i is exponential with mean 10/6. The distribution of the sum of 8 exponential randoms variable each with mean 10/6 which is an Erlang (gamma) random variable with a mean of 13 1/3 and a variance of $8*(10/6)^2 = 22\ 2/9$. Thus, T_8 is Erlang distributed with an offset of 105.

2-9.   If an analytic result can be incorporated into a simulation model, the Rao-Blackwell theorem provides the basis for the statement that the variance for the derived model will be less than or equal to the variance for the model without the analytic result. For example, if a portion of the model is replaced by a set of queueing equations and these equations are solved to obtain the waiting time given the current status of the system, then the variance of the time in the system under these calculation circumstances will be less than or equal to the variance of the time in the system if direct sampling of the waiting times is performed.

2-10.   A system variable relates to a characteristic of the system. If the system variable has random fluctuations, then it is characterized as a random variable which has an underlying distribution. For a system variable, the only procedure for specifying the underlying distribution is to establish assumptions concerning the system variable and then deriving the distribution function of the system variable based on the assumptions. The distribution function would be a model of the system variable. The mean of a system variable is a theoretical construct which relates to its central tendency. The mean of a model of a system variable is a parameter of the distribution function, that is, a parameter of the model of the system variable. The expected value of a variable in a model is a mathematical computation which weights each possible value of the variable by the probability that the value will occur. The average of a model variable is a weighted sum of samples of the model variable. Before the samples are taken, it is a random variable describing a central tendency of the samples (the samples are also random variables). After a sample is taken and an observation obtained, the average is computed from the values of the observations of each of the samples of the model variable. An average of the system variable is the same as the average of a model variable with the difference that the random variable under consideration is from the system and not the model. A confidence interval attempts to make a statement about the probability that a sample interval includes the mean value, that is, a 95 percent confidence interval indicates that out of 100 sample confidence intervals, 95 should include the mean value. If the confidence interval is taken from observations of the system then the confidence interval is drawn for the mean of the system variable. If the observations are taken from a model of the system variable, then the confidence interval is for the mean of the model. Coverage relates to the fraction of sample confidence intervals that include the mean. Tolerance intervals relate to the number of intervals that include the average. A tolerance interval can also be developed to specify the probability that a sample interval will include a sample value.

2-11.   Let $Y = \min [X_1, X_2, ..., X_N]$

and by definition $P[X \leq t] = F_X(t)$.
For Y to be the minimum then $Y > t$ requires $X_i > t$ for all i.
Thus,

$$P[Y > t] = P[X_1 > t, X_2 > t, ..., X_N > t]$$

By independence of X_i

$$P[Y > t] = P[X_1 > t] \, P[X_2 > t]... P[X_N > t]$$

Since the X_i are identically distributed, the subscript i can be omitted and

$$P[Y > t] = (P[X > t])^N = (1 - P[X \leq t])^N$$

$$= (1 - F_X(t))^N$$

The cumulative distribution function is

$$F_Y(t) = P[Y \leq t] = 1 - P[Y > t] = 1 - (1 - F_X(t))^N.$$

For the exponential distribution $F_X(t) = 1 - e^{-t/\mu}$

and $F_Y(t) = 1 - (e^{-t/\mu})^N = 1 - e^{-Nt/\mu} = 1 - e^{-t/(\mu/N)}$

which we recognize as the form for the exponential distribution with a mean of μ/N.

2-12.   For the exponential distribution, $P[B \geq t] = e^{-\lambda t}$

Let $t_2 = t_1 + \Delta t$.

From the definition of conditional probability

$$P[B > t_2 ; B > t_1] = \frac{P[B>_2 ; B>T_1]}{P[B > T_1]}$$

Since $t_2 > t_1$, $P[B > t_2; B > t_1] = P[B > t_2]$

and $P[B > t_2 ; B > t_1] = \dfrac{P[B> t_2]}{P[B > t_1]} = \dfrac{e^{-\lambda t_2}}{e^{-\lambda t_1}} = e^{-\lambda(t_2-t_1)}$

$$= e^{-\lambda \Delta t}.$$

This result is referred to as the forgetfulness property and only holds for the exponential distribution. It specifies that if an event has not occurred by time t_1, then the probability of the event occurring in the next Δt time units is the same as if the t_1 time units had not expired.

This result can be used in conjunction with the result of Exercise 2-11 in the following manner. Suppose that we have n servers working in parallel with each server having an exponential service time with mean μ. The time until the next end-of-service according to the result of Exercise 2-11 will be exponentially distributed with a mean time of μ/n. A sample from this derived distribution could be taken. At the end of the completion time, the time remaining for the completion of the n-1 services in progress is the same as if no time expired. Thus, the time to the next end-of-service for the n-1 servers in operation is a sample from an exponential distribution with a mean of $\mu/(n-1)$. Note that at each start-of-service event, a cancellation of the time until the next end-of-service event would be necessary (if one exists) and a new end-of-service time would have to be generated since there is one more server working.

CHAPTER 3

Simulation Modeling Perspectives

3-1. In considering the operation of a physician's office, the boundaries cannot be set unless a purpose is given for modeling. The physician's relation to other physicians, to medical centers, and to the hospital would need to be established. Furthermore, it must be established whether it is the business operations of the physician's office or the health care delivery that is of concern. Once the purpose for modeling is established, the entities, attributes, relationships,and activities can be described. The following references provide information that could be useful in answering this exercise. Kasanof, D., "Upgrading Productivity: A Computer Analysis", Patient Care, 1973, pp. 96-120. 1973, pp. 96-120. Carlson, R.C., J.C. Hershey, and D.H. Kropp, "Use of Optimization and Simulation Models to Analyze Outpatient Health Care Settings",Decision Sciences, Vol.10, No. 3, 1979, pp. 412-433.

3-2. Examples of activities whose duration is based on the status of the system are numerous. Whether or not they must be modeled using state events depends on the situation. Good examples are: processing of a job on a computer in a multiprocessing environment; the completion of a batch which is determined based on pressure and temperature considerations where temperature and pressure are dependent on energy usage which is shared by several processes; the activity of driving to Chicago's O'Hare Airport; and the length of a Ph.D. qualifying or preliminary oral examination.

3-3. In the event orientation, there are four events: arrival to preparation area; completion of preparation; arrival to spraying machine; and end of spraying. This assumes that there is a time delay between completion of preparation and arrival to a spraying machine. If there is no time delay then these events could be combined. It also assumes that the event, arrival to the preparation area, and the assignment of a worker to the preparation occur at the same time. If this is not the case, then separate events must be used to model these changes in system status.

The process orientation would involve the flow of the job through the activities of preparation and spraying and would have to include concepts for waiting for preparation and waiting for spraying and, if necessary, activities for traveling from preparation to spraying. Thus, the process orientation would visualize an entity going through the following sequence of activities: waiting for preparation, preparation, travel to spraying, waiting for spraying, and spraying.

3-4. The description of a residential heating and cooling control system depends on the purpose for modeling. Here we will consider the temperature in the residence as the main state variable. Possible time events are: resetting of the thermostat; turning on the heating system; turning on the cooling system; breakdown of the compressor or furnace; blowing a circuit breaker; and starting an appliance. A few of the state events are: temperature crossing the thermostat setting, amount of fuel crossing a reorder point level for coal or oil system, and the moisture in the house crossing a threshold that turns off a humidifier.

3-5. A good description of the modeling of an elevator system is contained in Sweet, A.L. and S.D. Duket, "A Simulation Study of Energy Consumption by Elevators in Tall Buildings", Computing and Industrial Engineering, Vol. 1, 1976, pp. 3-11.

3-6. Illustrations of simulation models using FORTRAN are contained in Schmidt, J.W. and R.E. Taylor, Simulation and Analysis of Industrial Systems, Homewood, Illinois: Richard D. Irwin, Inc., 1970, pp. 459-470 (Inventory Simulators); Emshoff, J.R. and R.L. Sission, Design and Use of Computer Simulation Models, Macmillan, 1970, p. 94 (Plastic Shop Simulation); Bratley, P., B.L. Fox, and L.E. Schrage, A Guide to Simulation, Springer-Verlag, 1983; Banks, J. and J.S. Carson, II, Discrete-Event System Simulation, Prentice-Hall, 1984; and, Law, A.M. and W.D. Kelton, Simulation Modeling and Analysis, McGraw-Hill, 1982.

CHAPTER 4

Applications of Simulation

4-1. The solution to this problem is contained in the papers that describe each application. See also solutions to Exercises 4-2 and 4-3.

4-2. The table in D4-2.1 describes the application in terms of the class of sponsor, model type, and modeling approach. Other categories for describing simulation modesl are: performance measures employed; level of detail (aggregation); decision types involved in the model; size of the model; and worth of the model.

4-3. The table in D4-3.1 gives a ranking for each application based on the difficulty associated with: the modeling effort; obtaining data; and applying the results. A rank of 1 indicates the easiest or least difficult. The benefits from each application are also ranked with a rank of 1 indicating highest benefit. The rankings were done on a subjective basis and comments to support the ranking are included in the table.

4-4. The concepts presented in Chapter 16 provide a basis for making a specification of a network simulation language to evaluate materials- handling equipment.

4-5. A simulation course is a good integrating mechanism as it provides a means for modeling a wide class of problems. It also provides an opportunity to use analysis tools and methods learned in many other courses such as regression techniques, data acquisition, distribution and curve fitting, tests of hypothesis, design of experiments, plant layout, linear programming, engineering economic analysis, work measurement, human factors, and project planning and scheduling.

4-6. Developing performance measures for the Department of Defense is an extremely difficult task. A level of analysis must be selected and performance measures oriented to that level of analysis. At a high level, performance measures would relate to the probability of being able to defend the country, the fraction of GNP allocated to defense, total manpower involved in defense and defense related research, technology transfer from defense activities to commercial activities, and the training of personnel. At lower levels, performance measures would be necessary to assess operational readiness, the cost of delivering a given level of fire power, the number of strategic and tactical weapons, the value of information, the ability to initiate new programs, and the logistical support capabilities of the military. The diverse nature of the goals of the Department of Defense (maintaining peace, technology development and transfer, operational readiness, logistical support, etc.) make the establishment of performance measures a major task in the military and defense environments.

Application	Class of Sponsor	Model Type	Modeling Approach
4.3 Submarine Purchase	Industry	Discrete Event	Economic Comparison of Alternatives
4.4 FMS	Industry	Network	Design Evaluation
4.5 Corn Syrup Refinery	Industry	Combined	Design Evaluation
4.6 Pipeline Construction	Industry/Government	Network	Project Planning
4.7 Sporting Event	Government	Discrete Event	Manpower Allocation and Routing
4.8 MLRS	Industry/Military	Discrete Event	Productive Capacity Evaluation
4.9 Ingot Mould	Industry	Discrete Event	Forecasting, Inventory Analysis, and Design Evaluation
4.10 Work Flow Analysis	Service Industry	Network	Bottleneck Analysis
4.11 Chemical Manufacturing	Industry	Discrete Event	Batch Scheduling
4.12 Automated Warehouse	Industry	Combined	Operational Planning
4.13 Air Terminal Cargo	Military	Network	Capacity Planning
4.14 Machine Allocation	Industry	Discrete Event	Labor Requirements
4.15 Refinery Offsite	Industry	Combined	Design and Operational Planning

D4-2.1

Application	Difficulty of Modeling Effort		Difficulty In Obtaining Data		Difficulty In Applying Results		Benefits	
	Rank	Time for Initial Model	Rank		Rank		Rank	
4.3 Submarine Purchase	7	4 weeks	3	Low—Standard Data	2	Direct Use	3	Capital Cost Avoidance
4.4 FMS	2	1 week	1	Low—Assumed Values	1	Direct Use	6	Capital Cost Avoidance
4.5 Corn Syrup Refinery	5	2 weeks	7	Average – Flow Rates Known Operations Assumed	6	Direct Use	1	Improved Productivity
4.6 Pipeline Construction	4	2 weeks	12	High – Controversial	11	Indirect Use – Legal Issues	13	Avoided Potential Project Overruns
4.7 Sporting Event	6	4 weeks	4	Low – Good Records Available	10	Direct for Model No Organization/ Decision Methods for Future Applications	10	Staff Reductions
4.8 MLRS	3	4 weeks	8	Average– Operation Times Standard – Procedures Hypothesized	7	Direct Use in Changing & Evaluating Design	5	Better Planning Redesign Avoidance
4.9 Ingot Mould	12	6 weeks	11	High– Processing Times Standard – Demand Characterization Difficult – High Volumes Required	13	Require Difficult Operational & Budgeting Changes	7	Improved Provisioning Procedures. Capital Cost Avoidance
4.10 Work Flow Analysis	1	2 weeks	2	Low – Standard Data	3	Direct Use	11	Understanding of Operations
4.11 Chemical Manufacturing	10	6 weeks	10	High – Batch Times Available – Schedules Difficult to Forecast	12	Difficult – Organization Jurisdiction	12	Understanding of Scheduling Difficulties
4.12 Automated Warehouse	11	6 weeks	5	Average – Good Company Records	4	Direct Use	4	Productivity Improvement
4.13 Air Terminal Cargo	8	6 weeks	6	Average– Many Assumptions Necessary	8	Indirect Use For Planning	9	Productivity Planning Improvement
4.14 Machine Allocation	9	6 weeks	9	Average–Machining Times Known – Human Characteristics Difficult	5	Direct Use For Setting Labor Requirements	8	Labor Savings
4.15 Refinery Offsite	13	12 weeks	13	Difficult– Many New Operations & Procedures	9	Continuous Use In Design & Planning	2	Design aid

*   A rank of 1 indicates the least difficult or the highest benefit.

CHAPTER 5

Basic Network Modeling

5-1. The SLAM II network segment which sets the time to the minimum if the random sample is less than the minimum, or to the maximum when the sample is greater than the maximum, is shown in D5-1.1. This type of truncation of a sampling process is referred to as sampling from a mixed distribution.

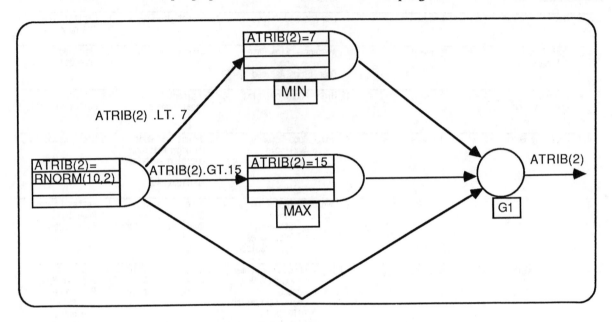

D5-1.1

The SLAM II network segment for sampling from a truncated normal distribution in which the probabilities in the tails of the distribution are redistributed is shown in D5-1.2. This is called sampling from a truncated distribution and is achieved by rejecting samples outside the desired range. It is a form of the acceptance/rejection method described in Chapter 18.

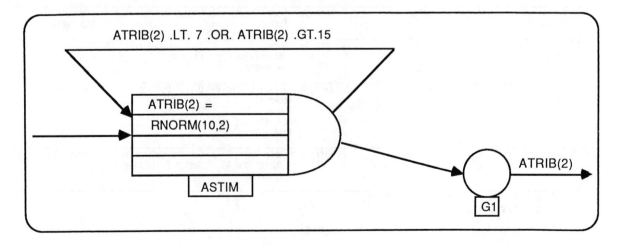

D5-1.2

5-2.   The tables requested are shown in D5-2.1 and D5-2.2.

A graphical representation is provided in D5-2.3 with statistical estimates for the quantities of interest in D5-2.4. The statements and SLAM II output for this example are then given in D5-2.5 and D5-2.6 respectively.

CUSTOMER NUMBER	ARRIVAL TIME	START ACT 1	FIN ACT 1	ENTER Q2	TIME BLOCKED	START ACT 2	FIN ACT 2	TIME IN SYSTEM
1	0	---	---	---	---	0	2	2
2	0	---	---	0	---	2	4	4
3	0	---	---	0	---	4	6	6
4	0	0	1	2	1	6	8	8
5	0	2	3	4	1	8	10	10
6	0	4	5	6	1	10	12	12
7	0	6	7	8	1	12	14	14
8	0	8	9	10	1	14	16	16
9	4	10	11	12	1	16	18	14
10	8	12	13	14	1	18	20	12
11	12	14	15	16	1	20	22	10
12	16	16	17	18	1	22	24	8
13	20	20	21	21	---	24	26	6
14	24	24	25	25	---	26	28	4
15	28	28	---	---	---	---	---	---
					$\overline{9}$			$\overline{126}$

D5-2.1

EVENT TIME	CUST NO	EVENT TYPE	NO. IN Q1	NO. IN Q2	SERV. 1 STATUS	IDLE TIME	SERV.2 STATUS
0	8	arrival enter Q1	4	2	Busy		Busy
1	4	finish act 1	4	2	Block		Busy
2	1	finish act 2 leave system					
	2	start act 2					
	4	enter Q2					
	5	start act 1	3	2	Busy		Busy
3	5	finish act 1	3	2	Block		Busy
4	9	arrival enter Q1					
	2	finish act 2 leave system					
	3	start act 2					
	5	enter Q2					
	6	start act 1	3	2	Busy		Busy
5	6	finish act	3	2	Block		Busy

D5-2.2

EVENT TIME	CUST NO	EVENT TYPE	NO. IN Q1	NO. IN Q2	SERV. 1 STATUS	IDLE TIME	SERV.2 STATUS
6	3	finish act 2 leave system					
	4	start act 2					
	6	enter Q2					
	7	start act 1	2	2	Busy		Busy
7	7	finish act 1	2	2	Block		Busy
8	10	arrival enter Q1					
	4	finish act 2 leave system					
	5	start act 2					
	7	enter Q2					
	8	start act 1	2	2	Busy		Busy
9	8	finish act 1	2	2	Block		Busy
10	5	finish act 2 leave system					
	6	start act 2					
	8	enter Q2					
	9	start act 1	1	2	Busy		Busy
11	9	finish act 1	1	2	Block		Busy
12	11	arrival enter Q1					
	6	finish act 2 leave system					
	7	start act 2					
	9	enter Q2					
	10	start act 1	1	2	Busy		Busy
13	10	finish act 1	1	2	Block		Busy
14	7	finish act 2 leave system					
	8	start act 2					
	10	enter Q2					
	11	start act 1	0	2	Busy		Busy
15	11	finish act 1	0	2	Block		Busy
16	12	arrival enter Q1					
	8	finish act 2 leave system					
	9	start act 2					
	11	enter Q2					
	12	start act 1	0	2	Busy		Busy
17	12	finish act 1	0	2	Block		Busy
18	9	finish act 2 leave system					
	10	start act 2					
	12	enter Q2	0	2	Idle		Busy

D5-2.2 (continued)

EVENT TIME	CUST NO	EVENT TYPE	NO. IN Q1	NO. IN Q2	SERV. 1 STATUS	IDLE TIME	SERV.2 STATUS
20	13	arrival					
		start act 1					
	10	finish act 2					
		leave					
	11	start act 2	0	1	Busy	2	Busy
21	13	finish act 1					
		enter Q2	0	2	Idle		Busy
22	11	finish act 2					
		leave system					
	12	start act 2	0	1	Idle		Busy
24	14	arrival					
		start act 1					
	12	finish act 2					
		leave system					
	13	start act 2	0	0	Busy	3	Busy
25	14	finish act 1					
		enter Q2	0	1	Idle		Busy
26	13	finish act 2					
		leave system					
	14	start act 2	0	0	Idle		Busy
28	15	arrival					
		start act 1					
	14	finish act 2					
		leave system	0	0	Busy	3	Idle

D5-2.2 (concluded)

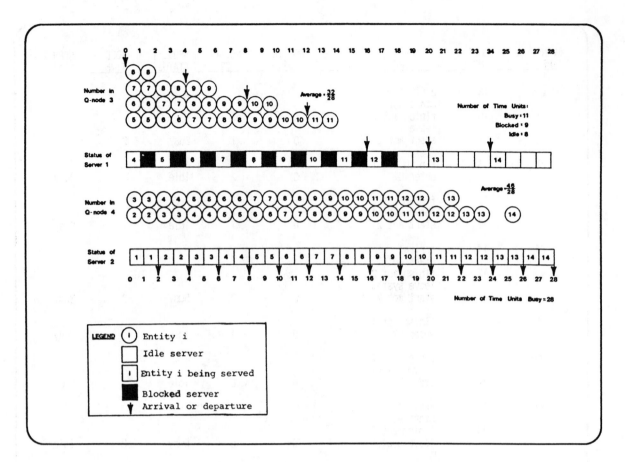

D5-2.3

Average number in Q-node 3   = 32/28 = 1.1429
Maximum number in Q-node 3   = 4
Average number in Q-node 4   = 46/28 = 1.6429
Maximum number in Q-node 4   = 2
Average server 1 utilization   = 11/28 = 0.3929
Average time in Q-node 3   = 32/11 = 2.9091
Average time in Q-node 4   = 46/13 = 3.5385
Longest period idle for server 1 = 3
Longest period busy for server 1 = 1
Blocked time per unit time for
          server 1   = 9/28 = 0.3214
Longest blocked time for server 1= 1
Average server 2 utilization   = 28/28 = 1
Longest period idle for server 2 = 0
Longest period busy for server 2 = 28

D5-2.4

```
 1   GEN,ROLSTON,PROBLEM 5.2,7/21/80,1;
 2   LIMITS,2,1,20;
 3   NETWORK;
 4           CREATE,4,,1;
 5           QUEUE(1),3;
 6           ACT(1)/1,1.0;
 7           QUEUE(2),2,2,BLOCK;
 8           ACT(1)/2,2.0;
 9           COLCT,INT(1),TIME IN SYSTEM;
10           END;
11   INIT,0,28;
12   FIN;
```

D5-2.5

SLAM II SUMMARY REPORT

SIMULATION PROJECT PROBLEM 5.2   BY ROLSTON

DATE 7/21/1980   RUN NUMBER 1 OF 1

CURRENT TIME   0.2800E+02
STATISTICAL ARRAYS CLEARED AT TIME 0.0000E+00

STATISTICS FOR VARIABLES BASED ON OBSERVATION

	MEAN VALUE	STANDARD DEVIATION	COEFF. OF VARIATION	MINIMUM VALUE	MAXIMUM VALUE	NUMBER OF OBSERVATIONS
TIME IN SYSTEM	0.9000E+01	0.4279E+01	0.4754E+00	0.2000E+01	0.1600E+02	14

FILE STATISTICS

FILE NUMBER	LABEL/TYPE	AVERAGE LENGTH	STANDARD DEVIATION	MAXIMUM LENGTH	CURRENT LENGTH	AVERAGE WAITING TIME
1	QUEUE	1.1429	1.3553	4	0	2.9091
2	QUEUE	1.6429	0.6662	2	0	3.5385
3	CALENDAR	2.3929	0.4884	4	2	1.0308

SERVICE ACTIVITY STATISTICS

ACTIVITY INDEX	START NODE OR ACTIVITY LABEL	SERVER CAPACITY	AVERAGE UTILIZATION	STANDARD DEVIATION	CURRENT UTILIZATION	AVERAGE BLOCKAGE	MAXIMUM IDLE TIME/SERVERS	MAXIMUM BUSY TIME/SERVERS	ENTITY COUNT
1	QUEUE	1	0.3929	0.4884	1	0.3214	3.0000	1.0000	11
2	QUEUE	1	1.0000	0.0000	0	0.0000	0.0000	28.0000	14

D5-2.6

5-3.   The network and control statements for this exercise are shown in D5-3.1 and D5-3.2. There are two CREATE nodes which generate entities representing each part type. ASSIGN nodes give a value to ATRIB(1) which is the part type number. The model consists of three QUEUE nodes in series representing the waiting areas for the drill, straightening and finishing operations. Following the drill activity, parts are routed to QUE2 if they are part type 1 or to QUE3 if they are part type 2. Following the finishing operation, the parts are segregated by part type in order to collect time-in-the-system statistics by part type. The summary report obtained for the statements in D5-3.2 is shown in D5-3.3. From the output, it is seen that there is hardly any waiting in the queues. The system is underutilized as each of the service activities has a utilization of less than 50% per server. These low utilization statistics are to be expected as the cycle time for part type 1 is 30 minutes of which only 10 minutes is for drilling, 15 minutes for straightening and 5 minutes for finishing on the average. Similarly for part type 2 the cycle time is 20 minutes with 10 minutes for drilling and 5 minutes for finishing on the average. An expected value analysis for the drills would indicate that in a 60 minute cycle there would be 50 minutes of processing time for the two drills. This corresponds to a utilization of 5/12ths or 0.416. The average utilization shown in D5-3.3 is 0.418. For a system with low utilizations, an expected value analysis can provide good estimates of the performance measures.

5-3,Embellishment(a).With the added resources, there should be no queueing time and the expected time in the system should be the sum of the expected processing times plus the traveling times. For part type 1, this is 32 minutes and 25 minutes for part type 2. The SLAM II Summary Report for this run is shown in D5-3.4 where the estimates of the mean time in the system are 32.13 and 24.96 minutes, respectively. Note that only 79 and 119 observations were obtained for part types 1 and 2 as the 80th and 120th part arrival did not complete processing by the end of the simulation.

The estimates of the variance of the time in the system for each part is the sum of the variance of each activity performed on a part. For part type 1 the variance is 226 minutes^2 and for part type 2, the variance is 1 minute^2.

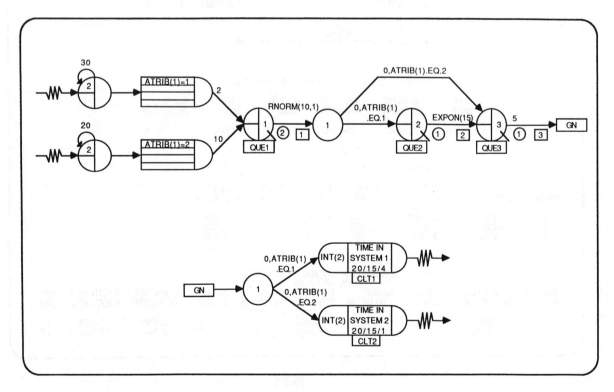

D5-3.1

```
 1  GEN,FLOSS,PROBLEM 5.3,6/10/86,1;
 2  LIMITS,3,2,100;
 3  NETWORK;
 4          CREATE,30,,2;                              CREATE PART TYPE 1
 5          ASSIGN,ATRIB(1)=1;                         ASSIGN PART TYPE
 6          ACTIVITY,2,,QUE1;                          TRAVEL TO DRILL AREA
 7          CREATE,20,,2;                              CREATE PART TYPE 2
 8          ASSIGN,ATRIB(1)=2;                         ASSIGN PART TYPE
 9          ACTIVITY,10;                               TRAVEL TO DRILL AREA
10  QUE1    QUEUE(1);                                  WAIT FOR DRILLS
11          ACTIVITY(2)/1,RNORM(10,1);                 DRILL PARTS
12          GOON,1;                                    ROUTE PARTS BY PART TYPE
13          ACTIVITY,,ATRIB(1).EQ.1,QUE2;              IF TYPE 1 BRANCH TO STRAIGHTENNG
14          ACTIVITY,,ATRIB(1).EQ.2,QUE3;              IF TYPE 2 BRANCH TO FINISHING
15  QUE2    QUEUE(2);                                  WAIT FOR STRAIGHTENER
16          ACTIVITY(1)/2,EXPON(15);                   STRAIGHTEN PARTS
17  QUE3    QUEUE(3);                                  WAIT FOR FINISHING
18          ACTIVITY(1)/3,5;                           FINISH PART
19          GOON,1;                                    COLLECT STATISTICS BY PART TYPE
20          ACTIVITY,,ATRIB(1).EQ.1,CLT1;              IF TYPE 1 BRANCH TO COLLECT NODE
21          ACTIVITY,,ATRIB(1).EQ.2,CLT2;              IF TYPE 2 BRANCH TO COLLECT NODE
22  CLT1    COLCT,INT(2),TIME IN SYSTEM 1,20/15/4;     COLLECT TIME IN SYSTEM
23          TERMINATE;                                 DEPART FROM SYSTEM
24  CLT2    COLCT,INT(2),TIME IN SYSTEM 2,15/20/1;     COLLECT TIME IN SYSTEM
25          TERMINATE;                                 DEPART FROM SYSTEM
26          END;
27  INIT,0,2400;
28  FIN;
```

D5-3.2

```
                    S L A M   I I   S U M M A R Y   R E P O R T


          SIMULATION PROJECT PROBLEM 5.3              BY FLOSS

          DATE  6/10/1986                             RUN NUMBER   1 OF   1


          CURRENT TIME   0.2400E+04
          STATISTICAL ARRAYS CLEARED AT TIME  0.0000E+00
```

STATISTICS FOR VARIABLES BASED ON OBSERVATION

	MEAN VALUE	STANDARD DEVIATION	COEFF. OF VARIATION	MINIMUM VALUE	MAXIMUM VALUE	NUMBER OF OBSERVATIONS
TIME IN SYSTEM 1	0.3387E+02	0.1238E+02	0.3655E+00	0.1606E+02	0.6537E+02	80
TIME IN SYSTEM 2	0.2530E+02	0.1422E+01	0.5622E-01	0.2254E+02	0.3037E+02	119

FILE STATISTICS

FILE NUMBER	LABEL/TYPE	AVERAGE LENGTH	STANDARD DEVIATION	MAXIMUM LENGTH	CURRENT LENGTH	AVERAGE WAITING TIME
1	QUE1 QUEUE	0.0000	0.0000	0	0	0.0000
2	QUE2 QUEUE	0.0364	0.1872	1	0	1.0914
3	QUE3 QUEUE	0.0392	0.1957	2	0	0.4703
4	CALENDAR	4.3160	0.6714	7	5	7.7474

SERVICE ACTIVITY STATISTICS

ACTIVITY INDEX	START NODE OR ACTIVITY LABEL	SERVER CAPACITY	AVERAGE UTILIZATION	STANDARD DEVIATION	CURRENT UTILIZATION	AVERAGE BLOCKAGE	MAXIMUM IDLE TIME/SERVERS	MAXIMUM BUSY TIME/SERVERS	ENTITY COUNT
1	DRILL PARTS	2	0.8365	0.6898	0	0.0000	2.0000	2.0000	200
2	STRAIGHTEN P	1	0.4978	0.5000	0	0.0000	31.6767	111.2866	80
3	FINISH PART	1	0.4150	0.4927	1	0.0000	20.3989	15.0000	199

D5-3.3

```
                    S L A M   I I   S U M M A R Y   R E P O R T

        SIMULATION PROJECT PROBLEM 5.3A              BY FLOSS

        DATE  6/10/1986                              RUN NUMBER    1 OF    1


        CURRENT TIME   0.2400E+04
        STATISTICAL ARRAYS CLEARED AT TIME  0.0000E+00


        **STATISTICS FOR VARIABLES BASED ON OBSERVATION**

                        MEAN      STANDARD    COEFF. OF    MINIMUM    MAXIMUM    NUMBER OF
                        VALUE     DEVIATION   VARIATION    VALUE      VALUE      OBSERVATIONS

TIME IN SYSTEM 1    0.3213E+02  0.1177E+02  0.3662E+00  0.1606E+02  0.6537E+02      79
TIME IN SYSTEM 2    0.2496E+02  0.1127E+01  0.4515E-01  0.2254E+02  0.2853E+02      119


                        **FILE STATISTICS**

FILE                   AVERAGE    STANDARD    MAXIMUM    CURRENT    AVERAGE
NUMBER  LABEL/TYPE     LENGTH     DEVIATION   LENGTH     LENGTH     WAITING TIME

   1    QUE1 QUEUE     0.0000     0.0000        0          0        0.0000
   2    QUE2 QUEUE     0.0000     0.0000        0          0        0.0000
   3    QUE3 QUEUE     0.0000     0.0000        0          0        0.0000
   4         CALENDAR  4.3161     0.7117        6          6        8.0737


                        **SERVICE ACTIVITY STATISTICS**

ACTIVITY  START NODE OR   SERVER    AVERAGE      STANDARD    CURRENT      AVERAGE    MAXIMUM IDLE    MAXIMUM BUSY   ENTITY
INDEX     ACTIVITY LABEL  CAPACITY  UTILIZATION  DEVIATION   UTILIZATION  BLOCKAGE   TIME/SERVERS    TIME/SERVERS   COUNT

   1      DRILL PARTS     20        0.8361       0.6878         0         0.0000     20.0000         2.0000         200
   2      STRAIGHTEN P    10        0.5004       0.5362         1         0.0000     10.0000         2.0000         79
   3      FINISH PART     15        0.4129       0.5574         1         0.0000     15.0000         2.0000         198
```

D5-3.4

5-3,Embellishment(b). Since the utilizations in the original model are low, a guess at the expected time in the system for each part type is the sum of the expected values, that is, 32 minutes and 25 minutes.

5-3,Embellishment(c). The revised model with 10% of the parts requiring refinishing is shown in D5-3.5. If we consider no waiting time for finishing, then the expected time for the finishing operation including the time to refinish 10% of the parts is 5 divided by 0.9 or 5.555. This increase in the amount of finishing time should increase the queueing time and add to the congestion of the system. Thus, each time the part is recycled, there is a chance that it encounters some waiting time. From this discussion we estimate that the expected values will be higher.

5-3,Embellishment(d). The SLAM II statement model showing the branching back to straightening for part type 1 when refinishing is required a second time is shown in D5-3.6. The fraction of part type 1's that will need to be restraightened is estimated at 0.1 * 0.1 or 0.01.

```
 1  GEN,FLOSS,PROBLEM 5.3C,6/10/86,1;
 2  LIMITS,3,2,100;
 3  NETWORK;
 4          CREATE,30,,2;                              CREATE PART TYPE 1
 5          ASSIGN,ATRIB(1)=1;                         ASSIGN PART TYPE
 6          ACTIVITY,2,,QUE1;                          TRAVEL TO DRILL AREA
 7          CREATE,20,,2;                              CREATE PART TYPE 2
 8          ASSIGN,ATRIB(1)=2;                         ASSIGN PART TYPE
 9          ACTIVITY,10;                               TRAVEL TO DRILL AREA
10  QUE1    QUEUE(1);                                  WAIT FOR DRILLS
11          ACTIVITY(2)/1,RNORM(10,1);                 DRILL PARTS
12          GOON,1;                                    ROUTE PARTS BY PART TYPE
13          ACTIVITY,,ATRIB(1).EQ.1,QUE2;              IF TYPE 1 BRANCH TO STRAIGHTENNG
14          ACTIVITY,,ATRIB(1).EQ.2,QUE3;              IF TYPE 2 BRANCH TO FINISHING
15  QUE2    QUEUE(2);                                  WAIT FOR STRAIGHTENER
16          ACTIVITY(1)/2,EXPON(15);                   STRAIGHTEN PARTS
17  QUE3    QUEUE(3);                                  WAIT FOR FINISHING
18          ACTIVITY(1)/3,5;                           FINISH PART
19          GOON,1;                                    ROUTE PARTS TO BE REFINISHED
20          ACTIVITY,,0.1,QUE3;                        10% OF PARTS REFINISHED
21          ACTIVITY,,0.9;                             90% OF PARTS CONTINUE
22          GOON,1;                                    COLLECT STATISTICS BY PART TYPE
23          ACTIVITY,,ATRIB(1).EQ.1,CLT1;              IF TYPE 1 BRANCH TO COLLECT NODE
24          ACTIVITY,,ATRIB(1).EQ.2,CLT2;              IF TYPE 2 BRANCH TO COLLECT NODE
25  CLT1    COLCT,INT(2),TIME IN SYSTEM 1,20/15/4;     COLLECT TIME IN SYSTEM
26          TERMINATE;                                 DEPART FROM SYSTEM
27  CLT2    COLCT,INT(2),TIME IN SYSTEM 2,15/20/1;     COLLECT TIME IN SYSTEM
28          TERMINATE;                                 DEPART FROM SYSTEM
29          END;
30  INIT,0,2400;
31  FIN;
```

D5-3.5

```
 1  GEN,FLOSS,PROBLEM 5.3D,6/10/86,1;
 2  LIMITS,3,3,100;
 3  NETWORK;
 4          CREATE,30,,2;                              CREATE PART TYPE 1
 5          ASSIGN,ATRIB(1)=1;                         ASSIGN PART TYPE
 6          ACTIVITY,2,,QUE1;                          TRAVEL TO DRILL AREA
 7          CREATE,20,,2;                              CREATE PART TYPE 2
 8          ASSIGN,ATRIB(1)=2;                         ASSIGN PART TYPE
 9          ACTIVITY,10;                               TRAVEL TO DRILL AREA
10  QUE1    QUEUE(1);                                  WAIT FOR DRILLS
11          ACTIVITY(2)/1,RNORM(10,1);                 DRILL PARTS
12          GOON,1;                                    ROUTE PARTS BY PART TYPE
13          ACTIVITY,,ATRIB(1).EQ.1,QUE2;              IF TYPE 1 BRANCH TO STRAIGHTENNG
14          ACTIVITY,,ATRIB(1).EQ.2,QUE3;              IF TYPE 2 BRANCH TO FINISHING
15  QUE2    QUEUE(2);                                  WAIT FOR STRAIGHTENER
16          ACTIVITY(1)/2,EXPON(15);                   STRAIGHTEN PARTS
17  QUE3    QUEUE(3);                                  WAIT FOR FINISHING
18          ACTIVITY(1)/3,5;                           FINISH PART
19          ASSIGN,ATRIB(3)=ATRIB(3)+1,1;              ASSIGN NUMBER OF TIMES FINISHED
20          ACTIVITY,,0.1,DEC1;                        10% OF PARTS REQUIRE REFINISHING
21          ACTIVITY,,0.9;                             90% OF PARTS CONTINUE
22          GOON,1;                                    COLLECT STATISTICS BY PART TYPE
23          ACTIVITY,,ATRIB(1).EQ.1,CLT1;              IF TYPE 1 BRANCH TO COLLECT NODE
24          ACTIVITY,,ATRIB(1).EQ.2,CLT2;              IF TYPE 2 BRANCH TO COLLECT NODE
25  CLT1    COLCT,INT(2),TIME IN SYSTEM 1,20/15/4;     COLLECT TIME IN SYSTEM
26          TERMINATE;                                 DEPART FROM SYSTEM
27  CLT2    COLCT,INT(2),TIME IN SYSTEM 2,15/20/1;     COLLECT TIME IN SYSTEM
28          TERMINATE;                                 DEPART FROM SYSTEM
29  DEC1    GOON,1;                                    ROUTE PARTS TO MACHINING AREA
30          ACTIVITY,,ATRIB(3).GT.1.AND.ATRIB(1).EQ.1,QUE2;  REQUIRES RESTRAIGHTENING
31          ACTIVITY,,,QUE3;                           REQUIRES REFINISHING
32          END;
33  INIT,0,2400;
34  FIN;
```

D5-3.6

5-4.   The network and control statements for this exercise are shown in D5-4.1. In D5-4.2, the summary report for the 10th run is given. Items of note are: the number of observations is 502 which is the number of arrivals plus the one entity that started in QUEUE node QUEZ and the one entity that was in service. Recall if there is an entity in a queue, all following servers are busy. There is a large amount of variability in this problem as seen by the estimates for the standard deviations. This is due in part to the balking which in some of the runs is excessive. Under file statistics, note that the current length is 0 as the simulation ended when there were no entitities to be processed. The average waiting time is computed using Little's formula where the average length is divided by the actual arrival rate to the queue and includes zero values for those arrivals that did not wait. For file 1, the arrival rate is 500 divided by 2423 and, for file 2, it is 501 divided by 2423.

```
 1   GEN,ROLSTON,PROBLEM 5.4,8/28/80,10;
 2   LIMITS,2,1,50;
 3   NETWORK;
 4        CREATE,EXPON(5),1,1,500;
 5        QUEUE(1);
 6        ACT,RNORM(4,1);
 7   QUEZ QUEUE(2),1,2,BALK(BUFF);
 8        ACT,EXPON(4.);
 9        COLCT,INT(1),TIME IN SYSTEM;
10        TERM;
11   BUFF COLCT,BET,TIME BET BALKS;
12        ACT,3,,QUEZ;
13        END;
14   FIN;
```

D5-4.1

SLAM II SUMMARY REPORT

SIMULATION PROJECT PROBLEM 5.4   BY ROLSTON

DATE 8/28/1980   RUN NUMBER 10 OF 10

CURRENT TIME   0.2673E+04
STATISTICAL ARRAYS CLEARED AT TIME 0.0000E+00

STATISTICS FOR VARIABLES BASED ON OBSERVATION

	MEAN VALUE	STANDARD DEVIATION	COEFF. OF VARIATION	MINIMUM VALUE	MAXIMUM VALUE	NUMBER OF OBSERVATIONS
TIME IN SYSTEM	0.1807E+02	0.1041E+02	0.5760E+00	0.3234E+01	0.6110E+02	502
TIME BET BALKS	0.1134E+02	0.3993E+02	0.3521E+01	0.1733E-01	0.3050E+03	230

FILE STATISTICS

FILE NUMBER	LABEL/TYPE	AVERAGE LENGTH	STANDARD DEVIATION	MAXIMUM LENGTH	CURRENT LENGTH	AVERAGE WAITING TIME
1	QUEUE	1.0268	1.4727	8	0	5.4889
2	QUEZ QUEUE	0.6291	0.7956	2	0	3.3563
3	CALENDAR	2.7267	1.0653	9	0	2.7647

SERVICE ACTIVITY STATISTICS

ACTIVITY INDEX	START NODE OR ACTIVITY LABEL	SERVER CAPACITY	AVERAGE UTILIZATION	STANDARD DEVIATION	CURRENT UTILIZATION	AVERAGE BLOCKAGE	MAXIMUM IDLE TIME/SERVERS	MAXIMUM BUSY TIME/SERVERS	ENTITY COUNT
0	QUEZ QUEUE	1	0.7323	0.4427	0	0.0000	21.8590	95.3770	
0	QUEUE	1	0.7468	0.4349	0	0.0000	25.8877	128.5726	

D5-4.2

5-5.   Include in the input statements for Exercise 5-4, the following:

      TERM,500;

      .
      .
      .

      INIT,0,2500;
      MONTR,CLEAR,200;
      SIMULATE;
      MONTR,CLEAR,200;
      SIMULATE;

      .
      .
      .

      FIN;

Repeated use of the MONTR statement is required as MONTR statements only apply for a single run. To avoid the repeated use of MONTR statements, calls to subroutine CLEAR in a user written FORTRAN routine at time 200 could be performed as discussed in Chapter 12.

5-6.   The network for this exercise is shown in D5-6.1.

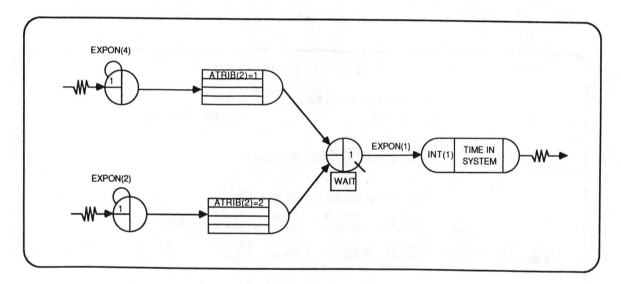

D5-6.1

There are two types of entities in the system which are identified by their second attribute. Type 1 entities have an interarrival time that is a sample from an exponential distribution with a mean of 4. Type 2 entities have an interarrival time that is a sample from an exponential distribution with a mean of 2. The arrival time for the entities is stored in the first attribute. All entities are routed to the QUEUE node labeled WAIT, where they wait for service if the one server is busy. The entities are ranked in the queue based on the value of their second attribute; those having lower values of this attribute are given priority. The service time is exponentially distributed with a mean of 1. After service, an entity leaves the system and statistics are collected on its time in the system.

5-7.   Since workstation 2 has the larger mean processing time, more storage spaces should be assigned before workstation 2. An R&D program for decreasing the time requirements at workstation 2 seems in order. A profit or cost structure to evaluate improvements or additional investments should be considered.

5-8.   The model for this abstracted system is a single server queue with a limited queue capacity. The SLAM II statements are shown in D5-8.1. The Summary Report is shown in D5-8.2. The results indicate that the same average utilization for workstation 2 is obtained but that there is less balking since there are 7 items allowed in the queue before balking begins whereas there was only a capacity of 4 for the first workstation.

```
1   GEN,FLOSS,PROBLEM 5.8,6/20/86,1;
2   LIMITS,2,1,50;
3   NETWORK;
4         CREATE,EXPON(.4,1),,1;              CREATE ARRIVALS
5         QUEUE(2),0,7,BALK(SUB);             WAIT FOR STATION 2
6         ACT/2,EXPON(.50,3);                 SERVICE STATION 2
7         COLCT,INT(1),TIME IN SYSTEM,20/0/.4; COLLECT STATISTICS
8         TERM;                               DEPART SYSTEM
9   SUB   COLCT,BET,TIME BET. BALKS;          COLLECT STATISTICS
10        TERM;                               DEPART SYSTEM
11        END
12  INIT,0,300;
13  FIN;
```

D5-8.1

```
                    S L A M   I I   S U M M A R Y   R E P O R T

         SIMULATION PROJECT PROBLEM 5.8          BY FLOSS

         DATE  6/20/1986                         RUN NUMBER    1 OF    1


         CURRENT TIME   0.3000E+03
         STATISTICAL ARRAYS CLEARED AT TIME  0.0000E+00


                    **STATISTICS FOR VARIABLES BASED ON OBSERVATION**
```

	MEAN VALUE	STANDARD DEVIATION	COEFF. OF VARIATION	MINIMUM VALUE	MAXIMUM VALUE	NUMBER OF OBSERVATIONS
TIME IN SYSTEM	0.2961E+01	0.1674E+01	0.5654E+00	0.1068E-02	0.7110E+01	561
TIME BET. BALKS	0.1228E+01	0.3382E+01	0.2755E+01	0.2485E-02	0.3327E+02	234

FILE STATISTICS

FILE NUMBER	LABEL/TYPE	AVERAGE LENGTH	STANDARD DEVIATION	MAXIMUM LENGTH	CURRENT LENGTH	AVERAGE WAITING TIME
1		0.0000	0.0000	0	0	0.0000
2	QUEUE	4.6077	2.2149	7	7	2.4294
3	CALENDAR	1.9768	0.1507	3	2	0.2774

SERVICE ACTIVITY STATISTICS

ACTIVITY INDEX	START NODE OR ACTIVITY LABEL	SERVER CAPACITY	AVERAGE UTILIZATION	STANDARD DEVIATION	CURRENT UTILIZATION	AVERAGE BLOCKAGE	MAXIMUM IDLE TIME/SERVERS	MAXIMUM BUSY TIME/SERVERS	ENTITY COUNT
2	SERVICE STAT	1	0.9768	0.1507	1	0.0000	0.8625	115.2183	561

D5-8.2

5-8,Embellishment. From queueing theory, the probability of having 8 units in the system (7 in the queue and 1 in service) is given by the following formula

$$\pi_8 = \frac{(1-\rho)\,\rho^8}{1-\rho^9}$$

where ρ is equal to 1.25. Since the arrivals to the system are exponentially distributed, the probability of a unit balking is the same as the probability of 8 units being in the system. This gives a balking probability of 0.231. The estimate from D5-8.2 is 129 divided by 737 or 0.175. An estimate of ρ for the run can be made by computing the observed arrival rate and service rate. The arrival rate is 737/300 or 2.46. The service rate is (.9491)(300)/608 or approximately 2.14. This gives a ρ of 1.15. The estimate of the probability of balking with $\rho = 1.15$ is 0.182.

5-9,Embellishment(a). Add a second activity between the CREATE node and the QUEUE node INSP. Note that this change doubles the arrival rate so that it exceeds the service rate for the inspectors. Thus, the system will be unstable. The value of MNTRY on the LIMITS statement must be increased to run this exercise. In general, conditional branching is used to generate multiple entities based on a single entity arrival. By indexing a variable by 1 and then recycling if the variable has not reached a specified value, multiple entities can be generated.

5-9,Embellishment(b). The model is changed to have two activities emanating from QUEUE node ADJT both representing the adjustor. The changes are shown in the statement model given in D5-9.1.

5-9,Embellishment(c). Add 5 time units to the inspection activity and change the probabilities of routing entities to the adjustor and departure nodes accordingly. The statement model including these changes and a request for a trace is given in D5-9.2.

```
 1   GEN,ROLSTON,PROBLEM 5.9B,7/25/80,1;
 2   LIMITS,2,2,50;
 3   NETWORK;
 4          CREATE,UNFRM(3.5,7.5),,1;
 5   INSP   QUEUE(1);
 6          ACT(2)/1,UNFRM(6,12);
 7          GOON;
 8          ACT,,.85,DPRT;
 9          ACT,,.15,ADJT;
10   ADJT   QUEUE(2);
11          ACT/2,UNFRM(20,40),.60,INSP;
12          ACT/2,UNFRM(20,40),.40,DPRT;
13   DPRT   COLCT,INT(1),TIME IN SYSTEM;
14          TERM;
15          END;
16   INIT,0,480;
17   FIN;
```

D5-9.1

```
 1    GEN,ROLSTON,PROBLEM 5.9C,7/25/80,1;
 2    LIMITS,2,2,50;
 3    NETWORK;
 4            CREATE,UNFRM(3.5,7.5),,1;
 5    INSP    QUEUE(1);
 6            ACT(2)/1,UNFRM(6,12)+5;
 7            GOON;
 8            ACT,,.90,DPRT;
 9            ACT,,.10,ADJT;
10    ADJT    QUEUE(2);
11            ACT/2,UNFRM(20,40),,INSP;
12    DPRT    COLCT,INT(1),TIME IN SYSTEM;
13            TERM;
14            END;
15    INIT,0,480;
16    MONTR,TRACE,0,60,1;
17    FIN;
```

D5-9.2

5-10.   The network model and statement model are shown in D5- 10.1 and D5-10.2. Through the use of attributes the new features for the model are included. The parameters of the uniform distribution for the adjustment time are related to the entities being adjusted. In other situations, the parameters could be related to the adjustor's experience in which case the parameters could be specified by resetting the global variables XX(I). In this way, learning effects could be included in the model.

Since the probability of requiring an adjustment will now change depending on the entity being processed fixed probability cannot be assigned to a branch. In this case, a random number is generated and tested against the probability of requiring adjustment. DRAND is used to obtain the random number and ATRIB(3) is used as the attribute defining the probability of requiring adjustment. Note that on the branch to node DPRT, no condition is specified. It is not correct to specify the condition DRAND.GE.ATRIB(3) for this branch as a new random number would be generated if DRAND is specified. An incorrect result would be obtained if such a condition is specified. When more than two branches require variable probabilities then an ASSIGN node is required to set the random number value in an XX variable and to test XX against the different probabilities. The SLAM II Summary Report for this modified example is shown in D5-10.3.

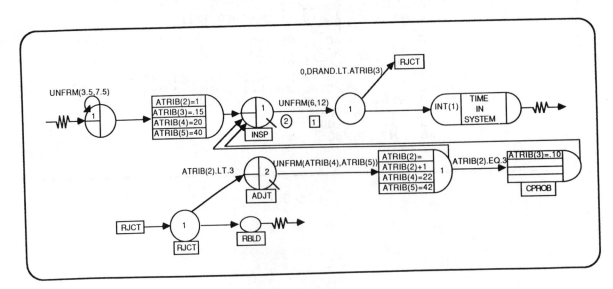

D5-10.1

```
 1     GEN,ROLSTON,PROBLEM 5.10,8/28/80,1;
 2     LIMITS,2,5,100;
 3     NETWORK;
 4            CREATE,UNFRM(3.5,7.5),,1;
 5            ASSIGN,ATRIB(2)=1,
 6                   ATRIB(3)=.15,
 7                   ATRIB(4)=20,
 8                   ATRIB(5)=40;
 9     INSP   QUEUE(1);
10            ACT(2)/1,UNFRM(6,12);
11            GOON,1;
12                ACT,0,DRAND.LT.ATRIB(3),RJCT;
13            ACT,,,DPRT;
14     DPRT   COLCT,INT(1),TIME IN SYSTEM;
15            TERM;
16     RJCT   GOON,1;
17                ACT,0,ATRIB(2).LT.3,ADJT;
18            ACT,,,RBLD;
19     RBLD   TERM;
20     ADJT   QUEUE(2);
21                ACT/2,UNFRM(ATRIB(4),ATRIB(5));
22            ASSIGN,ATRIB(2)=ATRIB(2)+1,
23                   ATRIB(4)=22,
24                   ATRIB(5)=42,1;
25                ACT,0,ATRIB(2).EQ.3,CPROB
26            ACT,,,INSP;
27     CPROB  ASSIGN,ATRIB(3)=.10;
28                ACT,,,INSP;
29            END;
30     INIT,0,480;
31     FIN;
```

D5-10.2

SLAM II SUMMARY REPORT

SIMULATION PROJECT PROBLEM 5.10   BY ROLSTON

DATE 8/28/1980   RUN NUMBER   1 OF   1

CURRENT TIME   0.4800E+03
STATISTICAL ARRAYS CLEARED AT TIME   0.0000E+00

STATISTICS FOR VARIABLES BASED ON OBSERVATION

	MEAN VALUE	STANDARD DEVIATION	COEFF. OF VARIATION	MINIMUM VALUE	MAXIMUM VALUE	NUMBER OF OBSERVATIONS
TIME IN SYSTEM	0.1338E+02	0.1255E+02	0.9378E+00	0.6068E+01	0.9918E+02	82

FILE STATISTICS

FILE NUMBER	LABEL/TYPE	AVERAGE LENGTH	STANDARD DEVIATION	MAXIMUM LENGTH	CURRENT LENGTH	AVERAGE WAITING TIME
1	INSP QUEUE	0.3516	0.5924	3	2	1.7579
2	ADJT QUEUE	0.0185	0.1766	2	2	0.8878
3	CALENDAR	3.2362	0.6082	5	4	4.3881

SERVICE ACTIVITY STATISTICS

ACTIVITY INDEX	START NODE OR ACTIVITY LABEL	SERVER CAPACITY	AVERAGE UTILIZATION	STANDARD DEVIATION	CURRENT UTILIZATION	AVERAGE BLOCKAGE	MAXIMUM IDLE TIME/SERVERS	MAXIMUM BUSY TIME/SERVERS	ENTITY COUNT
1	INSP QUEUE	2	1.7627	0.4326	2	0.0000	2.0000	2.0000	92
2	ADJT QUEUE	1	0.4735	0.4993	1	0.0000	104.6160	37.9087	7

D5-10.3

5-11.   The SLAM II program for the single-server single-queue system in which the service time is a probability
mass function is shown in D5-11.1. This exercise illustrates the use of DPROBN and ARRAY statements.

```
 1   GEN,FLOSS,PROBLEM 5.11,6/25/1986,1;
 2   LIMITS,1,0,100;
 3   ARRAY(1,4)/0.2,0.5,0.6,1;
 4   ARRAY(2,4)/4,6,7,10;
 5   NETWORK;
 6         CREATE,EXPON(7.75);              CREATE ARRIVALS
 7         QUEUE(1);                        WAIT FOR SERVICE
 8         ACTIVITY(1)/1,DPROBN(1,2);       SERVICE ACTIVITY
 9         TERMINATE;                       DEPART SYSTEM
10         END;
11   INIT,0,480;
12   FIN;
```

D5-11.1

CHAPTER 6

Resources and Gates

6-1.   The network and input statements are given in D6-1.1 and D6-1.2.

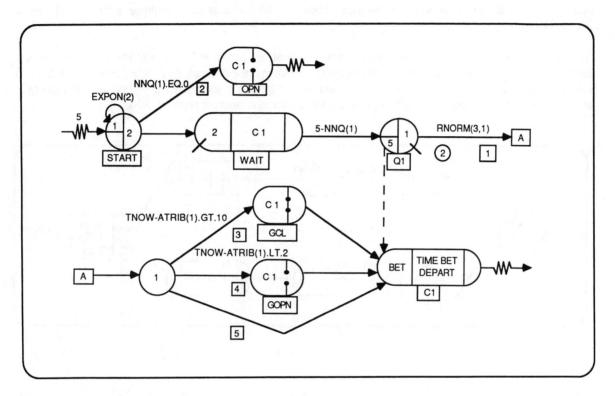

D6-1.1

```
 1    GEN,ROLSTON,PROBLEM 6.1,7/21/80,1;
 2    LIMITS,2,1,50;
 3    NETWORK;
 4            GATE/G1,OPEN,2;
 5    START   CREATE,EXPON(2),5,1,,2;
 6            ACT/2,,NNQ(1).EQ.0,OPN;
 7            ACT,,,WAIT;
 8    OPN     OPEN,G1;
 9            TERM;
10    WAIT    AWAIT(2),G1;
11            ACT,5-NNQ(1);
12    Q1      QUEUE(1),,5,BALK(C1);
13            ACT(2)/1,RNORM(3,1);
14            GOON,1;
15            ACT/3,,TNOW-ATRIB(1).GT.10,GCL;
16            ACT/4,,TNOW-ATRIB(1).LT.2,GOPN;
17            ACT/5,,,C1;
18    GCL     CLOSE,G1;
19    C1      COLCT,BET,TIM BET DEPART;
20            TERM;
21    GOPN    OPEN,G1;
22            ACT,,,C1;
23            ENDNETWORK;
24    INIT,0,100;
25    FIN;
```

D6-1.2

29

The problem statement does not clearly specify the disposition of entities balking from QUEUE node, Q1. The solution presumes that such entities are viewed as departures from the system. Attribute 1 is the arrival time of an entity. The gate is specified as OPEN at the beginning of the simulation and no entitites are initially in the queue. To obtain statistics on the frequency with which different branches are taken, activity numbers are given to branches. As a further exercise, it would be interesting to ask the students to estimate qualitatively the statistical quantitites of interest. There is sufficient interaction in this exercise to make this a nontrivial task.

6-2.   The SLAM II network and statement models for this situation are shown in D6-2.1 and D6-2.2. These models are analogous to the professor model of Illustration 6-2. Alternative network models are given in D6-2.3 and D6-2.4 to illustrate how a single server can be represented by two branches when probabilistic branching is employed and an attribute can be used to provide a service for two different types of service.

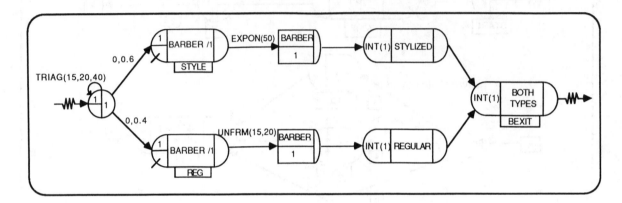

D6-2.1

```
 1    GEN,PRITSKER,PROBLEM 6.2,6/21/86,1;
 2    LIMITS,1,1,50;
 3    NETWORK;
 4          RESOURCE,BARBER,1;
 5          CREATE,TRIAG(15,20,40),,1,,1;
 6          ACT,,0.6,STYLE;                    CLASSIFY 60% AS STYLE
 7          ACT,,0.4,REG;                      CLASSIFY 40% AS REGULAR
 8    STYLE AWAIT(1),BARBER/1;                 REQUEST BARBER
 9          ACT,EXPON(20);                     PERFORM STYLE CUT
10          FREE,BARBER/1;                     RELEASE BARBER
11          COLCT,INT(1),STYLIZED;             COLLECT TIME IN SYSTEM
12          ACT,,,BEXIT;                       ROUTE TO EXIT
13    REG   AWAIT(1),BARBER/1;                 REQUEST BARBER
14          ACT,UNFRM(15,20);                  PERFORM REG CUT
15          FREE,BARBER/1;                     RELEASE BARBER
16          COLCT,INT(1),REGULAR;              COLLECT TIME IN SYSTEM
17    BEXIT COLCT,INT(1),BOTH TYPES;           COLLECT TIS FOR ALL
18          ENDNETWORK;
19    INIT,0,2000;
20    FIN;
```

D6-2.2

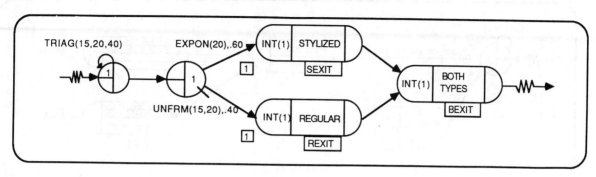

D6-2.3

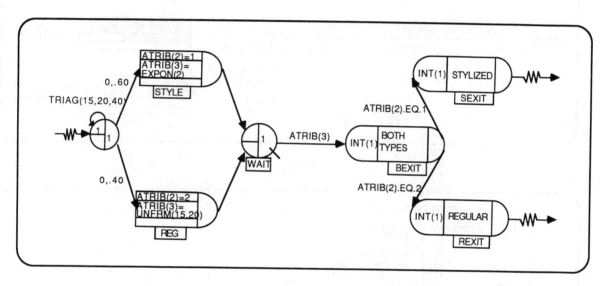

D6-2.4

6-3.   In the inventory example, the resource radio is used to keep track of the stock-on-hand and the number of units in file 1 represents the backorder quantity. To remove resources from the model, two new variables are needed which will be called STOCK_ON_H and BACKORDER_Q to represent the stock-on-hand and backorder quantity. These variables are equivalenced to XX(4) and XX(5), respectively. The STOCK_ON_H replaces NNRSC(RADIO) in Example 6-1 and it is necessary to include the logic in the model to satisfy backorders when an order is received. The network and statement models for this situation are shown in D6-3.1 and D6-3.2. The output statistics from this model are identical to those presented in Example 6-1. This exercise is a good one to illustrate the internal quantities that are maintained by SLAM II with regard to resources.

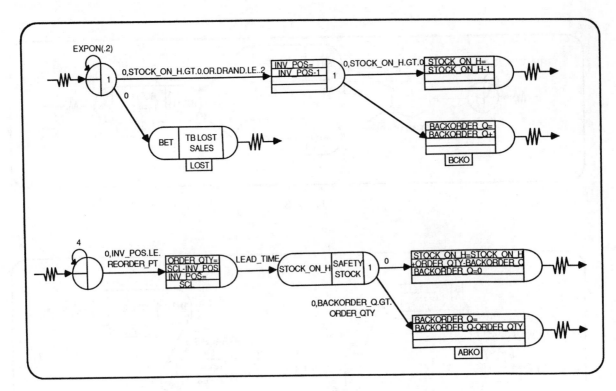

D6-3.1

```
 1    GEN,FLOSS,PROBLEM 6.3,6/11/86,1,Y,Y,Y/N;
 2    LIMITS,0,2,30;
 3    EQUIVALENCE/XX(1),INV_POS/
 4              XX(2),REORDER_PT/
 5              XX(3),SCL/
 6              XX(4),STOCK_ON_H/
 7              XX(5),BACKORDER_Q/
 8              ATRIB(1),ORDER_QTY/
 9              3,LEAD_TIME/;
10    INTLC,INV_POS=72,REORDER_PT=18,SCL=72,STOCK_ON_H=72,BACKORDER_Q=0;
11    TIMST,INV_POS,INV. POSITION;
12    TIMST,STOCK_ON_H,STOCK ON HAND;
13    NETWORK;
14    ;
15    ;     CUSTOMER ARRIVAL PROCESS
16    ;     ------------------------
17    ;
18          CREATE,EXPON(.2),,,1;                         CREATE ARRIVAL ENTITIES
19          ACT,,STOCK_ON_H.GT.0.OR.DRAND.LE..2;          CONTINUE IF RADIO AVAIL OR P=.2
20          ACT,,,LOST;                                   ELSE BRANCH TO LOST SALE
21          ASSIGN,INV_POS=INV_POS-1,1;                   DECREMENT INVENTORY POSITION
22          ACT,,STOCK_ON_H.GT.0;                         CONTINUE IF RADIO AVAIL
23          ACT,,,BCKO;                                   ELSE BRANCH TO BACKORDER
24          ASSIGN,STOCK_ON_H=STOCK_ON_H-1;               DECREMENT STOCK ON HAND
25          TERM;                                         DEPART THE SYSTEM
26    BCKO  ASSIGN,BACKORDER_Q=BACKORDER_Q+1;             INCREMENT BACKORDER QUANTITY
27          TERM;                                         DEPART THE SYSTEM
28    LOST  COLCT,BET,TB LOST SALES;                      COLLECT BET STATS ON LOST SALES
29          TERM;                                         DEPART THE SYSTEM
30    ;
31    ;     INVENTORY REVIEW PROCESS
32    ;     ------------------------
33    ;
34          CREATE,4;                                     CREATE A REVIEW ENTITY
35          ACT,0,INV_POS.LE.REORDER_PT;                  IF POSITION IS BELOW REODER PT.
36          ASSIGN,ORDER_QTY = SCL - INV_POS,
37                 INV_POS = SCL;                         ORDER UP TO STOCK CONTROL LEVEL
38          ACT,LEAD_TIME;                                DELAY RECEIPT BY 3 WEEK LEAD_TI
39          COLCT,STOCK_ON_H,SAFETY STOCK,,1;             COLLECT STATS ON SAFETY STOCK
40          ACT,,BACKORDER_Q.GT.ORDER_QTY,ABKO;           IF BACKORDERS GT ORDER BRANCH
41          ACT;                                          ELSE CONTINUE
42          ASSIGN,STOCK_ON_H=STOCK_ON_H+ORDER_QTY-BACKORDER_Q,
43                 BACKORDER_Q=0;                         ADJUST STOCK ON HAND
44          TERM;                                         END REVIEW
45    ABKO  ASSIGN,BACKORDER_Q=BACKORDER_Q-ORDER_QTY;     ADJUST BACKORDER QUANTITY
46          TERM;                                         END REVIEW
47          END;
48    INIT,0,312;
49    MONTR,CLEAR,52;
50    FIN;
```

D6-3.2

6-4.   The network and statement models are shown in D6-4.1 and D6-4.2. Interesting features of this model are: Collection of statistics on ATRIB variables; conditions to represent probabilistic branching; and, specification of histograms at COLCT nodes. Note also that the mean and variance for the Erlang distribution are specified in the problem statement. These must be converted to the mean of the corresponding exponential distribution and the number of exponential samples being summed. The problem statement specifies μ and σ^2 from which we compute $XMN = \sigma^2/\mu$ and $XK = \mu/XMN$. (Recall the Erlang distribution is a special case of the gamma distribution. XMN corresponds to the beta parameter of the gamma distribution and XK corresponds to the alpha parameter. The summary report for this exercise is shown in D6-4.3.

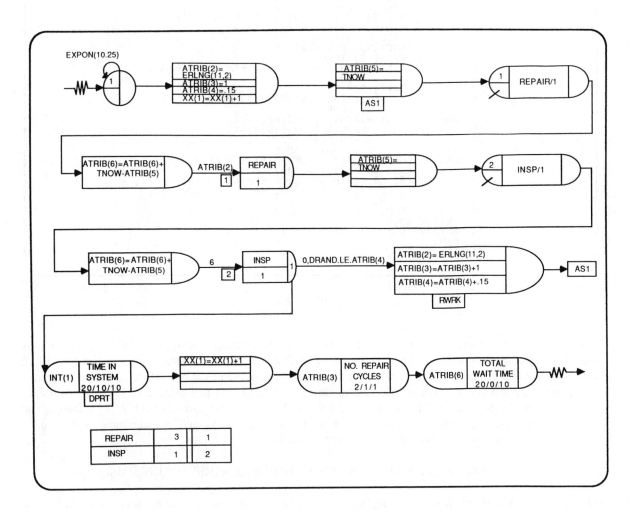

D6-4.1

```
 1          GEN,PRITSKER,PROBLEM 6.4,8/13/86,1;
 2          LIMITS,2,6,200;
 3          PRIORITY/1,LVF(2);
 4          TIMST,XX(1),NUMBER IN SYSTEM;
 5          INTLC,XX(1)=2;
 6          NETWORK;
 7                    RESOURCE,REPAIR(3),1;
 8                    RESOURCE,INSP(1),2;
 9          ;
10                    CREATE,EXPON(10.25),,1;
11                    ASSIGN,ATRIB(2)=ERLNG(11,2),
12                        ATRIB(3)=1,
13                        ATRIB(4)=.15,
14                        XX(1)=XX(1)+1;
15   ASGN     ASSIGN,ATRIB(5)=TNOW;
16                    AWAIT(1),REPAIR/1;
17                    ASSIGN,ATRIB(6)=ATRIB(6)+TNOW-ATRIB(5);
18                ACT/1,ATRIB(2);
19                    FREE,REPAIR/1;
20                    ASSIGN,ATRIB(5)=TNOW;
21                    AWAIT(2),INSP/1;
22                    ASSIGN,ATRIB(6)=ATRIB(6)+TNOW-ATRIB(5);
23                ACT/2,6;
24                    FREE,INSP/1,1;
25                    ACT,,DRAND.LE.ATRIB(4),RWRK;
26                    ACT,,,DPRT;
27   RWRK     ASSIGN,ATRIB(2)=ERLNG(11,2),
28                        ATRIB(3)=ATRIB(3)+1,
29                        ATRIB(4)=ATRIB(4)*.15;
30                        ACT,,,ASGN;
31   DPRT     COLCT,INT(1),TIME IN SYSTEM,20/10/10;
32                    ASSIGN,XX(1)=XX(1)-1;
33                    COLCT,ATRIB(3),NO OF REPAIR CYCLES,2/1/1;
34                    COLCT,ATRIB(6),TOTAL WAIT TIME,20/0/10;
35                    TERM;
36                    END;
37       INIT,0,2000;
38       ENTRY/1,0,1.0,1,.15,0/1,0,1.5,1,.15,0;
39       FIN;
```

D6-4.2

6-4,Embellishment(a). By defining a new attribute, we can achieve the complex ranking procedure specified in this embellishment. Specify attribute 7 to be the ranking attribute for file 1 and rank entities in file 1 based on HVF(7). For new arrivals, attribute 7 will be set equal to the negative of the repair time so that less negative values will be given priority. For recycled items, attribute 7 will be set equal to the time in the system minus the repair time for the item. In this way, recycled items will have a higher value of attribute 7. If there are no recycled items, the repair time will establish priority. For recycled items with the same time in the system, smaller repair times will determine which entity is processed first.

6-4,Embellishment(b). Change the assignment of attribute 2 in two locations to ATRIB(2) = UNFRM(0,48).

6-4,Embellishment(c). Modify the network by replacing the CREATE node with the subnetwork shown in D6-4.4. The branch leaving node G1 leads to an ASSIGN node where we set ATRIB(1) = TNOW. The use of a 1 in the MA field of the CREATE node will not work if this feedback branch is used. The INIT statement should be removed. Since there will be no inputs after time 2000, the run will eventually terminate when all entities are processed.

6-4,Embellishment(d). Change the inspection time as prescribed for activity 2 from 6 to RLOLGN (6,1.5).

```
                    S L A M   I I   S U M M A R Y   R E P O R T

              SIMULATION PROJECT PROBLEM 6.4              BY PRITSKER

              DATE  8/13/1986                             RUN NUMBER    1 OF    1


              CURRENT TIME   0.2000E+04
              STATISTICAL ARRAYS CLEARED AT TIME  0.0000E+00


                    **STATISTICS FOR VARIABLES BASED ON OBSERVATION**

                   MEAN         STANDARD      COEFF. OF      MINIMUM       MAXIMUM
                   VALUE        DEVIATION     VARIATION      VALUE         VALUE      NUMBER OF
                                                                                     OBSERVATIONS
TIME IN SYSTEM   0.4163E+02    0.2607E+02    0.6262E+00    0.7000E+01    0.1668E+03      172
NO OF REPAIR CYC 0.1163E+01    0.3703E+00    0.3184E+00    0.1000E+01    0.2000E+01      172
TOTAL WAIT TIME  0.9741E+01    0.1167E+02    0.1198E+01    0.0000E+00    0.7875E+02      172


                    **STATISTICS FOR TIME-PERSISTENT VARIABLES**

                   MEAN         STANDARD      MINIMUM       MAXIMUM       TIME          CURRENT
                   VALUE        DEVIATION     VALUE         VALUE         INTERVAL      VALUE

NUMBER IN SYSTEM 0.3590E+01    0.1975E+01    0.0000E+00    0.9000E+01    0.2000E+04    0.1000E+01


                    **FILE STATISTICS**

FILE                     AVERAGE       STANDARD      MAXIMUM   CURRENT    AVERAGE
NUMBER  LABEL/TYPE       LENGTH        DEVIATION     LENGTH    LENGTH     WAITING TIME

   1       AWAIT         0.5249        0.9893           5        0        5.2231
   2       AWAIT         0.3128        0.5874           3        0        3.1283
   3       CALENDAR      3.7519        1.1153           6        2        4.6781


                    **REGULAR ACTIVITY STATISTICS**

ACTIVITY         AVERAGE       STANDARD      MAXIMUM CURRENT   ENTITY
INDEX/LABEL      UTILIZATION   DEVIATION     UTIL    UTIL      COUNT

   1             2.1519        1.0107           3      1        200
   2             0.6000        0.4899           1      0        200


                    **RESOURCE STATISTICS**

RESOURCE  RESOURCE   CURRENT   AVERAGE        STANDARD      MAXIMUM       CURRENT
NUMBER    LABEL      CAPACITY  UTILIZATION    DEVIATION     UTILIZATION   UTILIZATION

   1      REPAIR       3       2.1519         1.0107           3             1
   2      INSP         1       0.6000         0.4899           1             0


RESOURCE  RESOURCE   CURRENT   AVERAGE        MINIMUM       MAXIMUM
NUMBER    LABEL      AVAILABLE AVAILABLE      AVAILABLE     AVAILABLE

   1      REPAIR       2       0.8481            0             3
   2      INSP         1       0.4000            0             1
```

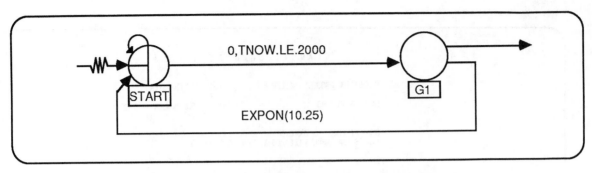

D6-4.4

6-5.   The model of Illustration 6-3 is changed to allow the number of resources available of each type of mill to be altered within the model. This is shown in D6-5.1 where the initial capacity of each mill resource is set at 0 and a disjoint network segment consisting of a series of ALTER nodes is used to alter the capacity in accordance with an XX variable. The value of the XX variable is set in an INTLC statement and new values are set by making multiple runs and including a new INTLC statement for each run. In D6-5.1 seven runs are indicated so that the number of castings produced for seven different alternative FMS configurations can be obtained. A summary of the seven runs is shown in D6-5.2. Note: after first printing the problem specifies a production of 173 castings in 80 hours. A discussion related to this exercise is included in references 41 and 66 of Chapter 4. The allocation condition to the flexible machine in this solution is naive which results in a negative assessment of the need for a flexible mill.

6-6.   The solution to the basic problem involves multiplying the PROCESS TIME by 1.1 when the flexible mill is assigned to perform an operation. The statement model for this situation is shown in D6-6.1.

6-6,Embellishment. To include a setup time when the flexible mill changes from one operation type to another, an array of setup times is defined in ARRAY with each row representing a previous operation type and each column the current operation type. The model for this situation is shown in D6-6.2. At line 29, the activity representing the flexible mill processing time is shown with the time to perform an operation equal to PROCESS_TIME + SETUP_TIME. SETUP_TIME is equivalenced to a value obtained from ARRAY. As discussed above, the row number for ARRAY is defined as the previous operation number, PREV_OPER. Since there is only one flexible mill only one previous operation number needs to be maintained. In line 3, 4, and 5 the setup times as prescribed by the problem statement are shown when the previous operation is 10, 20, and 30 respectively. Note the correspondence between operation 10 and row 1 of ARRAY. Also note that a zero value is prescribed along the diagonal of ARRAY to indicate that no setup time is required if the operation number has not changed.

6-7.   There are two interesting alternatives in modeling this problem. First, we will consider the aircraft as resources and use the global variable, XX(1), to count the number of unit loads waiting for aircraft. Thus, the unit loads (cargo) represent the entities that will be flowing through the network. ATRIB(1) represents the arrival time of a unit load while it is waiting for a plane. Once a plane has been allocated, ATRIB(1) is used for the flying time. ATRIB(2) is the resource type (aircraft type) allocated, and ATRIB(3) is the number of units requiring space. The model is described below in terms of the statement model shown in D6-7.1 starting at line 7.

```
 1   GEN,FLOSS,PROBLEM 6.5,6/16/86,7,,N,,N;
 2   LIMITS,4,3,500;
 3   ARRAY(1,3)/120,40,56;
 4   EQUIVALENCE/ATRIB(1),OPERATION/
 5              ATRIB(2),MILL/
 6              ARRAY(1,OPERATION),PROCESS_TIME;
 7   INTLC,XX(1)=5,XX(2)=2,XX(3)=3,XX(4)=0;
 8   NETWORK;
 9          RESOURCE/1,MILL1(0),1/2,MILL2(0),2;        DEFINE 10 MILLS
10          RESOURCE/3,MILL3(0),3/4,MILLF(0),4;        AS 4 RESOURCE TYPES
11   ;
12          CREATE;
13          ALTER,MILL1,XX(1);                         DEFINE NUMBER OF MACH 1
14          ALTER,MILL2,XX(2);                         DEFINE NUMBER OF MACH 2
15          ALTER,MILL3,XX(3);                         DEFINE NUMBER OF MACH 3
16          ALTER,MILLF,XX(4);                         DEFINE NUMBER OF FLEX M
17          TERM;
18   ;
19          CREATE,22,,3;
20   ;
21   SETA   ASSIGN,OPERATION = OPERATION + 1,          INCREMENT OPERATION
22              II=OPERATION,1;                        NUMBER AND
23   ;                                                 SET II TO PROPOSED MILL
24   ;                                                 CONDITIONS FOR
25          ACT,0,NNRSC(II).GT.0.OR.NNRSC(4).EQ.0,SETM; DEDICATED MILL
26   ;
27          ACT;
28          ASSIGN,MILL=4;                             ASSIGN FLEXIBLE MILL
29          ACT,,,AMILL;
30   SETM   ASSIGN,MILL = OPERATION;                   ASSIGN DEDICATED MILL
31   AMILL  AWAIT(MILL= 1,4),MILL/1;                   WAIT FOR MILL
32          ACT,PROCESS_TIME;                          MILL PROCESSING ACTIVITY
33          FREE,MILL/1,1;                             FREE MILL
34          ACT,0,OPERATION.LT.3,SETA;                 CHECK FOR ANOTHER
35   ;                                                 OPERATION
36          ACT;                                       MILL WORK COMPLETE
37          COLCT,INT(3),TIME IN SYSTEM;               COLLECT TIME FOR
38   ;                                                 MILL OPERATIONS
39          TERM;
40          END;
41   INIT,0,4800;
42   SIMULATE;
43   INTLC,XX(1)=6,XX(2)=2,XX(3)=2,XX(4)=0;
44   SIMULATE;
45   INTLC,XX(1)=6,XX(2)=1,XX(3)=3,XX(4)=0;
46   SIMULATE;
47   INTLC,XX(1)=5,XX(2)=1,XX(3)=3,XX(4)=1;
48   SIMULATE
49   INTLC,XX(1)=6,XX(2)=1,XX(3)=2,XX(4)=1;
50   SIMULATE;
51   INTLC,XX(1)=5,XX(2)=2,XX(3)=2,XX(4)=1;
52   SIMULATE;
53   INTLC,XX(1)=5,XX(2)=1,XX(3)=2,XX(4)=2;
54   FIN;
```

D6-5.1

Number of Mills	Castings Produced
5,2,3,0	192
6,1,3,0	115
5,1,3,1	157
6,1,2,1	169
5,2,2,1	164
5,1,2,2	193

D6-5.2

```
 1   GEN,FLOSS,PROBLEM 6.6,6/16/86;
 2   LIMITS,4,3,500;
 3   ARRAY(1,3)/120,40,56;
 4   EQUIVALENCE/ATRIB(1),OPERATION/
 5             ATRIB(2),MILL/
 6             ARRAY(1,OPERATION),PROCESS_TIME;
 7   NETWORK;
 8         RESOURCE/1,MILL1(5),1/2,MILL2(1),2;          DEFINE 10 MILLS
 9         RESOURCE/3,MILL3(2),3/4,MILLF(2),4;          AS 4 RESOURCE TYPES
10         CREATE,22,,3;
11   ;
12   SETA  ASSIGN,OPERATION = OPERATION + 1,
13             II=OPERATION,1;                          INCREMENT OPERATION
14   ;                                                  NUMBER AND
15   ;                                                  SET II TO PROPOSED MILL
16         ACT,0,NNRSC(II).GT.O.OR.NNRSC(MILLF).EQ.0,SETM; CONDITIONS FOR
17   ;                                                  DEDICATED MILL
18         ACT;
19         ASSIGN,MILL=4;                               ASSIGN FLEXIBLE MILL
20         ACT,,,AMILL;
21   SETM  ASSIGN,MILL = OPERATION;                     ASSIGN DEDICATED MILL
22   AMILL AWAIT(MILL= 1,4),MILL/1,,1;                  WAIT FOR MILL
23         ACT,PROCESS_TIME*1.1,MILL.EQ.4,FREE;         FLEX MILL PROC ACTIVITY
24         ACT,PROCESS_TIME;                            MILL PROCESSING ACTIVITY
25   FREE  FREE,MILL/1,1;                               FREE MILL
26         ACT,0,OPERATION.LT.3,SETA;                   CHECK FOR ANOTHER
27   ;                                                  OPERATION
28         ACT;                                         MILL WORK COMPLETE
29         COLCT,INT(3),TIME IN SYSTEM;                 COLLECT TIME FOR
30   ;                                                  MILL OPERATIONS
31         TERM;
32         END;
33   INIT,0,2400;
34   FIN;
```

D6-6.1

```
 1   GEN,FLOSS,PROBLEM 6.6A,6/16/86;
 2   LIMITS,4,3,500;
 3   ARRAY(1,3)/0,5,7;
 4   ARRAY(2,3)/10,0,8;
 5   ARRAY(3,3)/8,10,0;
 6   ARRAY(4,3)/120,40,56;
 7   EQUIVALENCE/ATRIB(1),OPERATION/
 8             ATRIB(2),MILL/
 9             XX(1),PREV_OPER/
10             ARRAY(PREV_OPER,OPERATION),SETUP_TIME/
11             ARRAY(4,OPERATION),PROCESS_TIME;
12   INTLC,PREV_OPER=1;
13   NETWORK;
14         RESOURCE/1,MILL1(5),1/2,MILL2(1),2;          DEFINE 10 MILLS
15         RESOURCE/3,MILL3(2),3/4,MILLF(2),4;          AS 4 RESOURCE TYPES
16         CREATE,22,,3;
17   ;
18   SETA  ASSIGN,OPERATION = OPERATION + 1,
19             II=OPERATION,1;                          INCREMENT OPERATION
20   ;                                                  NUMBER AND
21   ;                                                  SET II TO PROPOSED MILL
22         ACT,0,NNRSC(II).GT.O.OR.NNRSC(MILLF).EQ.0,SETM; CONDITIONS FOR
23   ;                                                  DEDICATED MILL
24         ACT;
25         ASSIGN,MILL=4;                               ASSIGN FLEXIBLE MILL
26         ACT,,,AMILL;
27   SETM  ASSIGN,MILL = OPERATION;                     ASSIGN DEDICATED MILL
28   AMILL AWAIT(MILL= 1,4),MILL/1,,1;                  WAIT FOR MILL
29         ACT,PROCESS_TIME+SETUP_TIME,MILL.EQ.4,APRV;  FLEX MILL PROC ACTIVITY
30         ACT,PROCESS_TIME,,FREE;                      MILL PROCESSING ACTIVITY
31   APRV  ASSIGN,PREV_OPER=OPERATION;                  ASSIGN PREVIOUS OPERATION
32   FREE  FREE,MILL/1,1;                               FREE MILL
33         ACT,0,OPERATION.LT.3,SETA;                   CHECK FOR ANOTHER
34   ;                                                  OPERATION
35         ACT;                                         MILL WORK COMPLETE
36         COLCT,INT(3),TIME IN SYSTEM;                 COLLECT TIME FOR
37   ;                                                  MILL OPERATIONS
38         TERM;
39         END;
40   INIT,0,2400;
41   FIN;
```

D6-6.2

```
 1    GEN,OREILLY,PROBLEM 6.7,1/30/85,1;
 2    LIM,2,3,200;
 3    TIMST,XX(1),NUMBER IN SYSTEM;
 4    NETWORK;
 5         RESOURCE/P80(3),1/P140(2),1;
 6         RESOURCE/SPACE(0),2;
 7         CREATE,1.,,1;
 8         ASSIGN,XX(1)=XX(1)+2,2;
 9           ACT,,NNRSC(P80).GT.0.AND.XX(1).GE.80,P1F;
10           ACT,,NNRSC(P140).GT.0.AND.XX(1).GE.140,P2F;
11           ACT;
12         AWAIT(2),SPACE/2;
13         TERM;
14    ;
15    P1F   ASSIGN,XX(1)=XX(1)-80,ATRIB(2)=1,ATRIB(3)=80;
16           ACT,,,WAIT;
17    P2F   ASSIGN,XX(1)=XX(1)-140,ATRIB(2)=2,ATRIB(3)=140;
18    WAIT  AWAIT(1),ATRIB(2);
19         ALTER,SPACE/ATRIB(3);
20         ASSIGN,ATRIB(1)=RNORM(180,60),1;
21           ACT,120,ATRIB(1).LT.120;
22           ACT,240,ATRIB(1).GT.240;
23           ACT,ATRIB(1);
24         FREE,ATRIB(2);
25           ACT,,XX(1).GE.80.AND.ATRIB(2).EQ.1,P1F;
26           ACT,,XX(1).GE.140.AND.ATRIB(2).EQ.2,P2F;
27         END;
28    INIT,0,6000;
29    FIN;
```

D6-7.1

7	Time between creations is 1
8	Increase the number of units waiting by 2
9	Branch to node P1F if a P80 plane is available and there are 80 or more unit loads waiting
10	Branch to node P2F if a P140 plane is waiting and there are 140 or more unit loads waiting
12	Delay unit loads in an AWAIT file in order to obtain waiting time statistics
15	Decrease unit loads waiting by 80 as a P80 plane is allocated and set ATRIB(2) to the P80 aircraft type
16	Branch to node WAIT
17	Decrease unit loads waiting by 140 as a P140 plane is allocated and set ATRIB(2) to the P140 aircraft type
18	Seize a plane
19	Alter the SPACE resource to allow loads to exit the first AWAIT node
20	Assign a sample of the flying time for the plane to ATRIB(1)
21	Set the minimum flying time to 120
22	Set the maximum flying time to 240
23	Let the flying time be ATRIB(1) if within the proper range
24	Free 1 unit of the resource defined in ATRIB(2) after the flying time
25	If 80 unit loads are waiting and a P80 plane has just been freed, branch to P1F so that the plane is seized again
26	If 140 unit loads are waiting and a P140 plane has justbeen freed, branch to P2F.

If neither condition is true, the plane will remain idle until enough unit loads are accumulated for allocation. This model is interesting as several SLAM II constructs are being used to provide variables on which conditions are based. This is the case for activity 1 and the two resources. Another interesting feature is that unit loads are being created but they are not the entities flowing through the system. The entities flowing through the system are planefuls of unit loads.

6-7,Embellishment. Adding a one minute per unit loading time can be accomplished by inserting 80 minute and 140 minute time activities after the planes are seized and again before they are freed up. This includes both loading and unloading times. This modification causes a large change in the statistics collected for this model.

6-7,Alternative Approach. An interesting way to view this problem is to view the cargo as the resource and the planes as the entities. This view is depicted in the statement model of D6-7.2. As unit loads arrive they alter the capacity of the cargo resource by 2. They then are delayed until cargo capacity is enough to allow a plane to leave the AWAIT node labeled WAIT. Three P80 aircraft are created at time zero, assigned a capacity of 80 loads in ATRIB(2), and routed to the AWAIT node to seize ATRIB(2) units of the CARGO resource. Two P140 aircraft are created at time zero and ATRIB(2) is assigned a value of 140. Units of seized cargo are never freed; the capacity reported at the end of the simulation will be the number of unit loads generated, and the average number of units available will be the average number of unit loads waiting for a plane.

```
 1    GEN,OREILLY,PROBLEM 6.7 ALT,1/30/85,1;
 2    LIM,2,3,150;
 3    PRIORITY/1,LVF(2);
 4    NETWORK;
 5         RESOURCE/CARGO(0),1;
 6         RESOURCE/SPACE(0),2;
 7         CREATE,1.,0,1;
 8         ALTER,CARGO/2;
 9         AWAIT(2),SPACE/2;
10         TERM;
11    ;
12         CREATE,0,,,3;
13         ASSIGN,ATRIB(2)=80;
14         ACT,,,WAIT;
15    ;
16         CREATE,0,,,2;
17         ASSIGN,ATRIB(2)=140;
18 WAIT  AWAIT(1),CARGO/ATRIB(2);
19         ALTER,SPACE/ATRIB(2);
20         ASSIGN,ATRIB(1)=RNORM(180,60),1;
21            ACT,120,ATRIB(1).LT.120,WAIT;
22            ACT,240,ATRIB(1).GT.240,WAIT;
23            ACT,ATRIB(1),,WAIT;
24         END;
25    INIT,0,6000;
26    FIN;
```

D6-7.2

6-8.   To change the model so that the repair time of the machine tool is not counted as utilization time for the machine tool, the capacity of the machine tool for its normal operation should be changed when it fails. This is accomplished by decreasing the number of tools available for processing parts. The model for this situation is shown in D6-8.1. When the tool fails, it is preempted at line 15. At that time the capacity of the tool is decreased by 1 (line 16) using an ALTER node. The tool is then freed which causes the capacity change to be satisfied and the number of tools in use is changed to 0. After the repair time, a second ALTER node is used to increase the capacity of the tool which will cause the tool to be reassigned if a part is waiting to be processed. The utilization of the tool will then be calculated as the total time that it is processing parts divided by the amount of time that it is available to process parts. The output from the five runs produce statistics as recorded in Table 6-1 of the textbook.

```
 1  GEN,FLOSS,PROBLEM 6.8,6/11/86,5,,,,NO;
 2  LIMITS,2,1,50;
 3  NETWORK;
 4       RESOURCE/TOOL(1),2,1;                      DEFINE RESOURCE
 5       CREATE,EXPON(1.),0,1;                      CREATE ARRIVALS
 6       AWAIT(1),TOOL/1;                           AWAIT THE TOOL
 7       ACT/1,UNFRM(.2,.5);                        SET UP
 8       GOON,1;                                    CONTINUE
 9       ACT/2,RNORM(.5,.1);                        PROCESSING
10       FREE,TOOL/1;                               FREE THE TOOL
11       COLCT,INT(1),TIME IN SYSTEM;               COLLECT STATISTICS
12       TERM;                                      DEPART SYSTEM
13  ;
14       CREATE,,20,,1;                             CREATE 1ST BREAKDOWN
15  DOWN PREEMPT(2),TOOL;                           PREEMPT THE TOOL
16       ALTER,TOOL/-1;                             DECREASE TOOLS AVAIL
17       FREE,TOOL/1;                               RELEASE TOOL
18       ACT/3,ERLNG(.75,3.);                       DOWN TIME
19       ALTER,TOOL/1;                              INCREASE TOOLS AVAIL
20       ACT,RNORM(20.,2.),,DOWN;                   TIME BETWEEN FAILURES
21       END;
22  INIT,0,500;
23  FIN;
```

D6-8.1

6-9.   The SLAM II network and statement model are shown in D6- 9.1 and D6-9.2. There are no special features associated with this model. Comments are included to make it readable. This model is a good one to embellish. (See Exercises 9-9 and 10-8.) The summary report for this example is given in D6-9.3.

6-9,Embellishment(a). This embellishment involves preventative maintenance which is performed periodically at a given time but only starts when the machine is free. The schedule for preventative maintenance is initiated using a CREATE node (line 20) which is scheduled for the first time at 4 hours (240 minutes) and every 8 hours thereafter. In D6-9.4, the ALTER node either reduces the capacity of the machine by 1 or requests that the machine capacity be reduced by 1 when it is freed. The entity arriving to the ALTER node continues in the network and activity 2 starts immediately independent of the status of the machine. Thirty minutes later the entity arrives to a second ALTER node which increases the capacity of the machine by 1. Thus, the preventative maintenance time is reduced so that the machine always returns to service at the end of the scheduled maintenance period. In this model, some preventative maintenance will always be performed because the processing time on the machine is uniformly distributed between 15 and 20 minutes. If the processing time were longer, then it would have been possible for preventative maintenance not to have been performed at all, that is, a preventative maintenance cycle could be missed. The summary report for this situation is shown in D6-9.5.

6-9,Embellishment(b). The network model to include the scrapping of a part if it fails inspection more than once is shown in D6-9.6 and the statement model in D6-9.7. The SLAM II summary report is shown in D6-9.8.

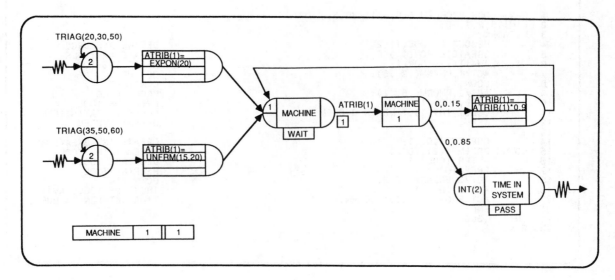

D6-9.1

```
 1    GEN,OREILLY,PROBLEM 6.9,6/11/86,1;
 2    LIMITS,1,2,100;
 3    NETWORK;
 4          RESOURCE/MACHINE(1),1;              DEFINE RESOURCE
 5          CREATE,TRIAG(20,30,50),,2;          CREATE TYPE 1 PARTS
 6          ASSIGN,ATRIB(1)=EXPON(20);          ASSIGN PROCESS TIME
 7          ACTIVITY,,WAIT;                     ROUTE TO MACHINE TOOL
 8          CREATE,TRIAG(35,50,60),,2;          CREATE TYPE 2 PARTS
 9          ASSIGN,ATRIB(1)=UNFRM(15,20);       ASSIGN PROCESS TIME
10    WAIT  AWAIT(1),MACHINE;                   WAIT FOR MACHINE
11          ACTIVITY/1,ATRIB(1);               PROCESS PART
12          FREE,MACHINE,1;                     RELEASE MACHINE
13          ACTIVITY,,0.85,PASS;                IF PASS INSP ROUTE TO PASS
14          ACTIVITY,,0.15;                     ELSE CONTINUE IF FAIL INSP
15          ASSIGN,ATRIB(1)=ATRIB(1)*0.9;       ASSIGN REWORK TIME
16          ACTIVITY,,,WAIT;                    ROUTE TO MACHINE TOOL
17    PASS  COLCT,INT(2),TIME IN SYSTEM;        COLLECT TIME IN SYS STATISTICS
18          TERMINATE;                          DEPART FROM SYSTEM
19          END;
20    INIT,0,2400;
21    FIN;
```

D6-9.2

```
              S L A M   I I   S U M M A R Y   R E P O R T

      SIMULATION PROJECT PROBLEM 6.9          BY OREILLY

      DATE  6/11/1986                         RUN NUMBER   1 OF   1


      CURRENT TIME   0.2400E+04
      STATISTICAL ARRAYS CLEARED AT TIME  0.0000E+00


              **STATISTICS FOR VARIABLES BASED ON OBSERVATION**

                MEAN       STANDARD      COEFF. OF    MINIMUM     MAXIMUM     NUMBER OF
                VALUE      DEVIATION     VARIATION    VALUE       VALUE       OBSERVATIONS

TIME IN SYSTEM  0.1033E+03  0.9424E+02   0.9127E+00  0.1277E+01  0.3442E+03    103


                          **FILE STATISTICS**

FILE                     AVERAGE      STANDARD     MAXIMUM   CURRENT    AVERAGE
NUMBER  LABEL/TYPE       LENGTH       DEVIATION    LENGTH    LENGTH     WAITING TIME

  1     WAIT AWAIT        4.8844       4.8685        17        17        84.3345
  2          CALENDAR     2.9514       0.2150         4         3        10.1482


                     **REGULAR ACTIVITY STATISTICS**

ACTIVITY           AVERAGE      STANDARD     MAXIMUM CURRENT   ENTITY
INDEX/LABEL        UTILIZATION  DEVIATION    UTIL    UTIL      COUNT

  1 PROCESS PART    0.9514       0.2150        1      .1        121


                        **RESOURCE STATISTICS**

RESOURCE  RESOURCE  CURRENT   AVERAGE       STANDARD    MAXIMUM      CURRENT
NUMBER    LABEL     CAPACITY  UTILIZATION   DEVIATION   UTILIZATION  UTILIZATION

  1       MACHINE      1        0.9514        0.2150        1            1


RESOURCE  RESOURCE  CURRENT    AVERAGE      MINIMUM     MAXIMUM
NUMBER    LABEL     AVAILABLE  AVAILABLE    AVAILABLE   AVAILABLE

  1       MACHINE      0        0.0486         0           1
```

D6-9.3

```
 1  GEN,OREILLY,PROBLEM 6.9A,6/26/86,1;
 2  LIMITS,1,2,100;
 3  NETWORK;
 4        RESOURCE/MACHINE(1),1;              DEFINE RESOURCE
 5        CREATE,TRIAG(20,30,50),,2;          CREATE TYPE 1 PARTS
 6        ASSIGN,ATRIB(1)=EXPON(20);          ASSIGN PROCESS TIME
 7        ACTIVITY,,WAIT;                     ROUTE TO MACHINE TOOL
 8        CREATE,TRIAG(35,50,60),,2;          CREATE TYPE 2 PARTS
 9        ASSIGN,ATRIB(1)=UNFRM(15,20);       ASSIGN PROCESS TIME
10  WAIT  AWAIT(1),MACHINE;                   WAIT FOR MACHINE
11        ACTIVITY/1,ATRIB(1);               PROCESS PART
12        FREE,MACHINE,1;                     RELEASE MACHINE
13        ACTIVITY,,0.85,PASS;                IF PASS INSP ROUTE TO PASS
14        ACTIVITY,,0.15;                     ELSE CONTINUE IF FAIL INSP
15        ASSIGN,ATRIB(1)=ATRIB(1)*0.9;       ASSIGN REWORK TIME
16        ACTIVITY,,,WAIT;                    ROUTE TO MACHINE TOOL
17  PASS  COLCT,INT(2),TIME IN SYSTEM;        COLLECT TIME IN SYS STATISTICS
18        TERMINATE;                          DEPART FROM SYSTEM
19  ;
20        CREATE,480,240;                     CREATE MAINTENANCE DOWNTIME
21        ALTER,MACHINE,-1;                   SEIZE MACHINE
22        ACTIVITY/2,30;                      PERFORM MAINTENANCE
23        ALTER,MACHINE,1;                    RELEASE MACHINE
24        TERMINATE;                          DEPART FROM SYSTEM
25        END;
26  INIT,0,2400;
27  FIN;
```

D6-9.4

```
                         S L A M   I I   S U M M A R Y   R E P O R T

                 SIMULATION PROJECT PROBLEM 6.9A            BY OREILLY

                 DATE  6/26/1986                            RUN NUMBER   1 OF    1


                 CURRENT TIME   0.2400E+04
                 STATISTICAL ARRAYS CLEARED AT TIME  0.0000E+00


                    **STATISTICS FOR VARIABLES BASED ON OBSERVATION**

                      MEAN         STANDARD      COEFF. OF     MINIMUM       MAXIMUM       NUMBER OF
                      VALUE        DEVIATION     VARIATION     VALUE         VALUE         OBSERVATIONS

   TIME IN SYSTEM    0.1210E+03   0.1130E+03    0.9338E+00    0.1277E+01    0.4169E+03        101


                            **FILE STATISTICS**

   FILE                         AVERAGE      STANDARD     MAXIMUM    CURRENT    AVERAGE
   NUMBER  LABEL/TYPE           LENGTH       DEVIATION    LENGTH     LENGTH     WAITING TIME

     1    WAIT AWAIT            5.8495       5.6703         20        19       101.7307
     2          CALENDAR       3.9836       0.2071          6         4        13.5421


                         **REGULAR ACTIVITY STATISTICS**

   ACTIVITY        AVERAGE       STANDARD      MAXIMUM CURRENT   ENTITY
   INDEX/LABEL     UTILIZATION   DEVIATION     UTIL    UTIL      COUNT

     1 PROCESS PART   0.9211       0.2695         1      1        118
     2 PERFORM MAIN   0.0625       0.2421         1      0          5


                            **RESOURCE STATISTICS**

   RESOURCE  RESOURCE   CURRENT    AVERAGE       STANDARD     MAXIMUM       CURRENT
   NUMBER    LABEL      CAPACITY   UTILIZATION   DEVIATION    UTILIZATION   UTILIZATION

      1      MACHINE       1        0.9211        0.2695          1             1


   RESOURCE  RESOURCE   CURRENT    AVERAGE      MINIMUM      MAXIMUM
   NUMBER    LABEL      AVAILABLE  AVAILABLE    AVAILABLE    AVAILABLE

      1      MACHINE       0        0.0164         -1            1
```

D6-9.5

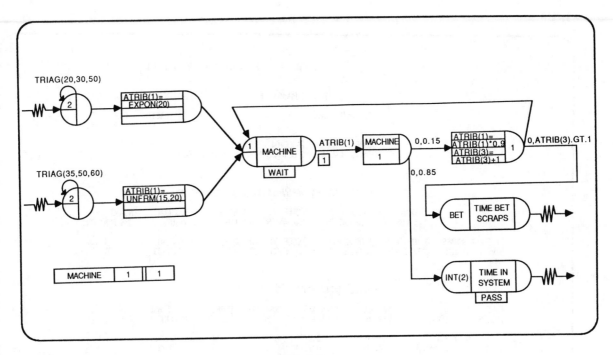

D6-9.6

```
 1   GEN,OREILLY,PROBLEM 6.9B,6/11/86,1;
 2   LIMITS,1,3,100;
 3   NETWORK;
 4        RESOURCE/MACHINE(1),1;              DEFINE RESOURCE
 5        CREATE,TRIAG(20,30,50),,2;          CREATE TYPE 1 PARTS
 6        ASSIGN,ATRIB(1)=EXPON(20);          ASSIGN PROCESS TIME
 7        ACTIVITY,,,WAIT;                    ROUTE TO MACHINE TOOL
 8        CREATE,TRIAG(35,50,60),,2;          CREATE TYPE 2 PARTS
 9        ASSIGN,ATRIB(1)=UNFRM(15,20);       ASSIGN PROCESS TIME
10   WAIT AWAIT(1),MACHINE;                   WAIT FOR MACHINE
11        ACTIVITY/1,ATRIB(1);               PROCESS PART
12        FREE,MACHINE,1;                    RELEASE MACHINE
13        ACTIVITY,,0.85,PASS;               IF PASS INSP ROUTE TO PASS
14        ACTIVITY,,0.15;                    ELSE CONTINUE IF FAIL INSP
15        ASSIGN,ATRIB(1)=ATRIB(1)*0.9,
16            ATRIB(3)=ATRIB(3)+1,1;         ASSIGN RWRK TIME AND # OF RWRKS
17        ACTIVITY,,ATRIB(3).GT.1;           IF FAILED TWICE ROUTE TO SCRAP
18        ACTIVITY,,,WAIT;                   ELSE ROUTE TO MACHINE TOOL
19        COLCT,BET,TIME BET SCRAPS;         COLLECT TIME BETWEEN SCRAP STAT
20        TERMINATE;                         DEPART FROM SYSTEM
21   PASS COLCT,INT(2),TIME IN SYSTEM;       COLLECT TIME IN SYS STATISTICS
22        TERMINATE;                         DEPART FROM SYSTEM
23        END;
24   INIT,0,2400;
25   FIN;
```

D6-9.7

```
                          S L A M   I I   S U M M A R Y   R E P O R T

              SIMULATION PROJECT PROBLEM 6.9B              BY OREILLY

              DATE  6/11/1986                              RUN NUMBER   1 OF   1


              CURRENT TIME   0.2400E+04
              STATISTICAL ARRAYS CLEARED AT TIME  0.0000E+00


                       **STATISTICS FOR VARIABLES BASED ON OBSERVATION**

                     MEAN        STANDARD     COEFF. OF    MINIMUM      MAXIMUM     NUMBER OF
                     VALUE       DEVIATION    VARIATION    VALUE        VALUE       OBSERVATIONS

      TIME BET SCRAPS  0.1099E+03  0.0000E+00   0.0000E+00   0.1099E+03   0.1099E+03       1
      TIME IN SYSTEM   0.1136E+03  0.1131E+03   0.9956E+00   0.1277E+01   0.3892E+03     101


                                   **FILE STATISTICS**

      FILE                  AVERAGE     STANDARD    MAXIMUM   CURRENT    AVERAGE
      NUMBER  LABEL/TYPE    LENGTH      DEVIATION   LENGTH    LENGTH     WAITING TIME

        1     WAIT AWAIT    5.2209      5.4721        19        17       92.1327
        2           CALENDAR 2.9478     0.2224         4         3       10.3281


                              **REGULAR ACTIVITY STATISTICS**

      ACTIVITY         AVERAGE      STANDARD    MAXIMUM CURRENT   ENTITY
      INDEX/LABEL      UTILIZATION  DEVIATION   UTIL    UTIL      COUNT

        1 PROCESS PART    0.9478      0.2224       1       1       118


                                 **RESOURCE STATISTICS**

      RESOURCE  RESOURCE  CURRENT   AVERAGE       STANDARD    MAXIMUM      CURRENT
      NUMBER    LABEL     CAPACITY  UTILIZATION   DEVIATION   UTILIZATION  UTILIZATION

        1       MACHINE      1        0.9478        0.2224        1            1


      RESOURCE  RESOURCE  CURRENT    AVERAGE      MINIMUM     MAXIMUM
      NUMBER    LABEL     AVAILABLE  AVAILABLE    AVAILABLE   AVAILABLE

        1       MACHINE      0         0.0522        0           1
```

D6-9.8

CHAPTER 7

Logic and Decision Nodes

7-1(a). An assembly operation is required which would be modeled by a SELECT node. Three queues would precede the SELECT node for storing steak, potatoes, and salad entities. When there is an entity in each queue, the assembly queue selection (ASM) rule combines them into a dinner entity.

7-1(b). There are several ways to delay the introduction of the steak entity into the oven until it is warmed up. For some problems, it is sufficient to delay the steak entity by 5 time units through the use of an activity. Another approach is to use an ACCUMULATE node which has incident to it the 5 minute delay activity and the activity associated with the steak entity. A first release requirement of 2 would cause the steak entity to wait. In this case, it would be possible to specify a subsequent number of releases for the ACCUMULATE node as 1 assuming that the oven has been warmed up. A third possiblity is to describe the oven as a resource which initially has a capacity of 0 until after it is warmed up.

7-1(c). An ACCUMULATE node can be used which requires that 4 people entities have arrived to it after completing the eating activity. When this occurs, the serve dessert activity would be allowed to start.

7-1(d). Attributes of entities would be employed to characterize the dinner that was ordered.

7-1(e). A gate could be used to prohibit the start of a dinner activity until the gate is opened by the pouring of wine for all diners.

7-2. The SLAM II network for this exercise is shown in D7-2.1. It is also assumed that the customer keeps driving around the block until a space is available in the queue. The time in queue statistic is an automatic output from the SLAM II program. SLAM II statements for this problem are given in D7-2.2. Note that concepts from Chapter 7 are not required to do this exercise or its embellishment.

7-2,Embellishment. The network segment and statement model for this embellishment are shown in D7-2.3 and D7-2.4 where a second queue is added which blocks the bank teller when five cars are waiting to enter the street. The time to service the cars waiting to go into the street is modeled by the time between gaps in the street traffic. Note that one car is in the "gap" activity and that four are allowed in the queue to provide a capacity of five cars that can wait before blocking the teller.

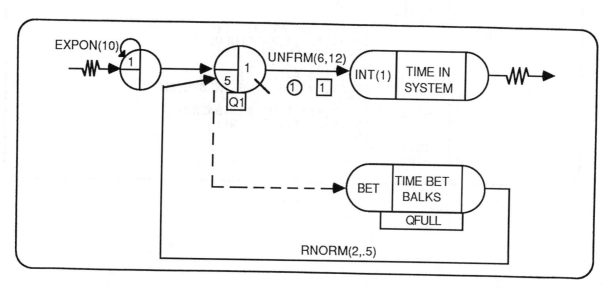

D7-2.1

```
1    GEN,ROLSTON,PROBLEM 7.2,7/22/80,1;
2    LIMITS,1,1,25;
3    NETWORK;
4              CREATE,EXPON(10),,1;
5    Q1        QUEUE(1),,5,BALK(QFULL);
6              ACT(1)/1,UNFRM(6,12);
7              COLCT,INT(1),TIME IN SYSTEM;
8              TERM;
9    QFULL     COLCT,BET,TIME BET BALKS;
10             ACT,RNORM(2,.5),,Q1;
11             END;
12   INIT,0,200;
13   FIN;
```

D7-2.2

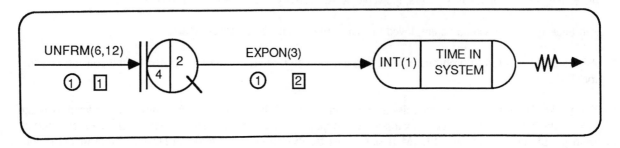

D7-2.3

```
1    GEN,ROLSTON,PROBLEM 7.2E,7/22/80,1;
2    LIMITS,2,1,50;
3    NETWORK;
4              CREATE,EXPON(10),,1;
5    Q1        QUEUE(1),,5,BALK(QFULL);
6              ACT(1)/1,UNFRM(6,12);
7              QUEUE(2),,4,BLOCK;
8              ACT(1)/2,EXPON(3);
9              COLCT,INT(1),TIME IN SYSTEM;
10             TERM;
11   QFULL     COLCT,BET,TIME BET BALKS;
12             ACT,RNORM(2,.5),,Q1;
13             END;
14   INIT,0,200;
15   FIN;
```

D7-2.4

7-3.   The network model of the two servers and two conveyors is shown in D7-3.1. Equivalence statements are used so that the attributes of the items on the conveyor are server number, SVR, the conveyor delay time, DLAY, and the service time, ST. Arriving items are routed to an ASSIGN node where the attributes of the item are prescribed. The item is then routed to a GOON node and placed in activity 3 which represents the first conveyor. The conveyor delivers the item to a QUEUE node which has zero capacity. At the queue the server number attribute SVR is used to determine if the server is idle or not. If the server is idle then the item proceeds into service with a time of ST. If the server requested is not available, the item balks to ASSIGN node B1 where DLAY is set to 12 and if the server number desired is 1, it returns to node G1 to continue on the first conveyor. If the server number is 2, then the item goes to GOON G2 for continuation on the second conveyor. Following service the item is routed to a GOON node where if service was just provided by server number 2, the item is routed to a collect node OUT where time in the system statistics are collected and the item is terminated. If the server number is 1 following the completion of service, the item is sent to an ASSIGN node where the server number is changed to 2, the delay is set equal to 5, and the service time obtained as a sample from the normal distribution is assigned. The item is then routed to node G2 where it traverses activity 4 which represents conveyor 2. Conveyor 2 takes the item back to QUEUE node Q1 which represents the queues before both servers. To collect statistics on the number of items on each conveyor, TIMST statements are used with variable NNACT(3) for collection of the number on conveyor 1 and NNACT(4) for collecting statistics on the number of items on conveyor 2. The statement model corresponding to the above network model is given in D7-3.2.

7-3,Embellishment(a). The network model and SLAM II input statements are shown in D7-3.3 and D7-3.4.

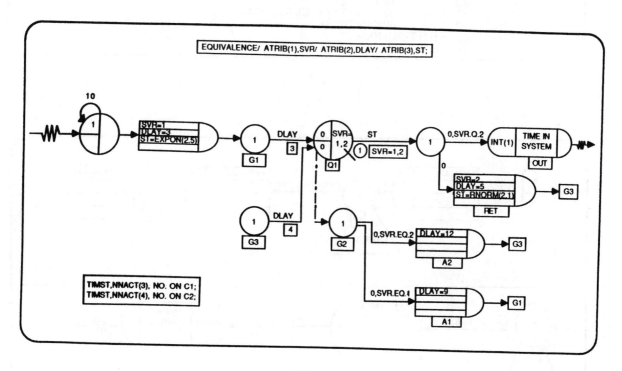

D7-3.1

```
 1    GEN,PRITSKER,PROBLEM 7.3,8/19/86;
 2    LIMITS,2,3,100;
 3    TIMST,NNACT(3),NO. ON C1;
 4    TIMST,NNACT(4),NO. ON C2;
 5    EQUIVALENCE/ATRIB(1),SVR/
 6                  ATRIB(2),DLAY/
 7                  ATRIB(3),ST;
 8    NETWORK;
 9          CREATE,10,,1;
10          ASSIGN,SVR=1,DLAY=3,ST=EXPON(2.5);
11    G1    GOON,1;
12          ACTIVITY/3,DLAY;
13    Q1    QUEUE(SVR=1,2),,0,BALK(G2);
14          ACTIVITY(1)/SVR=1,2,ST;
15          GOON,1;
16          ACTIVITY,,SVR.EQ.2,OUT;
17          ACTIVITY,,,RET;
18    OUT   COLCT,INT(1),TIME IN SYSTEM;
19          TERMINATE;
20    ;
21    RET   ASSIGN,SVR=2,DLAY=5,ST=RNORM(2,1);
22          ACTIVITY,,,G3;
23    ;
24    G2    GOON,1;
25          ACTIVITY,,SVR.EQ.1,A1;
26          ACTIVITY,,SVR.EQ.2,A2;
27    A1    ASSIGN,DLAY=9;
28          ACTIVITY,,,G1;
29    A2    ASSIGN,DLAY=12;
30    G3    GOON,1;
31          ACTIVITY/4,DLAY,,Q1;
32          ENDNETWORK;
33    INIT,0,400;
34    FIN;
```

D7-3.2

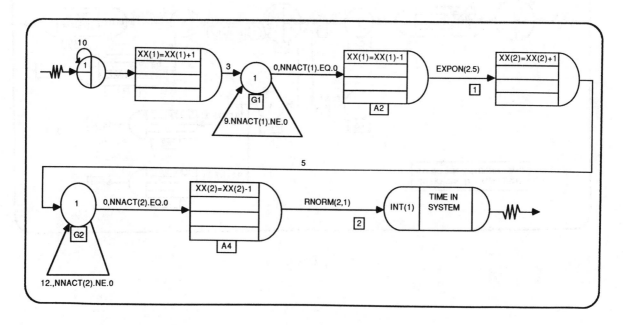

D7-3.3

```
 1   GEN,ROLSTON,PROBLEM 7.3A,7/22/80,1;
 2   LIMITS,0,1,50;
 3   TIMST,XX(1),CONVEYOR 1;
 4   TIMST,XX(2),CONVEYOR 2;
 5   NETWORK;
 6           CREATE,10,,1;
 7           ASSIGN,XX(1)=XX(1)+1;
 8           ACT,3;
 9   G1      GOON,1;
10           ACT,,NNACT(1).EQ.0,A2;
11           ACT,9,NNACT(1).NE.0,G1;
12   A2      ASSIGN,XX(1)=XX(1)-1;
13           ACT/1,EXPON(2.5);
14           ASSIGN,XX(2)=XX(2)+1;
15           ACT,5;
16   G2      GOON,1;
17           ACT,,NNACT(2).EQ.0,A4;
18           ACT,12,NNACT(2).NE.0,G2;
19   A4      ASSIGN,XX(2)=XX(2)-1;
20           ACT/2,RNORM(2,1);
21           COLCT,INT(1),TIME IN SYSTEM;
22           TERM;
23           END;
24   INIT,0,400;
25   INTLC,XX(1)=0,XX(2)=0;
26   FIN;
```

D7-3.4

7-4.   See solution to Exercise 9-7. By removing the resource requirements, the model is an example of a SLAM II network of a PERT model.

7-5(a). To model the situation in which all preceding activities must be completed before successor activities can be started, an ACCUMULATE node can be used with the release requirement equal to the number of preceding activities. This solution only holds if the modeler is sure that each of the preceding activities will be completed prior to the second completion of any of the activities. If this is not the case, a SELECT node with the assembly selection rule is used and each activity completion must be stored as an entity in a queue preceding the SELECT node.

7-5(b). If any one of the preceding activities can cause an activity to be started, a GOON node is used. The GOON node is a special case of the ACCUMULATE node with first and subsequent release requirements of 1.

7-5(c). To require 3 out of 5 of the preceding activities to be completed before a succeeding activity can be started, an ACCUMULATE node with a first and subsequent release requirement of 3 is used. The conditions described in Exercise 7-5(a) would also have to be satisfied for this situation.

7-6.   The SLAM II input statement and network models for this situation are shown in D7-6.1 and D7-6.2. The reciprocal of the time between arrivals to COLCT node LOST is the average number of customers per minute who cannot gain service from the ticket agents. The data values used in this example result in no queueing and very little utilization of the servers. It might be appropriate to ask the students to describe qualitatively the statistical characteristics of the system operation.

```
 1    GEN,ROLSTON,PROBLEM 7.6,7/22/80,1;
 2   LIMITS,5,1,100;
 3   NETWORK;
 4              CREATE,EXPON(5),,1;
 5              SELECT,SNQ,,BALK(LOST),Q1,Q2,Q3,Q4,Q5;
 6    Q1        QUEUE(1),,2;
 7              ACT/1,UNFRM(.5,1.5),,EXIT;
 8    Q2        QUEUE(2),,2;
 9              ACT/2,UNFRM(.5,1.5),,EXIT;
10    Q3        QUEUE(3),,2;
11              ACT/3,UNFRM(.5,1.5),,EXIT;
12    Q4        QUEUE(4),,2;
13              ACT/4,UNFRM(.5,1.5),,EXIT
14    Q5        QUEUE(5),,2;
15              ACT/5,UNFRM(.5,1.5);
16    EXIT      COLCT,INT(1),TIME IN SYSTEM;
17              TERM;
18    LOST      COLCT,BET,TIME BET BALKS;
19              TERM;
20              END;
21   INIT,0,300;
22   FIN;
```

D7-6.1

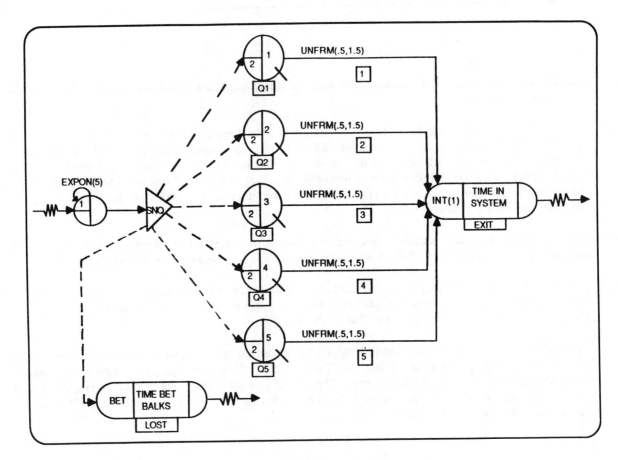

D7-6.2

7-6,Embellishment(a). The SLAM II model and the input statements for this embellishment are shown in D7-6.3 and D7-6.4. A simplification of the modeling occurs when a single queue is used with identical servers.

7-6,Embellishment(b). The network model and input statements are shown in D7-6.5 and D7-6.6, and indicate that adding functions to a network model is a straightforward process.

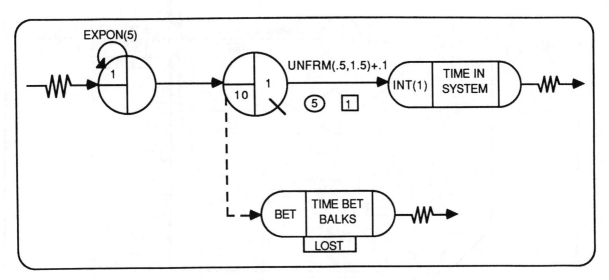

D7-6.3

```
 1      GEN,ROLSTON,PROBLEM 7.6A,7/22/80,1;
 2      LIMITS,1,1,100;
 3      NETWORK;
 4              CREATE,EXPON(5),,1;
 5              QUEUE(1),,10,BALK(LOST);
 6              ACT(5)/1,UNFRM(.5,1.5)+.1;
 7              COLCT,INT(1),TIME IN SYSTEM;
 8              TERM;
 9  LOST    COLCT,BET,TIME BET BALKS;
10              TERM;
11              END;
12      INIT,0,300;
13      FIN;
```

D7-6.4

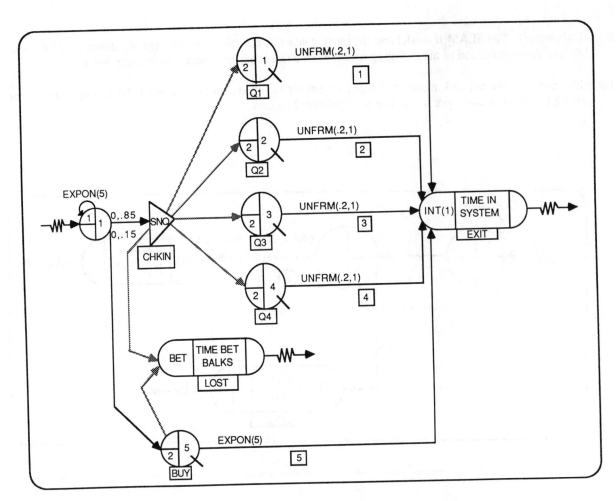

D7-6.5

```
1     GEN,ROLSTON,PROBLEM 7.6B,7/22/80,1;
2     LIMITS,5,1,100;
3     NETWORK;
4               CREATE,EXPON(5),,1,,1;
5               ACT,,.85,CHKIN;
6               ACT,,.15,BUY;
7     CHKIN     SELECT,SNQ,,BALK(LOST),Q1,Q2,Q3,Q4;
8     Q1        QUEUE(1),,2;
9               ACT/1,UNFRM(.2,1),,EXIT;
10    Q2        QUEUE(2),,2;
11              ACT/2,UNFRM(.2,1),,EXIT;
12    Q3        QUEUE(3),,2;
13              ACT/3,UNFRM(.2,1),,EXIT
14    Q4        QUEUE(4),,2;
15              ACT/4,UNFRM(.2,1),,EXIT;
16    BUY       QUEUE(5),,2,BALK(LOST);
17              ACT/5,EXPON(5),,EXIT;
18    EXIT      COLCT,INT(1),TIME IN SYSTEM;
19              TERM;
20    LOST      COLCT,BET,TIME BET BALKS;
21              TERM;
22              END;
23    INIT,0,300;
24    FIN;
```

D7-6.6

7-7.   In this exercise, shirt size and pant size for an order will be established by sampling form a discrete probability mass function. Attribute 6 will be used to maintain the shirt size for shirt portions of the order and pant size for pant's portions of the order. To satisfy the order, a match will be required between the size of a dozen shirts in inventory and the order request.

The revised statement model is shown in D7-7.1. In lines 12 through 15, the probability mass functions for the order sizes for shirts and pants are given. Lines 16 through 19 equivalence SHIRT to XX(1); PANT to XX(2); PREV_SHIRT to XX(3); and PREV_PANT to XX(4). Since there is a setup time penalty when switching sizes, it is necessary to maintain the previous size shirt or pants that was produced. When a new order is generated, the size of each desired commodity is also established. The production rule used in this solution is to continue to make the same size until an order is received with a different size.

At the CREATE node (line 25), an order is generated. The following two activities separate the order into an order for shirts and an order for pants. At ASSIGN node AS1, the shirt order characteristics are defined. If no shirt orders are outstanding, the shirt size SHIRT is set to ATRIB(6) and the entity is routed to an UNBATCH node where entities representing orders for one dozen shirts are placed in QUEUE node Q1 which is followed by a MATCH node ASM1. The MATCH node matches entities representing a dozen shirts with an order for a dozen shirts and when a match is made routes the combined entity to node BAT1. The manufacturing of shirts ends with the placement of an entity representing a dozen shirts into QUEUE node Q2. This is modeled by the statements in lines 41 to 46.

If shirt orders are outstanding, the number of shirt orders waiting to be released of size ATRIB(6) is increased by 1, that is, ARRAY(ATRIB(6),1)= ARRAY(ATRIB(6),1) + 1. The order is then placed in activity 6 which has an indefinite duration specified by STOPA(ATRIB(6)). The model coding for the order and manufacture of pants is similar to the above.

When a match is made, one entity is routed to BATCH node BAT1 which combines entities as was done in the original illustration. A second entity is routed from GOON node NEX1 to represent the production control portion of the model. Activities 6 and 7 hold entities representing orders for shirts and pants respectively. ATRIB(2) of an entity is equal to 1 for shirts and 2 for pants. Thus, by adding 5 to attribute 2 we have the activity number where orders for the appropriate commodity are held. If there are no orders outstanding no branching from NEX1 is performed and production control scheduling is not required. If orders of the same size as were produced are outstanding, that is, ARRAY(ATRIB(6),1).GT.0 then those orders in activity II with this size should be allowed to proceed. This is done by setting STOPA equal to ATRIB(6) and the number of outstanding orders with size ATRIB(6) is set to zero at ASSIGN node STPR. If an order with the same size is not outstanding, return to node NEX1. Note that ATRIB(6) is either 15, 16, or 17 for shirts or 30, 32, 34 or 36 for pants. Since these numbers are disjoint, entities will be released from either activity 6 or 7. The next size number is established as the ATRIB(6) value by adding ATRIB(2). Fortuitously, shirt sizes increase by a value of one and pant sizes by a value of two which is the code used for shirts and pants respectively. The next eight statements provide a means for searching through all the sizes that are prescribed in the problem statement. Since an order is waiting, the search stops when the order with the appropriate size is found and no further branching can be made from the ASSIGN node STPR. This completes the description of the model. An objective for evaluating production schedules would include both the value of in-process buffers for each commodity and the time required to fill an order.

```
 1    GEN,PRITSKER,PROBLEM 7.7,6/30/86,1,,,Y/N;
 2    LIMITS,6,6,350;
 3    ;
 4    ;    DEFINE PROBALISTIC ASSIGNMENT OF NUMBER OF DOZENS IN
 5    ;         EACH ORDER
 6    ;
 7    ARRAY(1,4)/ 1, 2, 3, 4;
 8    ARRAY(2,4)/.6,.8,.95,1.0;
 9    ;
10    ;    DEFINE PROBALISTIC ASSIGNMENT OF SIZE OF SHIRTS AND PANTS
11    ;
12    ARRAY(3,3)/.3,.8,1.0;
13    ARRAY(4,3)/15,16,17;
14    ARRAY(5,4)/.1,.3,.7,1.0;
15    ARRAY(6,4)/30,32,34,36;
16    ARRAY(15,1)/0;
17    ARRAY(16,1)/0;
18    ARRAY(17,1)/0;
19    ARRAY(30,1)/0;
20    ARRAY(32,1)/0;
21    ARRAY(34,1)/0;
22    ARRAY(36,1)/0;
23    EQUIVALENCE/XX(1),SHIRT;
24    EQUIVALENCE/XX(2),PANT;
25    EQUIVALENCE/XX(3),PREV_SHIRT;
26    EQUIVALENCE/XX(4),PREV_PANT;
27    INTLC,PREV_SHIRT=15,PREV_PANT=30;
28    NETWORK;
29    ;
30    ;    CREATE ORDERS
31    ;
32         CREATE,EXPON(30),,1,,2;
33         ACTIVITY,,,AS1;
34         ACTIVITY,,,AS3;
35    ;
36    ;    ORDER FOR SHIRTS
37    ;
38    AS1  ASSIGN,ATRIB(2)=1,ATRIB(3)=DPROBN(2,1),ATRIB(4)=8,
39             ATRIB(5)=60,ATRIB(6)=DPROBN(3,4),1;
40         ACTIVITY,,NNQ(1).EQ.0,AS5;
41         ACTIVITY;
42         ASSIGN,ARRAY(ATRIB(6),1)=ARRAY(ATRIB(6),1)+1;
43         ACTIVITY/6,STOPA(ATRIB(6)),,AS5;          WAIT SHIRT
44    AS5  ASSIGN,SHIRT=ATRIB(6);
45         UNBATCH,3;            SPLIT INTO ENTITIES FOR 1 DOZEN
46    Q1   QUEUE(1),,,,ASM1;
47    ;
48    ;    MANUFACTURE SHIRTS
49    ;
50         CREATE,TRIAG(11.,16.,22.);
51         ASSIGN,ATRIB(6)=SHIRT,1;
52         ACTIVITY/1,7,PREV_SHIRT.EQ.ATRIB(6),AS2;   MANF SHIRT
53         ACTIVITY/2,9;                              CHG SIZE
54    AS2  ASSIGN,PREV_SHIRT=ATRIB(6);
55    Q2   QUEUE(2),,,,ASM1;
56    ;
57    ;    ASSEMBLY OF A DOZEN SHIRTS WITH AN ORDER
58    ;
59    ASM1 MATCH,6,Q1/BAT1,Q2;
60    ;
61    ;    ORDER FOR PANTS
62    ;
63    AS3  ASSIGN,ATRIB(2)=2,ATRIB(3)=DPROBN(2,1),ATRIB(4)=12,
64             ATRIB(5)=102,ATRIB(6)=DPROBN(5,6),1;
65         ACTIVITY,,NNQ(3).EQ.0,AS6;
66         ACTIVITY;
67         ASSIGN,ARRAY(ATRIB(6),1)=ARRAY(ATRIB(6),1)+1;
68         ACTIVITY/7,STOPA(ATRIB(6)),,AS6;          WAIT PANT
69    AS6  ASSIGN,PANT=ATRIB(6);
70         UNBATCH,3;            SPLIT INTO ENTITIES FOR 1 DOZEN
71    Q3   QUEUE(3),,,,ASM2;
72    ;
73    ;    MANUFACTURE PANTS
74    ;
75         CREATE,TRIAG(11.,19.,22.);
76         ASSIGN,ATRIB(6)=PANT,1;
77         ACTIVITY/3,7,PREV_PANT.EQ.ATRIB(6),AS4;   MANF PANTS
78         ACTIVITY/4,9;                             CHG SIZE
79    AS4  ASSIGN,PREV_PANT=ATRIB(6);
80    Q4   QUEUE(4),,,,ASM2;
81    ;
82    ;    ASSEMBLY OF A DOZEN PANTS WITH AN ORDER
83    ;
84    ASM2 MATCH,6,Q3/BAT1,Q4;
85    ;
86    ;    BATCH NO. OF DOZ. REQUIRED FOR AN ORDER
87    ;
88    BAT1 BATCH,2/2,ATRIB(3),,FIRST/4,5,,2;      BATCH SHIRTS AND PANTS SEPARATELY
```

D7-7.1(1)

```
89            ACTIVITY,,ATRIB(2).EQ.1,Q5;
90            ACTIVITY,,ATRIB(2).EQ.2,Q6;
91            ACTIVITY,,NNQ(2.*ATRIB(2)-1.).EQ.0,NEX1;    QUEUE 1 OR 3 EMPTY
92       ;
93       ;    PRODUCTION CONTROL PORTION OF THE MODEL
94       ;
95  NEX1 ASSIGN,II=ATRIB(2)+5;
96            ACTIVITY,,NNACT(II).GT.0;
97            GOON,1;
98            ACTIVITY,,ARRAY(ATRIB(6),1).GT.0,STPR;
99            ACTIVITY,,CON;
100 STPR ASSIGN,ARRAY(ATRIB(6),1)=0,STOPA=ATRIB(6);
101          TERM;
102 CON  ASSIGN,ATRIB(6)=ATRIB(6)+ATRIB(2),1;
103          ACTIVITY,,ATRIB(2).EQ.1.AND.ATRIB(6).LE.17,STPR;
104          ACTIVITY,,ATRIB(2).EQ.2.AND.ATRIB(6).LE.36,STPR;
105          ACTIVITY,,ATRIB(2).EQ.1,RST1;
106          ACTIVITY,,ATRIB(2).EQ.2,RST2;
107 RST1 ASSIGN,ATRIB(6)=15;
108          ACTIVITY,,,STPR;
109 RST2 ASSIGN,ATRIB(6)=30;
110          ACTIVITY,,,STPR;
111      ;
112      ;    MATCH THE SHIRT AND PANTS ENTITIES
113      ;
114 Q5   QUEUE(5),,,,MAT1;
115 Q6   QUEUE(6),,,,MAT1;
116 MAT1 MATCH,1,Q5/GO1,Q6/GO1;
117 GO1  GOON,1;
118          ACTIVITY/5,8;      TRNSPRT TIME
119      ;
120      ;    BATCH SHIRTS AND PANTS INTO A CRATE ENTITY
121      ;
122          BATCH,,100,4,LOW(1)/4,5;
123      ;
124      ;    COLLECT STATISTICS ON THE TIME BETWEEN SHIPMENTS, THE VOLUME
125      ;    IN EACH CRATE, AND THE $ VALUE OF EACH CRATE
126      ;
127          COLCT,BET,TIME BET SHIPMT;
128          COLCT,ATRIB(4),VOL IN CRATE;
129          COLCT,ATRIB(5),$VAL OF CRATE;
130          TERM;
131          END;
132 INIT,0,2400;
133 FIN;
```

D7-7.1(2)

7-8.   Resources can be used in this exercise to maintain the number of shirts and pants in inventory. As a dozen shirts are produced, the number of shirts available is increased by 1 by changing the capacity of the resource using an ALTER node. A similar disjoint network would be used for the pants resource. The processing of an order is then accomplished through a third disjoint network where the order awaits for the number of shirts resource and pants resource as determined by the order quantities. A BATCH node is then used to put the completed orders into the crate. The network model for this exercise is shown in D7-8.1 and a statement model in D7-8.2. A summary report is provided in D7-8.3.

The use of resources to model the mail order clothes manufacturer employs resources named SHIRTS to model the inventory of dozens of shirts and PANTS to model the inventory of dozens of pants. The capacity of these resources is initially set to zero. The production of shirts and pants then involves creating an entity representing one unit of the resource which after a transportation delay increases the capacity of the resource by one. The number of resources available is the amount in stock. A disjoint network is then used to process the orders. The attributes of an order are: ATRIB(1), mark time; ATRIB(2), number of dozen shirts requested in the order; ATRIB(3), number of dozen pants requested in the order; ATRIB(4), the volume requirements of the order; and ATRIB(5), the dollar value of the order.

Following the creation of the order, the ordering entity awaits for ATRIB(2) units of the resource SHIRTS. It then waits for ATRIB(3) units of the resource PANTS. The order is then loaded into a crate and the entities representing the filled orders are batched in accordance with the specification that a hundred cubic feet constitutes a threshold for a crate being filled. The time between shipments, the volume in a crate and the dollar value of the orders in a crate are then collected.

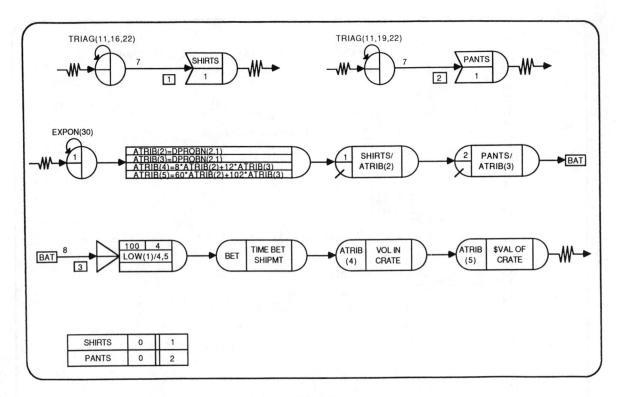

D7-8.1

```
1    GEN,PRITSKER,PROBLEM 7.8,6/26/86,,;
2    LIMITS,2,5,175;
3    ARRAY(1,4)/ 1, 2, 3, 4;
4    ARRAY(2,4)/.6,.8,.95,1.0;
5    NETWORK;
6          RESOURCE/SHIRTS(0),1;                          DEFINE RESOURCE SHIRTS
7          RESOURCE/PANTS(0),2;                           DEFINE RESOURCE PANTS
8          CREATE,TRIAG(11.,16.,22.);                     ARRIVAL OF 1 DOZ SHIRTS
9          ACT/1,7;                                       TRANSPORT SHIRTS
10         ALTER,SHIRTS,1;                                UPDATE INVENTORY
11         TERM;                                          DEPART FROM SYSTEM
12   ;
13         CREATE,TRIAG(11.,19.,22.);                     ARRIVAL OF 1 DOZ PANTS
14         ACT/2,7;                                       TRANSPORT PANTS
15         ALTER,PANTS,1;                                 UPDATE INVENTORY
16         TERM;                                          DEPART FROM SYSTEM
17   ;
18         CREATE,EXPON(30),,1;                           ARRIVAL OF ORDER
19         ASSIGN,ATRIB(2)=DPROBN(2,1),
20               ATRIB(3)=DPROBN(2,1),
21               ATRIB(4)=8*ATRIB(2)+12*ATRIB(3),
22               ATRIB(5)=60*ATRIB(2)+102*ATRIB(3);       ASSIGN ORDER INFO
23         AWAIT(1),SHIRTS/ATRIB(2);                      REMOVE SHIRTS FROM INV
24         AWAIT(2),PANTS/ATRIB(3);                       REMOVE PANTS FROM INV
25         ACT/3,8;                                       LOAD CRATES
26         BATCH,,100,4,LOW(1)/4,5;                       COMPLETE CRATE
27         COLCT,BET,TIME BET SHIPMT;                     COLLECT TIME BET SHIP
28         COLCT,ATRIB(4),VOL IN CRATE;                   COLLECT VOL IN CRATE
29         COLCT,ATRIB(5),$VAL OF CRATE;                  COLLECT DOLLAR VALUE
30         TERM;                                          DEPART FROM SYSTEM
31         END;
32   INIT,0,2400;
33   FIN;
```

D7-8.2

```
                        S L A M   I I   S U M M A R Y   R E P O R T

            SIMULATION PROJECT PROBLEM 7.8            BY PRITSKER

            DATE  6/26/1986                           RUN NUMBER   1 OF   1


            CURRENT TIME   0.2400E+04
            STATISTICAL ARRAYS CLEARED AT TIME  0.0000E+00


                    **STATISTICS FOR VARIABLES BASED ON OBSERVATION**

                        MEAN         STANDARD      COEFF. OF      MINIMUM       MAXIMUM      NUMBER OF
                        VALUE        DEVIATION     VARIATION      VALUE         VALUE        OBSERVATIONS

     TIME BET SHIPMT  0.1079E+03   0.2815E+02    0.2609E+00    0.5505E+02    0.1714E+03        20
     VOL IN CRATE     0.1173E+03   0.1520E+02    0.1295E+00    0.1000E+03    0.1440E+03        21
     $VAL OF CRATE    0.9531E+03   0.1239E+03    0.1300E+00    0.8100E+03    0.1176E+04        21


                              **FILE STATISTICS**

     FILE                   AVERAGE      STANDARD      MAXIMUM    CURRENT     AVERAGE
     NUMBER  LABEL/TYPE     LENGTH       DEVIATION     LENGTH     LENGTH      WAITING TIME

        1       AWAIT       0.8059       1.5475          7          0        23.0259
        2       AWAIT       0.5965       0.8617          3          0        17.0421
        3       CALENDAR    4.1071       0.7667          7          5         8.1328


                         **REGULAR ACTIVITY STATISTICS**

     ACTIVITY        AVERAGE       STANDARD     MAXIMUM CURRENT   ENTITY
     INDEX/LABEL     UTILIZATION   DEVIATION    UTIL    UTIL      COUNT

       1 TRANSPORT SH   0.4317      0.4953         1       0        148
       2 TRANSPORT PA   0.3967      0.4892         1       0        136
       3 LOAD CRATES    0.2787      0.4828         3       2         82


                              **RESOURCE STATISTICS**

     RESOURCE RESOURCE  CURRENT    AVERAGE       STANDARD      MAXIMUM       CURRENT
     NUMBER   LABEL     CAPACITY   UTILIZATION   DEVIATION     UTILIZATION   UTILIZATION

        1     SHIRTS      148      63.6587       34.0118        124           124
        2     PANTS       136      65.5304       37.8929        135           135


     RESOURCE RESOURCE  CURRENT     AVERAGE      MINIMUM      MAXIMUM
     NUMBER   LABEL     AVAILABLE   AVAILABLE    AVAILABLE    AVAILABLE

        1     SHIRTS       24       9.8501          0           27
        2     PANTS         1       1.9255          0            9
```

D7-8.3

7-9.   This exercise concerns an assembly line where the arrivals to the line are paced, although the line itself is not. Movement of units from one server to the next can occur if the current operation has been finished and the next server is idle. Each server is allowed 15 time units from the time it is received until it is to be completed. The network model is given in D7-9.1 and the corresponding statement model is in D7-9.2.

For I=1,3, ATRIB(I) is the time at station I, ATRIB(4) is the remaining processing time, and ATRIB(5) is the arrival time. The assignment of activity numbers causes SLAM II to print utilization statistics on the summary report for activities. For this example, this will provide utilization statistics on the stations and the total number of entities served by them. Note that the utilization of activity 1 must be added to the utilization of activity 2 to get the total utilization of the first server -- and similarly for the other 2 servers. By collecting statistics on XX(I) variables, the number of items routed offline from server I, an estimate of the number of entities diverted from each station is obtained. With the number served and the number diverted, an estimate of the fraction of entities diverted can be made.

In the statement model, D7-9.2, a PRIORITY statement is used to have the secondary ranking of the event calendar low-value-first based on the arrival time of an entity (ATRIB(5)). This secondary ranking is used so that events associated with entities already in the system are processed before new entities. In this way, new entities that might begin processing at exactly the same time as an entity completing processing are not counted as being in the activity concurrently. Although not specifying this secondary ranking for the event calendar does not change the average utilization, it would affect the maximum number of entities concurrently in a branch. A MONTR statement was used to obtain a trace of the events for the first 150 time units.

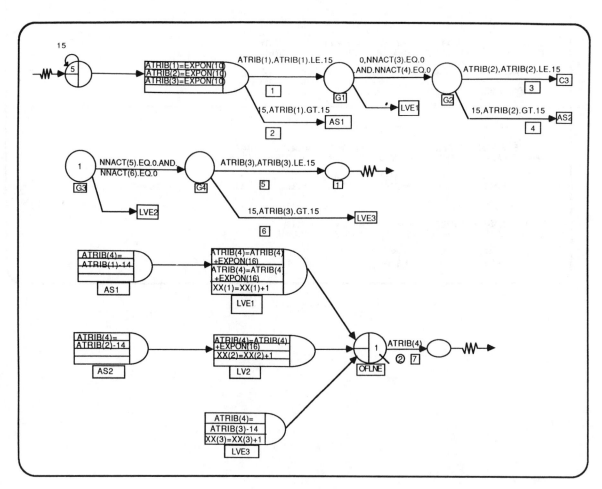

D7-9.1

```
 1    GEN,ROLSTON,PROBLEM 7.9,9/19/80,1;
 2    LIMITS,1,5,100;
 3    PRIORITY/NCLNR,LVF(5);
 4    TIMST,XX(1),FROM1;
 5    TIMST,XX(2),FROM2;
 6    TIMST,XX(3),FROM3;
 7    NETWORK;
 8          CREATE,15,,5;
 9          ASSIGN,ATRIB(1)=EXPON(10),
10                 ATRIB(2)=EXPON(10),
11                 ATRIB(3)=EXPON(10);
12          ACT/1,ATRIB(1),ATRIB(1).LE.15,G1;
13          ACT/2,15,ATRIB(1).GT.15,AS1;
14    G1    GOON,1;
15          ACT,,NNACT(3).EQ.0.AND.NNACT(4).EQ.0,G2;
16          ACT,,,LVE1;
17    G2    GOON;
18          ACT/3,ATRIB(2),ATRIB(2).LE.15,G3;
19          ACT/4,1,ATRIB(2).GT.15,AS2;
20    G3    GOON,1;
21          ACT,,NNACT(5).EQ.0.AND.NNACT(6).EQ.0,G4;
22          ACT,,,LVE2;
23    G4    GOON;
24          ACT/5,ATRIB(3),ATRIB(3).LE.15,T;
25          ACT/6,15,ATRIB(3).GT.15,LVE3;
26    T     TERM;
27    AS1   ASSIGN,ATRIB(4)=ATRIB(1)-14;
28    LVE1  ASSIGN,ATRIB(4)=ATRIB(4)+EXPON(16),
29                 ATRIB(4)=ATRIB(4)+EXPON(16),
30                 XX(1)=XX(1)+1;
31          ACT,,,OFLNE;
32    AS2   ASSIGN,ATRIB(4)=ATRIB(2)-14;
33    LVE2  ASSIGN,ATRIB(4)=ATRIB(4)+EXPON(16),
34                 XX(2)=XX(2)+1;
35          ACT,,,OFLNE;
36    LVE3  ASSIGN,ATRIB(4)=ATRIB(3)-14,
37                 XX(3)=XX(3)+1;
38    OFLNE QUEUE(1);
39          ACT(2)/7,ATRIB(4)
40          TERM;
41          END;
42    INIT,0,900;
43    MONTR,TRACE,0,150;
44    FIN;
```

D7-9.2

7-9,Embellishment(a). The solution is given in D7-9.3 and D7-9.4. In this case, the line itself is paced as well as the arrivals, that is, the movement of items only occurs at multiples of 15 minutes.

The solution uses ACCUMULATE nodes to control the movement of items. Upon reaching the server, an entity takes one of the first two branches to represent the actual processing. In addition, an entity is routed over a third branch which delays the arrival to the ACCUMULATE node by 15 minutes. No branch takes longer than 15 minutes, so movement occurs every 15 minutes when the entity routed over the third branch reaches the ACCUMULATE node.

7-9,Embellishment(b) In this case one entity is allowed to wait before each station on the assembly line. This change causes a new approach to be used as waiting areas and resources must now be included. The network and statement models are shown in D7-9.5 and D7-9.6.

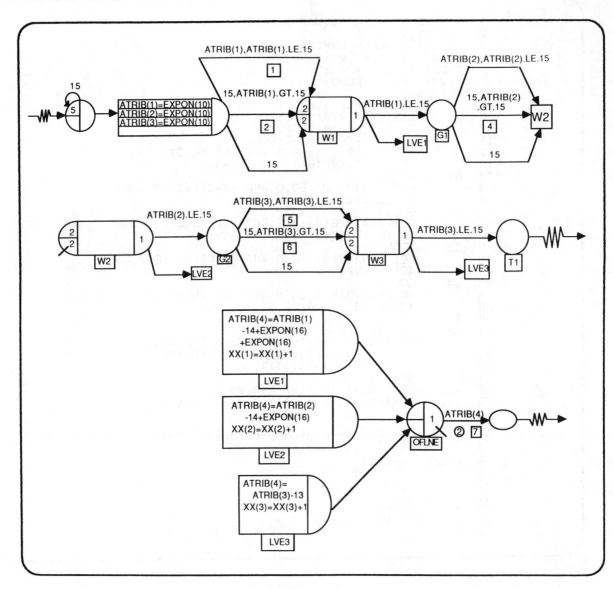

D7-9.3

```
1    GEN,ROLSTON,PROBLEM 7.9A,7/21/80,1;
2    LIMITS,1,5,50;
3    PRIORITY/NCLNR,LVF(5);
4    TIMST,XX(1),FROM1;
5    TIMST,XX(2),FROM2;
6    TIMST,XX(3),FROM3;
7    NETWORK;
8             CREATE,15,,5;
9             ASSIGN,ATRIB(1)=EXPON(10),
10                   ATRIB(2)=EXPON(10),
11                   ATRIB(3)=EXPON(10);
12             ACT/1,ATRIB(1),ATRIB(1).LE.15,W1;
13             ACT/2,15,ATRIB(1).GT.15,W1;
14             ACT,15,,W1;
15   W1        ACCUMULATE,2,2,,1;
16             ACT,,ATRIB(1).LE.15,G1;
17             ACT,,,LVE1;
18   G1        GOON;
19             ACT/3,ATRIB(2),ATRIB(2).LE.15,W2;
20             ACT/4,15,ATRIB(2).GT.15,W2;
21             ACT,15,,W2;
22   W2        ACCUMULATE,2,2,,1;
23             ACT,,ATRIB(2).LE.15,G2;
24             ACT,,,LVE2;
25   G2        GOON;
26             ACT/5,ATRIB(3),ATRIB(3).LE.15,W3;
27             ACT/6,15,ATRIB(3).GT.15,W3;
28             ACT,15,,W3;
29   W3        ACCUMULATE,2,2,,1;
30             ACT,,ATRIB(3).LE.15,T1;
31             ACT,,,LVE3;
32   T1        TERM;
33   LVE1      ASSIGN,ATRIB(4)=ATRIB(1)-15,
34                   ATRIB(4)=ATRIB(4)+1,
35                   ATRIB(4)=ATRIB(4)+EXPON(16),
36                   ATRIB(4)=ATRIB(4)+EXPON(16),
37                   XX(1)=XX(1)+1;
38             ACT,,,OFLNE;
39   LVE2      ASSIGN,ATRIB(4)=ATRIB(2)-15,
40                   ATRIB(4)=ATRIB(4)+1,
41                   ATRIB(4)=ATRIB(4)+EXPON(16),
42                   XX(2)=XX(2)+1;
43             ACT,,,OFLNE;
44   LVE3      ASSIGN,ATRIB(4)=ATRIB(3)-15,
45                   ATRIB(4)=ATRIB(4)+1,
46                   XX(3)=XX(3)+1;
47             ACT,,,OFLNE;
48   OFLNE     QUEUE(1);
49             ACT(2)/7,ATRIB(4);
50             TERM;
51             END;
52   INTLC,XX(1)=0,XX(2)=0,XX(3)=0;
53   INIT,0,900;
54   MONTR,TRACE,0,200;
55   FIN;
```

D7-9.4

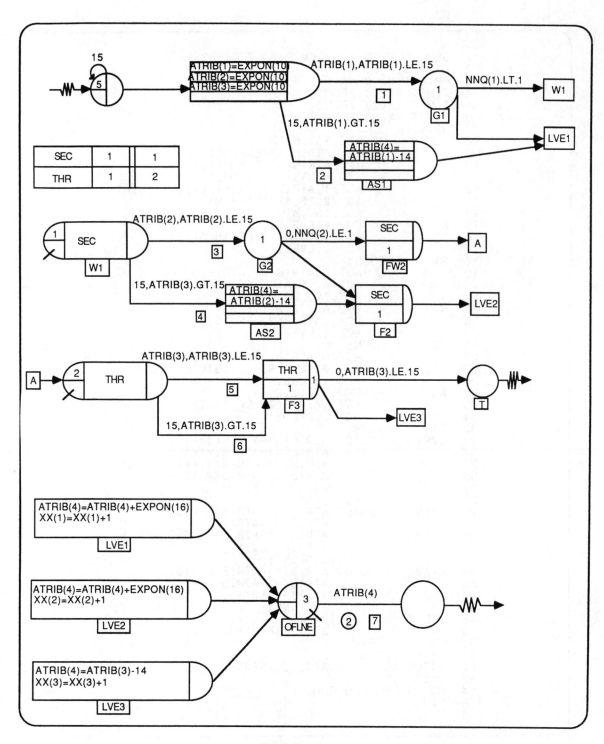

D7-9.5

```
 1    GEN,ROLSTON,PROBLEM 7.9B,9/19/80,1;
 2    LIMITS,3,5,100;
 3    PRIORITY/NCLNR,LVF(5);
 4    TIMST,XX(1),FROM1;
 5    TIMST,XX(2),FROM2;
 6    TIMST,XX(3),FROM3;
 7    NETWORK;
 8            RESOURCE/SEC(1),1;
 9            RESOURCE/THR(1),2;
10            CREATE,15,,5;
11            ASSIGN,ATRIB(1)=EXPON(10),
12                   ATRIB(2)=EXPON(10),
13                   ATRIB(3)=EXPON(10);
14            ACT/1,ATRIB(1),ATRIB(1).LE.15,G1;
15            ACT/2,15,ATRIB(1).GT.15,AS1;
16    G1      GOON,1;
17            ACT,,NNQ(1).LT.1,W1;
18            ACT,,,LVE1;
19    W1      AWAIT(1),SEC;
20            ACT/3,ATRIB(2),ATRIB(2).LE.15,G2;
21            ACT/4,15,ATRIB(2).GT.15,AS2;
22    G2      GOON,1;
23            ACT,,NNQ(2).LT.1,FW2;
24            ACT,,,F2;
25    FW2     FREE,SEC;
26            AWAIT(2),THR;
27            ACT/5,ATRIB(3),ATRIB(3).LE.15,F3;
28            ACT/6,15,ATRIB(3).GT.15,F3;
29    F3      FREE,THR,1;
30            ACT,,ATRIB(3).LE.15,T;
31            ACT,,,LVE3;
32    T       TERM;
33    AS1     ASSIGN,ATRIB(4)=ATRIB(1)-14;
34    LVE1    ASSIGN,ATRIB(4)=ATRIB(4)+EXPON(16),
35                   ATRIB(4)=ATRIB(4)+EXPON(16),
36                   XX(1)=XX(1)+1;
37            ACT,,,OFLNE;
38    AS2     ASSIGN,ATRIB(4)=ATRIB(2)-14;
39    F2      FREE,SEC;
40    LVE2    ASSIGN,ATRIB(4)=ATRIB(4)+EXPON(10),
41                   XX(2)=XX(2)+1;
42            ACT,,,OFLNE;
43    LVE3    ASSIGN,ATRIB(4)=ATRIB(3)-14,
44                   XX(3)=XX(3)+1;
45    OFLNE   QUEUE(3);
46            ACT(2)/7,ATRIB(4);
47            TERM;
48            END;
49    INIT,0,900;
50    MONTR,TRACE,0,150;
51    FIN;
```

D7-9.6

7-9,Embellishment(c). The network statements for the solution to embellishment(c) are shown in D7-9.7 and D7-9.8. This model allows units to be recycled back to the on-line servers if they are idle. An additional attribute, ATRIB(6), is set equal to the server number to which the unit should be recycled. The offline servers are modeled as resources to allow the use of conditional branching after the resource is allocated.

7-10.   For an analysis of this type of PERT network using simulation see pages 249, 254-256 of the GASP IV book (Pritsker, A.A.B., The GASP IV Simulation Language, John Wiley & Sons, Inc., 1974).

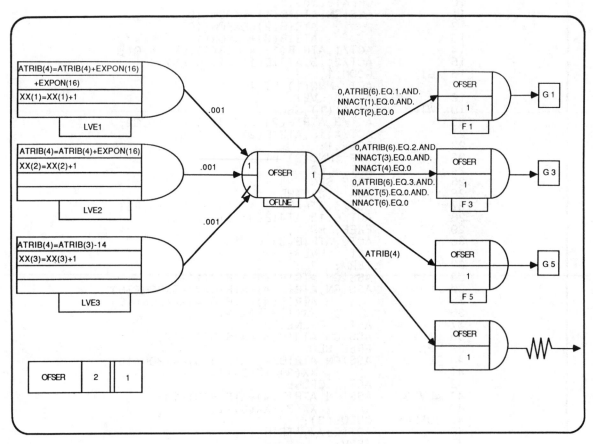

D7-9.7

```
 1    GEN,ROLSTON,PROBLEM 7.9C,9/19/80,1;
 2    LIMITS,1,6,100;
 3    PRIORITY/NCLNR,LVF(5);
 4    TIMST,XX(1),FROM1;
 5    TIMST,XX(2),FROM2;
 6    TIMST,XX(3),FROM3;
 7    NETWORK;
 8          RESOURCE/OFSER(2),1;
 9          CREATE,15,,5;
10          ASSIGN,ATRIB(1)=EXPON(10),
11                 ATRIB(2)=EXPON(10),
12                 ATRIB(3)=EXPON(10),1;
13          ACT,,NNACT(1).EQ.0.AND.NNACT(2).EQ.0,G1;
14          ACT,,,AS1;
15    G1    GOON;
16          ACT/1,ATRIB(1),ATRIB(1).LE.15,G2;
17          ACT/2,15,ATRIB(1).GT.15,AS2;
18    G2    GOON,1;
19          ACT,,NNACT(3).EQ.0.AND.NNACT(4).EQ.0,G3;
20          ACT,,,AS3;
21    G3    GOON;
22          ACT/3,ATRIB(2),ATRIB(2).LE.15,G4;
23          ACT/4,15,ATRIB(2).GT.15,AS4;
24    G4    GOON,1;
25          ACT,,NNACT(5).EQ.0.AND.NNACT(6).EQ.0,G5;
26          ACT,,,AS5;
27    G5    GOON;
28          ACT/5,ATRIB(3),ATRIB(3).LE.15,T;
29          ACT/6,15,ATRIB(3).GT.15,AS6;
30    T     TERM;
31    AS1   ASSIGN,ATRIB(4)=ATRIB(1)+1,
32                 ATRIB(6)=1;
33          ACT,,,LVE1;
34    AS2   ASSIGN,ATRIB(4)=ATRIB(1)-14,
35                 ATRIB(6)=1,
36                 ATRIB(1)=ATRIB(1)-15;
37          ACT,,,LVE1;
38    AS3   ASSIGN,ATRIB(6)=2;
39    LVE1  ASSIGN,ATRIB(4)=ATRIB(4)+EXPON(16)+EXPON(16),
40                 XX(1)=XX(1)+1;
41          ACT,.001,,OFLNE;
42    AS4   ASSIGN,ATRIB(4)=ATRIB(2)-14,
43                 ATRIB(6)=2,
44                 ATRIB(2)=ATRIB(2)-15;
45          ACT,,,LVE2;
46    AS5   ASSIGN,ATRIB(6)=3;
47    LVE2  ASSIGN,ATRIB(4)=ATRIB(4)+EXPON(16),
48                 XX(2)=XX(2)+1;
49          ACT,.001,,OFLNE;
50    AS6   ASSIGN,ATRIB(6)=3,
51                 ATRIB(3)=ATRIB(3)-15;
52    LVE3  ASSIGN,ATRIB(4)=ATRIB(3)-14,
53                 XX(3)=XX(3)+1;
54          ACT,.001,,OFLNE;
55    OFLNE AWAIT(1),OFSER,1;
56          ACT,,ATRIB(6).EQ.1.AND.NNACT(1).EQ.0.AND.NNACT(2).EQ.0,F1;
57          ACT,,ATRIB(6).EQ.2.AND.NNACT(3).EQ.0.AND.NNACT(4).EQ.0,F3;
58          ACT,,ATRIB(6).EQ.3.AND.NNACT(5).EQ.0.AND.NNACT(6).EQ.0,F5;
59          ACT,ATRIB(4);
60          FREE,OFSER;
61          TERM;
62    F1    FREE,OFSER;
63          ACT,,,G1;
64    F3    FREE,OFSER;
65          ACT,,,G3;
66    F5    FREE,OFSER;
67          ACT,,,G5;
68          END;
69    INIT,0,900;
70    MONTR,TRACE,0,150;
71    FIN;
```

7-11.   The network and statement models are shown in D7-11.1 and D7-11.2.

This example illustrates the use of a SELECT node for selecting from two parallel queues. In the model shown, it is assumed that customers for the inside of the bank do not arrive until 60 minutes after the start of operations. Also, the drive-in customers who balk before the opening of the inside of the bank, leave the system.

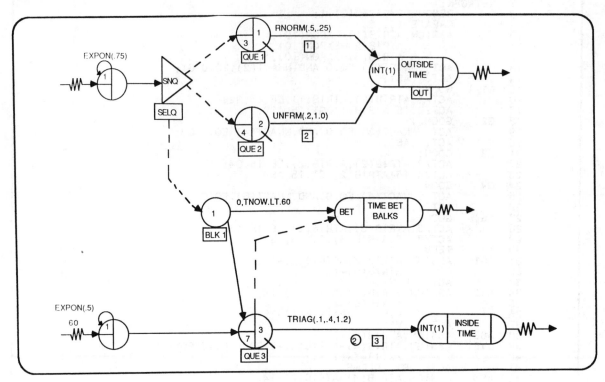

D7-11.1

```
 1     GEN,ROLSTON,PROBLEM 7.11,7/22/80,1;
 2     LIMITS,3,1,50;
 3     NETWORK;
 4             CREATE,EXPON(.75),,1;
 5     SELQ    SELECT,SNQ,,BALK(BLK1),QUE1,QUE2;
 6     QUE1    QUEUE(1),,3;
 7             ACT/1,RNORM(.5,.25),,OUT;
 8     QUE2    QUEUE(2),,4;
 9             ACT/2,UNFRM(.2,1.0);
10     OUT     COLCT,INT(1),OUTSIDE TIME;
11             TERM;
12     BLK1    GOON,1;
13             ACT,,TNOW.LT.60,LOST;
14             ACT,,,QUE3;
15     LOST    COLCT,BET,TIME BET BALKS;
16             TERM;
17             CREATE,EXPON(.5),60,1;
18     QUE3    QUEUE(3),,7,BALK(LOST);
19             ACT(2)/3,TRIAG(.1,.4,1.2);
20             COLCT,INT(1),INSIDE TIME;
21             TERM;
22             END;
23     INIT,0,480;
24     FIN;
```

D7-11.2

7-12.   The modifications to Exercise 7-11 for adding credit inquiries involve the introduction of a resource called MAN. Most of the changes to the model are straightforward and involve processing logic in accordance with the problem statement. The network changes and the complete statement model are shown in D7-12.1 and D7-12.2.

Attribute 2 represents the number of times a customer has seen the manager. It is set to 1 before a customer has seen the manager for the first time, and to 2 before he sees the manager for the second time. Ranking the AWAIT file HVF(2) fulfills the priority requirements. This attribute is also used for branching conditions. Statistics are collected on the time between negative responses to a customer's credit inquiry.

7-12,Embellishment. At ASSIGN node CHK include:

$ATRIB(3) = 1$

$ATRIB(4) = ATRIB(2) + ATRIB(3)$

In the ASSIGN node after the FREE node DONE include:

$ATRIB(4) = ATRIB(2) + ATRIB(3)$

Change the network starting at node G2 as shown in D7-12.3

The complete statement model is shown in D7-12.4. Note that AWAIT file 4 is ranked on HVF(4) and that attribute 3 represents the number of times the entity is processed through the credit inquiry process.

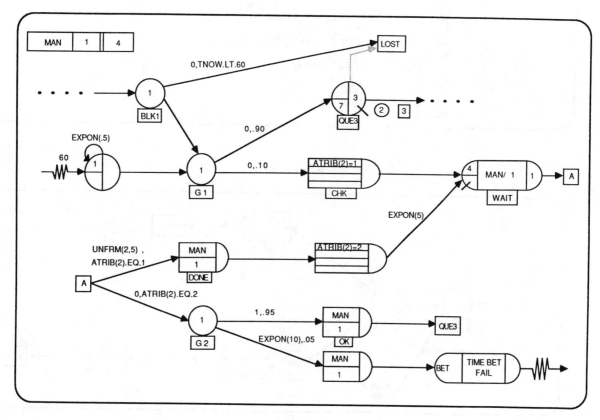

D7-12.1

```
 1     GEN,ROLSTON,PROBLEM 7.12,7/22/80,1;
 2     LIMITS,4,2,100;
 3     PRIORITY/4,HVF(2);
 4     NETWORK;
 5              RESOURCE/MAN(1),4;
 6              CREATE,EXPON(.75),,1;
 7     SELQ     SELECT,SNQ,,BALK(BLK1),QUE1,QUE2;
 8     QUE1     QUEUE(1),,3;
 9              ACT/1,RNORM(.5,.25),,OUT;
10     QUE2     QUEUE(2),,4;
11              ACT/2,UNFRM(.2,1.0);
12     OUT      COLCT,INT(1),OUTSIDE TIME;
13              TERM;
14     BLK1     GOON,1;
15              ACT,,TNOW.LT.60,LOST;
16              ACT,,,G1;
17     LOST     COLCT,BET,TIME BET BALKS;
18              TERM;
19              CREATE,EXPON(.5),60,1;
20     G1       GOON,1;
21              ACT,,.10,CHK;
22              ACT,,.90,QUE3;
23     QUE3     QUEUE(3),,7,BALK(LOST);
24              ACT(2)/3,TRIAG(.1,.4,1.2);
25              COLCT,INT(1),INSIDE TIME;
26              TERM;
27     CHK      ASSIGN,ATRIB(2)=1;
28     WAIT     AWAIT(4),MAN/1,1;
29              ACT,UNFRM(2,5),ATRIB(2).EQ.1,DONE;
30              ACT,,ATRIB(2).EQ.2,G2;
31     DONE     FREE,MAN/1;
32              ASSIGN,ATRIB(2)=2;
33              ACT,EXPON(5),,WAIT;
34     G2       GOON,1;
35              ACT,1,.95,OK;
36              ACT,EXPON(10),.05;
37              FREE,MAN/1;
38              COLCT,BET,TIME BET FAIL;
39              TERM;
40     OK       FREE,MAN/1;
41              ACT,,,QUE3;
42              END;
43     INIT,0,480;
44     FIN;
```

D7-12.2

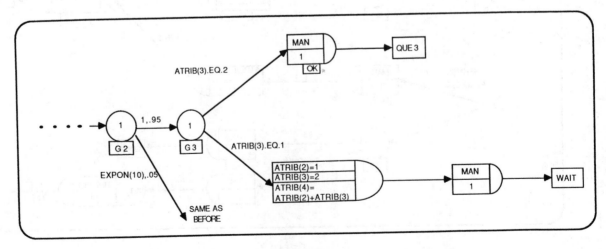

D7-12.3

```
 1         GEN,ROLSTON,PROBLEM 7.12E,7/22/80,1;
 2         LIMITS,4,4,100;
 3         PRIORITY/4,HVF(4);
 4         NETWORK;
 5                   RESOURCE/MAN(1),4;
 6                   CREATE,EXPON(.75),,1;
 7  SELQ     SELECT,SNQ,,BALK(BLK1),QUE1,QUE2;
 8  QUE1     QUEUE(1),,3;
 9                   ACT/1,RNORM(.5,.25),,OUT;
10  QUE2     QUEUE(2),,4;
11                   ACT/2,UNFRM(.2,1.0);
12  OUT      COLCT,INT(1),OUTSIDE TIME;
13           TERM;
14  BLK1     GOON,1;
15                   ACT,,TNOW.LT.60,LOST;
16                   ACT,,,G1;
17  LOST     COLCT,BET,TIME BET BALKS;
18           TERM;
19           CREATE,EXPON(.5),60,1;
20  G1       GOON,1;
21                   ACT,,.10,CHK;
22                   ACT,,.90,QUE3;
23  QUE3     QUEUE(3),,7,BALK(LOST);
24                   ACT(2)/3,TRIAG(.1,.4,1.2);
25           COLCT,INT(1),INSIDE TIME;
26           TERM;
27  CHK      ASSIGN,ATRIB(2)=1,
28                   ATRIB(3)=1,
29                   ATRIB(4)=ATRIB(2)+ATRIB(3);
30  WAIT     AWAIT(4),MAN/1,1;
31                   ACT,UNFRM(2,5),ATRIB(2).EQ.1,DONE;
32                   ACT,,ATRIB(2).EQ.2,G2;
33  DONE     FREE,MAN/1;
34           ASSIGN,ATRIB(2)=2,
35                   ATRIB(4)=ATRIB(2)+ATRIB(3);
36           ACT,EXPON(5),,WAIT;
37  G2       GOON,1;
38                   ACT,1,.95,G3;
39                   ACT,EXPON(10),.05;
40           FREE,MAN/1;
41           COLCT,BET,TIME BET FAIL;
42           TERM;
43  G3       GOON,1;
44                   ACT,,ATRIB(3).EQ.2,OK;
45                   ACT,,ATRIB(3).EQ.1;
46           ASSIGN,ATRIB(2)=1,
47                   ATRIB(3)=2,
48                   ATRIB(4)=ATRIB(2)+ATRIB(3);
49           FREE,MAN/1;
50                   ACT,,,WAIT
51  OK       FREE,MAN/1;
52                   ACT,,,QUE3;
53           END;
54  INIT,0,480;
55  FIN;
```

D7-12.4

7-13. The network model is given in D7-13.1 and the statement model in D7-13.2. The statement model is well documented and shows the flow of kits through the network. ATRIB(1) is used to attach a due date to each kit entity. This assembly occurs at activity 1 and part A is sent to QUE1 and part B to QUE2. Kits are cleaned in activities 2 and 3 and assembled at select node ASMB. The kits are routed to queue node QUE5 if the current time is within ten minutes or less of the kit's due date, otherwise the kit is routed to queue node QUE6. Select node SLT1 selects kits from queue node QUE5 first and, if none, from queue node QUE6. Select node SLT1 selects activity 4 in preference to activity 5 in that the first expert assembler is used when possible. Kits waiting in QUEUE node QUE6 are processed by server 6, the trainee assembler whenever the trainee assembler is available. The other assemblers will process kits in QUEUE node QUE6 when they are idle and there are no kits in QUEUE node QUE5.

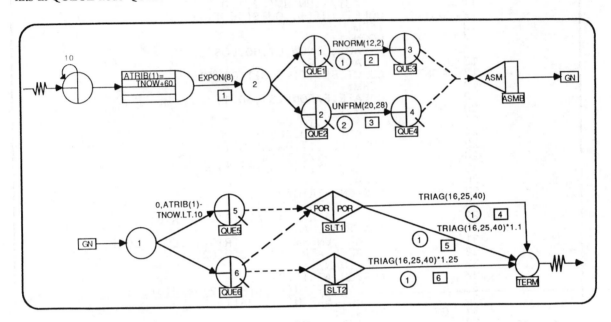

D7-13.1

```
 1     GEN,FLOSS,PROBLEM 7.13,6/30/86,1;
 2     LIMITS,6,1,175;
 3     NETWORK;
 4            CREATE,10;                              ARRIVAL OF KIT
 5            ASSIGN,ATRIB(1)=TNOW+60;                ASSIGN DUE DATE FOR PARTS
 6            ACTIVITY/1,EXPON(8);                    DISASSEMBLE KIT
 7            GOON,2;                                 ROUTE TO CLEANING STATIONS
 8            ACTIVITY,,,QUE1;                        PART A TO CLEANING STATION
 9            ACTIVITY,,,QUE2;                        PART B TO CLEANING STATION
10     QUE1   QUEUE(1);                               PART A WAIT FOR CLN MACHINE
11            ACTIVITY(1)/2,RNORM(12,2);              CLEAN PART A
12     QUE3   QUEUE(3),,,,ASMB;                       ROUTE TO ASSEMBLY
13     QUE2   QUEUE(2);                               PART B WAIT FOR CLN MACHINES
14            ACTIVITY(2)/3,UNFRM(20,28);             CLEAN PART B
15     QUE4   QUEUE(4),,,,ASMB;                       ROUTE TO ASSEMBLY
16     ASMB   SELECT,ASM,,,QUE4,QUE3;                 SELECT A PART A AND A PART B
17            ACTIVITY;                               ROUTE TO ASSEMBLERS
18            GOON,1;                                 CHOOSE ASSEMBLER
19            ACTIVITY,,ATRIB(1)-TNOW.LT.10,QUE5;     IF DUE IN LESS THAN 10 MIN
20            ACTIVITY,,,QUE6;                        ELSE CONTINUE
21     QUE5   QUEUE(5),,,,SLT1;                       WAIT FOR EXPERT ASSEMBLERS
22     QUE6   QUEUE(6),,,,SLT1,SLT2;                  WAIT FOR ANY ASSEMBLER
23     SLT1   SELECT,POR,POR,,QUE5,QUE6;              SELECT EXPERT ASSEMBLER
24            ACTIVITY(1)/4,TRIAG(16,25,40),,TERM;    ASSEMBLE
25            ACTIVITY(1)/5,TRIAG(16,25,40)*1.1,,TERM; ASSEMBLE
26     SLT2   SELECT,,,,QUE6;                         SELECT TRAINEE ASSEMBLER
27            ACTIVITY(1)/6,TRIAG(16,25,40)*1.25;     ASSEMBLE
28     TERM   TERM;                                   DEPART FROM SYSTEM
29            END;
30     INIT,0,2400;
31     FIN;
```

D7-13.2

The SLAM II Summary Report is given in D7-13.3 and shows that the expert assemblers are extremely busy (98% of the time) and the trainee is idle most of the time. This indicates that the due date setting procedure and rule for assignment to the expert assemblers needs to be investigated.

7-13,Embellishment. The requirement that parts from the same kit be assembled together can be accomplished by assigning a unique kit number to each arriving kit as an attribute and changing the select node ASMB to a match node that matches on the unique kit number. The assignment of a unique kit number could be made by replacing the GOON node that splits the kit into two parts by an ASSIGN node and assigning attribute 2 equal to NNCNT(1). Alternatively, a mark time could be assigned to a kit and a MATCH on mark time could be used.

SLAM II SUMMARY REPORT

SIMULATION PROJECT PROBLEM 7.13   BY FLOSS

DATE 6/30/1986   RUN NUMBER   1 OF   1

CURRENT TIME 0.2400E+04
STATISTICAL ARRAYS CLEARED AT TIME 0.0000E+00

FILE STATISTICS

FILE NUMBER	LABEL/TYPE	AVERAGE LENGTH	STANDARD DEVIATION	MAXIMUM LENGTH	CURRENT LENGTH	AVERAGE WAITING TIME
1	QUE1 QUEUE	18.4819	10.5354	39	37	185.5920
2	QUE2 QUEUE	18.5752	10.7837	38	36	186.5293
3	QUE3 QUEUE	1.1056	0.8403	3	0	13.2009
4	QUE4 QUEUE	0.0204	0.1412	1	0	0.2430
5	QUE5 QUEUE	12.3941	8.2440	27	27	156.5564
6	QUE6 QUEUE	0.0016	0.0402	1	0	0.3533
7	CALENDAR	6.7946	0.8139	10	8	6.5098

REGULAR ACTIVITY STATISTICS

ACTIVITY INDEX/LABEL	AVERAGE UTILIZATION	STANDARD DEVIATION	MAXIMUM UTIL	CURRENT UTIL	ENTITY COUNT
1 DISASSEMBLE	0.8008	0.6995	3	2	239

SERVICE ACTIVITY STATISTICS

ACTIVITY INDEX	START NODE OR ACTIVITY LABEL	SERVER CAPACITY	AVERAGE UTILIZATION	STANDARD DEVIATION	CURRENT UTILIZATION	AVERAGE BLOCKAGE	MAXIMUM IDLE TIME/SERVERS	MAXIMUM BUSY TIME/SERVERS	ENTITY COUNT
2	CLEAN PART A	1	0.9960	0.0630	1	0.0000	9.5610	2390.4390	201
3	CLEAN PART B	2	1.9879	0.1411	2	0.0000	2.0000	2.0000	201
0	ASMB SELECT	1	0.0000	0.0000	0	0.0000	32.5085	0.0000	
4	ASSEMBLE	1	0.9789	0.1436	1	0.0000	32.5085	2235.3643	89
5	ASSEMBLE	1	0.9782	0.1460	1	0.0000	46.0559	2184.4612	79
6	ASSEMBLE	1	0.0528	0.2236	0	0.0000	2172.0481	41.0184	4

D7-13.3

7-14.   A paper on alternative modeling views discusses this exercise in detail and shows various alternative models and the evolutionary modeling process associated with this exercise (Pritsker, A.A.B. - "Model Evolution: A Rotary Index Table Case History", 1986 WSC Proceedings). Exercise 13-2 presents a continuous modeling approach to computing the expected time to complete the eight concurrent operations. A derivation of the equation in Exercise 13-2 is given in the above cited paper. In this exercise, a few of the alternative models will be presented.

In D7-14.1, a model which uses batching and unbatching to perform the operations in parallel is shown. The estimate of the length of the production run is the time for the simulation to end which occurs after 50 parts have been produced. For this model, the elapsed time on the first run is 340.3 minutes.

A second model of the rotary table uses a QUEUE node and service activity to represent the operation times. Following the completion of each operation, the part produced is put into activity 9 with an indefinite duration as specified by REL(FINI). An ACCUMULATE node is given the node label FINI and when 8 entities are routed to node FINI, it is released which causes each of the entities in activity 9 to be routed to the end node of the activity. This model is shown in D7- 14.2. There is no difference in the time to produce the 50 parts on the first run of this model.

A third model accumulates the 8 entities into a single entity and then uses an M number of 9 to recreate 8 new entities to initiate the 8 operations for each rotation of the table. The ninth entity is sent to node TERM to count the number of completed parts. This model is shown in Figure D7-14.3

7-15.   The 99.7% confidence interval is $20.78 + 2.97 * 0.107$ which is (20.46, 21.10). The tolerance interval is $20.78 \pm 1.380 * 2.137/20 = (20.63, 20.93)$.

```
 1    GEN,PRITSKER,PROBLEM 7.14,6/17/86,1;
 2    LIMITS,0,3,100;
 3    NETWORK;
 4          CREATE,0,8,,8;                        LOAD THE SYSTEM
 5          ASSIGN,II=II+1,ATRIB(1)=II;           ASSIGN OPERATION NUMBER
 6    TIME  ASSIGN,ATRIB(2)=UNFRM(3,6),1;         ASSIGN OPERATION TIME
 7          ACTIVITY(1)/ATRIB(1)=1,8,ATRIB(2);    PERFORM OPERATIONS
 8          BATCH,,8,,,ALL(3),2;                  BATCH OPERATIONS AS THEY COMPLETE
 9          ACTIVITY,,,UNLD;                      ROUTE TO UNLOAD OPERATION
10          ACTIVITY,1;                           INDEX ROTARY TABLE
11          UNBATCH,3,1;                          UNBATCH THE PARTS
12          ACTIVITY,,,TIME;                      ROUTE TO NEXT OPERATION
13    UNLD  TERMINATE,50;                         COUNT NUMBER OF PARTS UNLOADED
14          END;
15    FIN;
```

D7-14.1

```
 1   GEN,FLOSS,PROBLEM 7.14A,6/17/86,1;
 2   LIMITS,8,1,100;
 3   NETWORK;
 4          CREATE,0,8,,8;                          LOAD ROTARY TABLE
 5          ASSIGN,II=II+1,ATRIB(1)=II;             ASSIGN OPERATION NUMBER
 6   QUE    QUEUE(ATRIB(1)=1,8);                     WAIT FOR OPERATIONS TO BEGIN
 7          ACTIVITY(1)/ATRIB(1)=1,8,UNFRM(3,6);    PERFORM OPERATIONS
 8          GOON,2;                                 CONTINUE
 9          ACTIVITY,,,FINI;                        ROUTE TO FINISHED PARTS
10          ACTIVITY/9,REL(FINI);                   WAIT FOR ALL PARTS TO FINISH
11          GOON,1;                                 CONTINUE
12          ACTIVITY/10,1,,QUE;                     INDEX TABLE
13   FINI   ACCUMULATE,8,8;                         UNLOAD ROTARY TABLE
14          TERMINATE,50;                           COUNT NUMBER OF COMPLETED PARTS
15          END;
16   FIN;
```

D7-14.2

```
 1   GEN,FLOSS,PROBLEM 7.14B,6/17/86,1;
 2   LIMITS,8,1,100;
 3   NETWORK;
 4          CREATE,0,8,,8;                          LOAD ROTARY TABLE
 5   ASGN   ASSIGN,II=II+1,ATRIB(1)=II;             ASSIGN OPERATION NUMBER
 6          QUEUE(ATRIB(1)=1,8);                     WAIT FOR OPERATION
 7          ACTIVITY(1)/ATRIB(1)=1,8,UNFRM(3,6);    PERFORM OPERATION
 8          ACCUMULATE,8,8;                         WAIT FOR ALL PARTS TO COMPLETE
 9          ASSIGN,II=0,9;                          INITIALIZE COUNTER
10          ACTIVITY,,,TERM;                        ROUTE TO TERM
11          ACTIVITY,1,,ASGN;                       INDEX ROTARY TABLE
12          ACTIVITY,1,,ASGN;                       INDEX ROTARY TABLE
13          ACTIVITY,1,,ASGN;                       INDEX ROTARY TABLE
14          ACTIVITY,1,,ASGN;                       INDEX ROTARY TABLE
15          ACTIVITY,1,,ASGN;                       INDEX ROTARY TABLE
16          ACTIVITY,1,,ASGN;                       INDEX ROTARY TABLE
17          ACTIVITY,1,,ASGN;                       INDEX ROTARY TABLE
18          ACTIVITY,1,,ASGN;                       INDEX ROTARY TABLE
19   TERM   TERMINATE,50;                           COUNT NUMBER OF COMPLETED PARTS
20          END;
21   FIN;
```

D7-14.3

CHAPTER 8

SLAM II Processor, Inputs and Outputs

8-1.   Change the model for Example 6-2 to have the PREEMPT node specify the saving of the remaining time in attribute 2. A trace for Example 6-2 which has attribute 1 and attribute 2 printed could be used to examine the remaining processing time for those jobs that were preempted from the operation. This manual approach would require a matching of attribute 1 values to determine if the entity was preempted and then analysis of the attribute 2 values for those that were preempted. The statement to achieve such a trace for the entire simulation is shown below.

MONTR,TRACE,0,500,ATRIB(1),ATRIB(2);

An alternative approach is to modify the problem so that entities that are preempted are sent to a send node with the remaining processing time saved in attribute 2. Attribute 3 would be used to indicate which activity was interrupted. If the interruption occurred for activity 2 the preempted entity would be sent to a collect node to collect statistics on the value of attribute 2 before returning the preempted entity back to the AWAIT node. Both statistics and a histogram would be requested at the COLCT node. The COLCT node would be given a label and a trace of all entities arriving to the COLCT node could then be requested.

8-2.   Errors in GEN statement: The month must be entered as an integer which, in this case, is 7, and the name cannot include a minus (-) sign.

Errors in LIMIT statement: MNTRY, the maximum number of concurrent entities in the network is specified as 0; MFIL, the largest file number used, is specified as 2, while the largest file number in the network is 1. Although this is not a fatal error, it will cause an additional line of output on the summary report; MATR, the maximum number of attributes per entity, is specified as 1. Since no attribute values are assigned, this could have been specified as 0, but, again, this is not a fatal error.

Error in PRIORITY statement: HVT should be HVF. The statement specifies the ranking criterion for file 2 which is the event file. When ranking is specified for the event file, it is the secondary ranking rule. The ranking criterion is specified as high value first, HVF, for attribute 2. The maximum number of attributes per entity is specified as 1 and the next two attributes are used to store the event code and event time for the event calendar. Thus, the PRIORITY statement used here specifies that events with the same time should be ordered with high numbered events first.

8-3.   The control statements are shown in D8-3.1. Although there are no changes between the second and third runs, a SIMULATE statement is required. However, no SIMULATE statement is required following the fourth run as the FIN statement specifies that all following runs should be made under the same conditions. Thus, the FIN statement is read on the fourth run and no further SLAM statements are expected after the fourth run.

8-4.   The network and statement models are shown in D8-4.1 and D8-4.2. The SLAM II histogram is shown in D8-4.3.

This version of the Thief of Baghdad problem can be solved analytically (see solution to Exercise 2-7). Through the use of the GERTE program, the analytic solution for the time for the thief to reach freedom is 12 with a variance of 14.81.

```
1    GEN,ROLSTON,PROBLEM 8.3,7/18/80,5;
2    LIMITS,1,1,150;
3    PRIORITY/1,LVF(1);
4    INTLC,XX(1)=0.0,XX(2)=1.0;
5    NETWORK;
6            CREATE,8,,,100;
7            ASSIGN,ATRIB(1)=EXPON(7);
8              ACTIVITY,,,QOFS;
9            CREATE,12,,,50;
10           ASSIGN,ATRIB(1)=EXPON(10);
11   QOFS    QUEUE(1);
12           ACTIVITY/1,ATRIB(1)+RNORM(XX(1),XX(2));
13           TERM;
14           ENDNETWORK;
15   INIT,0,1000;
16   SIMULATE;
17   INTLC,XX(2)=2.0;
18   SIMULATE;
19   SIMULATE;
20   PRIORITY/1,LIFO;
21   FIN;
```

D8-3.1

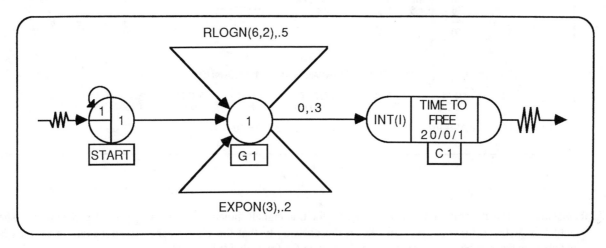

D8-4.1

```
1    GEN,PRITSKER,THIEF OF BAGHDAD 8.4,7/18/80,1000,,N,,N,Y/1000;
2    LIMITS,0,1,10;
3    INIT,,,N;
4    NETWORK;
5    START   CREATE,,,1;
6    G1        GOON,1;
7              ACT,RLOGN(6,2),.5,G1;
8              ACT,,.3,C1;
9              ACT,EXPON(3),.2,G1;
10   C1        COLCT,INT(1),TIME TO FREE,20/0/1;
11             TERM;
12             ENDNETWORK;
13   FIN;
```

D8-4.2

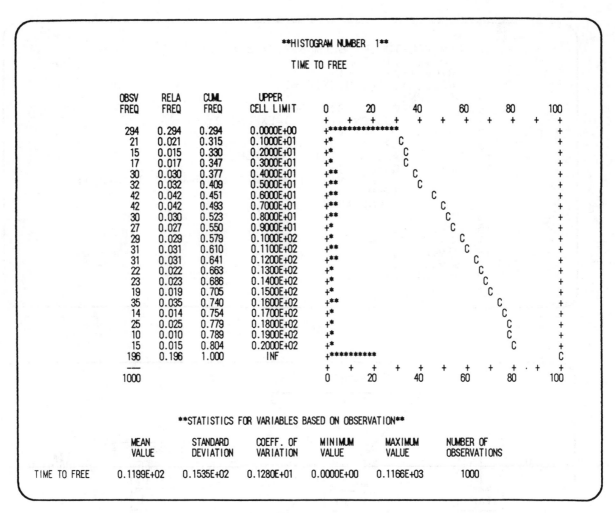

D8-4.3

8-4,Embellishment. The network and statement models for this embellishment are shown in D8-4.4 and D8-4.5. The output is given in D8-4.6. An estimate of the probability that the thief reaches freedom before he dies is obtained as the number of observations made at node C1 (594) divided by 1000.

8-5.   One entity is created at time .0001 and activity 1 and activity 2 are scheduled to end at time 0.0001. Since they both have 0 times, they will be taken from the event calendar in the order in which they are placed in the input statements. Thus, an entity arrives to node A and then an entity arrives to node B. Since a default value is indicated for the time between creations, no further entities are scheduled from the CREATE node. Also, a limit of 1 creation is specified.

The entity arrival to node A occurs at time 0.0001 and activities 3 and 5 are scheduled to end at time 1.0001. Assuming activity 3 precedes activity 5 on the input statements, the arrival to node C will precede the arrival to node D, both occurring at time 1.0001.

An entity arrives to node B at time 0.0001 and activities 4 and 6 are scheduled to end at time 1.0001. At this point, four events are on the event calendar representing completions of activities 3, 5, 4, and 6. The events are in the above order as the secondary ranking for the future events file is specified as FIFO in the problem statement. This would also be the default specification. Thus, the sequencing of activity completions for the network assuming a FIFO secondary ranking for events, is 1,2,3,5,4,6.

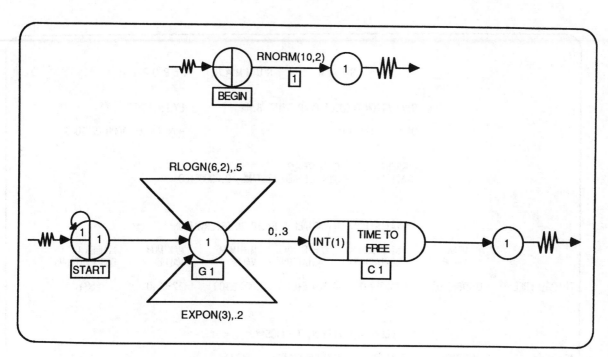

D8-4.4

```
 1   GEN,PRITSKER,DEAD THIEF 8.4A,7/18/80,1000,,N,,N,Y/1000;
 2   LIMITS,0,1,10;
 3   INIT,,,N;
 4   NETWORK;
 5   START    CREATE,,,1;
 6   G1       GOON,1;
 7            ACT,RLOGN(6,2),.5,G1;
 8            ACT,,.3,C1;
 9            ACT,EXPON(3),.2,G1;
10   C1       COLCT,INT(1),TIME TO FREE,20/0/1;
11            TERM,1;
12   BEGIN    CREATE;
13            ACT/1,RNORM(10,2);
14            TERM,1;
15            ENDNETWORK;
16   FIN;
```

D8-4.5

```
            S L A M   I I   S U M M A R Y   R E P O R T

         SIMULATION PROJECT DEAD THIEF 8.4A          BY PRITSKER

         DATE  7/18/1980                        RUN NUMBER 1000 OF 1000


         CURRENT TIME   0.1197E+02
         STATISTICAL ARRAYS CLEARED AT TIME  0.0000E+00



         **STATISTICS FOR VARIABLES BASED ON OBSERVATION**

            MEAN        STANDARD    COEFF. OF   MINIMUM     MAXIMUM     NUMBER OF
            VALUE       DEVIATION   VARIATION   VALUE       VALUE       OBSERVATIONS

TIME TO FREE  0.2786E+01  0.3458E+01  0.1241E+01  0.0000E+00  0.1238E+02    594


         **REGULAR ACTIVITY STATISTICS**

ACTIVITY      AVERAGE      STANDARD    MAXIMUM CURRENT  ENTITY
INDEX/LABEL   UTILIZATION  DEVIATION   UTIL    UTIL     COUNT

   1          0.7047       0.4562        1       1        0
```

D8-4.6

8-5,Embellishment. When a LIFO ranking is specified for the event calendar by the following statement:

PRIORITY/NCLNR,LIFO;

the activity completions occur in the following order 1,2,6,4,5,3. The secondary ranking rule for the event calendar does not alter the ranking of 0 time activities as they are performed in the order in which the activity statements appear on the input statements.

8-6.   The statement model and trace for the FIFO case are shown in D8- 6.1 and D8-6.2.

ASSIGN nodes were added so as to be able to follow the entity flow. (The ASSIGN nodes could have replaced the GOON nodes, however, by adding the ASSIGN nodes further activity sequencing is illustrated.)

Since a trace is requested, SLAM II requires that at least two attribute values be used and this is indicated on the LIMITS statement. The use of the small time for the first creation was to allow comparison with a trace obtained for the LIFO case where initial events occurring at time 0 would have been placed in front of the event that triggered the monitoring.

```
 1    GEN,ROLSTON,PROBLEM 8.6,7/18/80,2;
 2    LIMITS,0,3,10;
 3    NETWORK;
 4            CREATE,,.0001,,1;
 5                ACT/1,,,AS1;
 6                ACT/2,,,AS2;
 7    AS1     ASSIGN,ATRIB(1)=1;
 8                ACT,,,A;
 9    AS2     ASSIGN,ATRIB(1)=2;
10                ACT,,,B;
11    A       GOON;
12                ACT/3,1.0,,C;
13                ACT/5,1.0,,D;
14    B       GOON;
15                ACT/4,1.0,,C;
16                ACT/6,1.0,,D;
17    C       TERM;
18    D       TERM;
19            ENDNETWORK;
20    MONTR,TRACE;
21    SIMULATE;
22    PRIORITY/NCLNR,LIFO;
23    MONTR,TRACE;
24    FIN;
```

D8-6.1

SLAM II TRACE BEGINNING AT TNOW= 0.0000E+00

TNOW	JEVNT	NODE ARRIVAL		CURRENT VARIABLE BUFFER		
		LABEL	TYPE			
0.1000E-03			CREATE	0.0000E+00	0.0000E+00	0.0000E+00
		AS1	ASSIGN	0.0000E+00	0.0000E+00	0.0000E+00
		A	GOON	0.1000E+01	0.0000E+00	0.0000E+00
		AS2	ASSIGN	0.0000E+00	0.0000E+00	0.0000E+00
		B	GOON	0.2000E+01	0.0000E+00	0.0000E+00
0.1000E+01		C	TERM	0.1000E+01	0.0000E+00	0.0000E+00
		D	TERM	0.1000E+01	0.0000E+00	0.0000E+00
		C	TERM	0.2000E+01	0.0000E+00	0.0000E+00
		D	TERM	0.2000E+01	0.0000E+00	0.0000E+00

D8-6.2

8-7   A good problem to assign is the exponential interarrival, exponential service, single channel, single server queueing situation. Although there is a lot of variability in this problem situation due to the use of the exponential distributions, most students are familiar with the problem and it provides a good comparison of results. Using an arrival rate of 1 and a service rate of 1.25, the expected number in the system is 4, the expected number in the queue is 3.2, and the average waiting time in the queue is 3.2. The utilization of the server in this case is 0.8. If the simulation is run for 1000 times units then the expected number of arrivals is 1000.

8-8.   The listing of the input statements for this problem is shown in D8-8.1. Note that the STAT and CONTINUOUS statements are only briefly mentioned in Chapter 8.

```
 1    GEN,ROLSTON,PROBLEM 8.8,8/12/80,1;
 2    LIMITS,2,4,25;
 3    PRIORITY/1,HVF(4);
 4    STAT,1,HT;
 5    STAT,2,WT;
 6    STAT,3,TOL;
 7    STAT,4,SIZE;
 8    STAT,5,GRADE,15/10/5;
 9    TIMST,SS(1),BAL;
10    TIMST,SS(2),POL;
11    CONTINUOUS,3,2,.001,1.0,.0,,.0001,.00001;
12    RECORD,TNOW,TIME,0,B,2.0;
13    VAR,SS(3),X,XPOS,MIN(1),MAX(1);
14    VAR,SS(4),Y,YPOS,MIN(1),MAX(1);
15    VAR,SS(5),Z,ZPOS,MIN(1),MAX(1);
16    SEEDS,567471923;
17    INIT,10;
18    ENTRY/2,3,5;
19    ENTRY/3,0,0,0,0,2,11/3,0,0,0,0,3,225;
20    MONTR,TRACE,50,100;
21    MONTR,SUMRY,100,25;
22    FIN;
```

D8-8.1

CHAPTER 9

Network Modeling With User-Written Inserts

9-1. The network of Figure 6-9 is changed by adding an EVENT node after the PREEMPT node in order to remove from file 1 all but the last three entities to arrive. Note that the default value is used for the send node label for the unit that is preempted. When the default value is used, the entity preempted is returned to the AWAIT node at which it seized the resource. It is placed at the head of the queue waiting for the resource.

The removal and processing of all entities but the last three to arrive is done in subroutine EVENT(I). The code for the EVENT subroutine is shown in D9-1.1. Statistics are collected on the number of units subcontracted and the time between the subcontracting of units. The input statements for this model are shown in D9-1.2, and the output summary report is shown in D9-1.3. From D9-1.3, it is seen that the resource is utilized over 90 percent of the time and from statistics on activity 3, it is being repaired over 9 percent of the time. Also from the statistics on activity 3, we see that 22 repairs occurred during the 500 time unit simulation. Since the repair time on the average is 2.25 time units and the time between breakdowns is 20 time units, we expect $500/22.25=22.47$ breakdowns to occur. From statistics on the number of subcontractings performed, it is seen that 13 observations were made. Thus, in 13 of the 22 breakdowns there were more than three units waiting for the resource. When units were subcontracted, four units were sent out on the average and a minimum of one unit and maximum of 10 units were sent out. Note that the time in the system is printed after statistics obtained through calls to subroutine COLCT.

```
          SUBROUTINE EVENT(I)
          COMMON/SCOM1/ATRIB(100),DD(100),DDL(100),DTNOW,II,MFA,MSTOP,NCLNR
         1,NCRDR,NPRNT,NNRUN,NNSET,NTAPE,SS(100),SSL(100),TNEXT,TNOW,XX(100)
      C   CHECK NO. OF JOBS WAITING FOR TOOL
          IF(NNQ(1).LT.4) RETURN
      C   IF FOUR OR MORE, INCLUDING INTERRUPTED JOB,
      C   REMOVE ALL BUT LAST THREE TO ARRIVE FROM QUEUE
          NUM=NNQ(1)-3
          DO 10 J=1,NUM
          CALL RMOVE(1,1,ATRIB)
   10     CONTINUE
      C   COLLECT STATISTICS ON NUMBER REMOVED
      C   AND TIME BETWEEN REMOVALS
          XNUM=NUM
          CALL COLCT(XNUM,1)
          TBD=TNOW-XX(1)
          XX(1)=TNOW
          CALL COLCT(TBD,2)
          RETURN
          END
```

D9-1.1

```
    1   GEN,ROLSTON,PROBLEM 9.1,7/2/86,1;
    2   LIMITS,2,2,50;
    3   STAT,1,NUM SUBC;
    4   STAT,2,TIME BET SUBC;
    5   NETWORK;
    6           RESOURCE/TOOL,2,1;
    7           CREATE,EXPON(1.),,1;
    8           AWAIT(1),TOOL;
    9           ACT/1,UNFRM(.2,.5);
   10           GOON;
   11           ACT/2,RNORM(.5,.1);
   12           FREE,TOOL;
   13           COLCT,INT(1),TIME IN SYSTEM;
   14           TERM;
   15           CREATE,,20.,,1;
   16   DOWN    PREEMPT(2),TOOL;
   17           EVENT,1;
   18           ACT/3,ERLNG(.75,3.);
   19           FREE,TOOL;
   20           ACT,RNORM(20.,2.),,DOWN;
   21           END;
   22   INIT,0,500;
   23   INTLC,XX(1)=0;
   24   FIN;
```

D9-1.2

From the regular activity statistics section of the summary report, it is seen that the resource is busy over 33 percent of the time in setups, 47 percent of the time in processing, and 9 percent of the time being repaired. These three utilization values sum to the resource utilization of .9018. Also of interest is the information that nine more setups were performed than units processed through the system, that is, 9 units were preempted and routed to subcontracting because of failures.

```
                    S L A M   I I   S U M M A R Y   R E P O R T

         SIMULATION PROJECT PROBLEM 9.1              BY ROLSTON

         DATE  7/ 2/1986                             RUN NUMBER   1 OF    1


         CURRENT TIME   0.5000E+03
         STATISTICAL ARRAYS CLEARED AT TIME  0.0000E+00


              **STATISTICS FOR VARIABLES BASED ON OBSERVATION**

                   MEAN        STANDARD     COEFF. OF    MINIMUM      MAXIMUM     NUMBER OF
                   VALUE       DEVIATION    VARIATION    VALUE        VALUE       OBSERVATIONS

    NUM SUBC       0.4700E+01  0.2983E+01   0.6347E+00   0.1000E+01   0.9000E+01     10
    TIME BET SUBC  0.4839E+02  0.5693E+02   0.1176E+01   0.1925E+02   0.1613E+03     10
    TIME IN SYSTEM 0.4209E+01  0.2531E+01   0.6013E+00   0.6372E+00   0.1157E+02    464


                         **FILE STATISTICS**

    FILE                      AVERAGE     STANDARD    MAXIMUM    CURRENT    AVERAGE
    NUMBER  LABEL/TYPE        LENGTH      DEVIATION   LENGTH     LENGTH     WAITING TIME

      1            AWAIT      3.5996      3.2763        14         5        3.4812
      2     DOWN PREEMPT      0.0000      0.0000         1         0        0.0000
      3            CALENDAR   2.8044      0.3967         4         3        0.5555


                    **REGULAR ACTIVITY STATISTICS**

    ACTIVITY        AVERAGE       STANDARD    MAXIMUM CURRENT   ENTITY
    INDEX/LABEL     UTILIZATION   DEVIATION   UTIL    UTIL      COUNT

       1            0.3334        0.4714        1       0        471
       2            0.4710        0.4992        1       1        464
       3            0.0919        0.2889        1       0         22


                       **RESOURCE STATISTICS**

    RESOURCE  RESOURCE  CURRENT   AVERAGE      STANDARD    MAXIMUM      CURRENT
    NUMBER    LABEL     CAPACITY  UTILIZATION  DEVIATION   UTILIZATION  UTILIZATION

       1      TOOL         1      0.8963       0.3049        1            1


    RESOURCE  RESOURCE  CURRENT    AVERAGE     MINIMUM     MAXIMUM
    NUMBER    LABEL     AVAILABLE  AVAILABLE   AVAILABLE   AVAILABLE

       1      TOOL         0       0.1037         0           1
```

D9-1.3

9-2.   To model a breakdown of the repair process that causes an added delay due to a lack of a spare part, the disjoint network involving the PREEMPT node is modified as shown in D9-2.1. No user-written functions are necessary.

A disjoint network is used to preempt the repair process if it is ongoing. Attribute 1 is the mark time for the entities. It is always zero for the breakdown entity and greater than zero for the repair breakdown. The repair breakdown is given priority at the PREEMPT node by using HIGH(1) for its PR parameter.

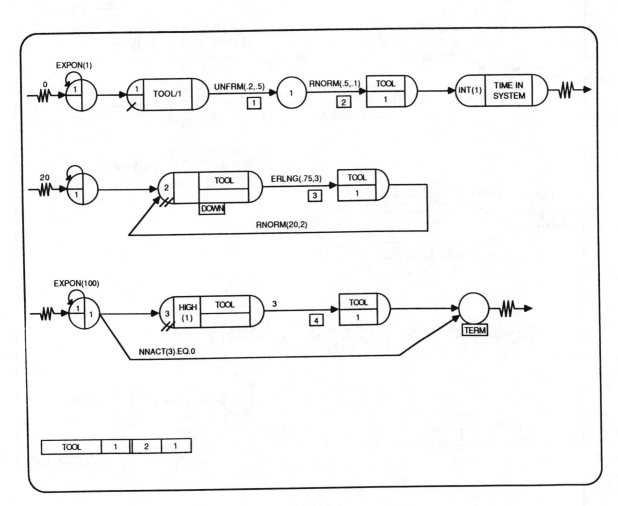

D9-2.1

9-3. The network and statement models are presented in D9-3.1 and D9-3.2. The SLAM II variables are defined below.

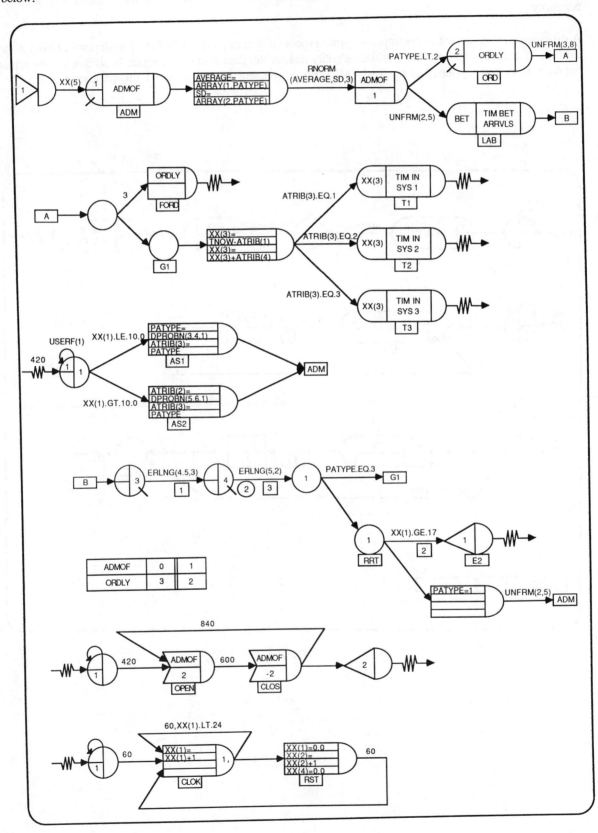

D9-3.1

```
 1      GEN,ROLSTON,PROBLEM 9.3,8/20/86,1;
 2      LIMITS,4,4,200;
 3      PRIORITY/1,LVF(2);
 4      STAT,1,NUM IN Q1;
 5      ARRAY(1,3)/15.0,40.0,30.0;
 6      ARRAY(2,3)/1.5,4.0,3.0;
 7      ARRAY(3,2)/0.9,1.0;
 8      ARRAY(4,2)/1.0,2.0;
 9      ARRAY(5,3)/0.5,0.6,1.0;
10      ARRAY(6,3)/1.0,2.0,3.0;
11      EQUIVALENCE/ATRIB(2),PATYPE/XX(6),AVERAGE/XX(7),SD;
12      NETWORK;
13            RESOURCE/ADMOF(0),1;
14            RESOURCE/ORDLY(3),2;
15            CREATE,,,,1;
16            ACT,60;
17  CLOK  ASSIGN,XX(1)=XX(1)+1,1;
18            ACT,60.0,XX(1).LT.24,CLOK;
19            ACT,,,RST;
20  RST   ASSIGN,XX(1)=0.0,
21                  XX(2)=XX(2)+1,
22                  XX(4)=0;
23            ACT,60,,CLOK;
24  ;
25            CREATE,,,,1;
26            ACT,420;
27  OPEN  ALTER,ADMOF/2;
28            ACT,600;
29  CLOS  ALTER,ADMOF/-2;
30            ACT,840,,OPEN;
31            ACT;
32            EVENT,2;
33            TERM;
34  ;
35            CREATE,USERF(1),420.0,1,,1;
36            ACT,,XX(1).LE.10.0,AS1;
37            ACT,,XX(1).GT.10.0,AS2;
38  AS1   ASSIGN,PATYPE=DPROBN(3,4,1),ATRIB(3)=PATYPE;
39            ACT,,,ADM;
40  AS2   ASSIGN,ATRIB(2)=DPROBN(5,6,1),ATRIB(3)=PATYPE;
41            ACT,,,ADM;
42  ;
43            ENTER,1;
44            ACT,XX(5);
45  ADM   AWAIT(1),ADMOF;
46            ASSIGN,AVERAGE=ARRAY(1,PATYPE),SD=ARRAY(2,PATYPE);
47            ACT,RNORM(AVERAGE,SD,3);
48            FREE,ADMOF,1;
49            ACT,,PATYPE.LT.2,ORD;
50            ACT,UNFRM(2,5),,LAB;
51  ORD   AWAIT(2),ORDLY;
52            ACT,UNFRM(3,8);
53            GOON;
54            ACT,3,,FORD;
55            ACT,,,G1;
56  FORD  FREE,ORDLY;
57            TERM;
58  LAB   COLCT,BET,TIM BET ARRVLS;
59            QUEUE(3);
60            ACT/1,ERLNG(4.5,3);
61            QUEUE(4);
62            ACT(2)/3,ERLNG(5.,2);
63            GOON,1;
64            ACT,,PATYPE.EQ.3,G1;
65            ACT;
66  RRT   GOON,1;
67            ACT/2,,XX(1).GE.17,E2;
68            ACT;
69            ASSIGN,PATYPE=1;
70            ACT,UNFRM(2,5),,ADM;
71  E2    EVENT,1;
72            TERM;
73  G1    GOON;
74            ASSIGN,XX(3)=TNOW-ATRIB(1),
75                  XX(3)=XX(3)+ATRIB(4);
76            ACT,,ATRIB(3).EQ.1,T1;
77            ACT,,ATRIB(3).EQ.2,T2;
78            ACT,,ATRIB(3).EQ.3,T3;
79  T1    COLCT,XX(3),TIM IN SYS 1;
80            TERM;
81  T2    COLCT,XX(3),TIM IN SYS 2;
82            TERM;
83  T3    COLCT,XX(3),TIM IN SYS 3;
84            TERM;
85            END;
86      INIT,0,14400;
87      FIN;
```

XX(1) Number of hours (used when assigning patient types)
XX(2) Number of days
XX(3) Time in system
XX(4) Number of next day arrivals
XX(5) Time delay until next day arrival
XX(6) Average
XX(7) Standard deviation
ATRIB(1) Arrival time
ATRIB(2) Patient type
ATRIB(3) Patient type (for collecting stats)
ATRIB(4) Accumulated time in system.

An entity is created at the beginning of the simulation and routed to ASSIGN node CLOK which causes the variable XX(1) to be incremented after 60 time units. XX(1) is used to keep track of time in hours on the network. XX(2) represents the time in days and is incremented every 24 hours. Both of these variables are used in the network and in the discrete event routines to stop and start the flow of entities. XX(4) is utilized in subroutine COMBCK to calculate interarrival times.

A disjoint network is used to model the arrival and departure of patients to the resource, admitting office. The initial capacity is zero; at time 420 (7:00 A.M.) it is increased by 2. After 600 time units (5:00 P.M.), the capacity is changed back to zero. A feedback loop repeats this process for the next day.

If the admission officer resource is utilized at the time the ALTER node is activated, the resource is altered when it is freed. The arrival and departure of other servers is not modeled. Entities which are finished with the admitting process when the admissions officers leave are allowed to continue through the network; that is, the system is allowed to empty out.

Patient entities are inserted into the network through the arrival subnetwork or at ENTER node 1 by a call to subroutine ENTER. Function DPROBN is used to assign the admitting time. Branching is based on patient type,PATYPE, which is equivalenced to ATRIB(2). At GOON node RRT, a test is made on the time variable XX(1). If it is before 5:00 P.M. (17.0), the patient entity is routed back to the admissions office after changing the PATYPE to 1 so that it will now be processed as a type 1 entity. If it is after 5:00 P.M., subroutine COMBCK is called in EVENT 1.

Subroutine EVENT and function USERF are shown in D9-3.3 and D9-3.4. Subroutine COMBCK is shown in D9-3.5. This event is called when a type 2 entity, which has finished in the lab, cannot return to the admissions office because it is after 5:00 P.M. The time in system is stored in ATRIB(4), the PATYPE is set to one so that the patient will be processed as a type 2, and an arrival is scheduled for the next day. The logic is set up so that the entities scheduled to arrive through this routine will do so with an exponential interarrival distribution.

Subroutine BOOT (D9-3.6) is called when the admissions office capacity is reset to zero by having an entity arrive to an EVENT node with a code of 2. This routine is used to remove any entities in the AWAIT node and reschedule their arrival on the next day. A summary report is given in D9-3.7.

```
                     SUBROUTINE EVENT(I)
                     GOTO (1,2),I
           1         CALL COMBCK
                     RETURN
           2         CALL BOOT
                     RETURN
                     END
```

```
      FUNCTION USERF(IFN)
      COMMON/SCOM1/ATRIB(100),DD(100),DDL(100),DTNOW,II,MFA,MSTOP,NCLNR
     1,NCRDR,NPRNT,NNRUN,NNSET,NTAPE,SS(100),SSL(100),TNEXT,TNOW,XX(100)
      DTIME=EXPON(15.0,2)
      TABS=TNOW+DTIME
      TLIM=XX(2)*1440.0+16.0*60.0
      IF (TLIM.LE.TABS) GOTO 10
      USERF=DTIME
      RETURN
   10 CONTINUE
      TMROW=(XX(2)+1.0)*1440.0+7.0*60.0
      USERF=TMROW+DTIME-TNOW
      RETURN
      END
```

D9-3.4

```
      SUBROUTINE COMBCK
      COMMON/SCOM1/ATRIB(100),DD(100),DDL(100),DTNOW,II,MFA,MSTOP,NCLNR
     1,NCRDR,NPRNT,NNRUN,NNSET,NTAPE,SS(100),SSL(100),TNEXT,TNOW,XX(100)
      COMMON/UCOM2/TARV
C     SAVE TIME IN SYSTEM
      TISYS=TNOW-ATRIB(1)
      ATRIB(4)=ATRIB(4)+TISYS
C     SET ATRIB(2) TO 1 TO PROCESS AS TYPE 1 PATIENT
      ATRIB(2)=1.0
C     SCHEDULE ARRIVAL TOMORROW AT EXPONENTIAL RATE
      TMROW=(XX(2)+1.0)*1440.0+7.0*60.0
      XX(4)=XX(4)+1
      IF(XX(4).EQ.1.0) TARV=0.0
      TARV=TARV+EXPON(15.0,2)
      DTIME=TMROW+TARV-TNOW
      ATRIB(1)=TNOW+DTIME
      XX(5)=DTIME
      CALL ENTER(1,ATRIB)
      RETURN
      END
```

D9-3.5

```
      SUBROUTINE BOOT
      COMMON/SCOM1/ATRIB(100),DD(100),DDL(100),DTNOW,II,MFA,MSTOP,NCLNR
     1,NCRDR,NPRNT,NNRUN,NNSET,NTAPE,SS(100),SSL(100),TNEXT,TNOW,XX(100)
C     COLLECT STATS ON NUMBER LEFT IN QUEUE WHEN
C     ADMISSIONS OFFICE CLOSES
      NPULL=NNQ(1)
      XPULL=NPULL
      CALL COLCT(XPULL,1)
      IF(NPULL.EQ.0) RETURN
C     SCHEDULE THEIR ARRIVAL TOMORROW AT EXPONENTIAL RATE
      DMROW=840.0
      DO 10 IFROM=1,NPULL
      CALL RMOVE(1,1,ATRIB)
      DMROW=DMROW+EXPON(15.0,2)
      TISYS=TNOW-ATRIB(1)
      ATRIB(4)=ATRIB(4)+TISYS
      XX(5)=DMROW
      CALL ENTER(1,ATRIB)
   10 CONTINUE
      RETURN
      END
```

D9-3.6

```
                          S L A M   I I   S U M M A R Y   R E P O R T

              SIMULATION PROJECT PROBLEM 9.3          BY ROLSTON

              DATE  8/20/1986                         RUN NUMBER    1 OF    1


              CURRENT TIME   0.1440E+05
              STATISTICAL ARRAYS CLEARED AT TIME  0.0000E+00


                    **STATISTICS FOR VARIABLES BASED ON OBSERVATION**

                      MEAN        STANDARD      COEFF. OF     MINIMUM       MAXIMUM      NUMBER OF
                      VALUE       DEVIATION     VARIATION     VALUE         VALUE        OBSERVATIONS

  NUM IN Q1           0.7000E+00  0.1494E+01    0.2135E+01    0.0000E+00    0.4000E+01        10
  TIM BET ARRVLS      0.1216E+03  0.2690E+03    0.2212E+01    0.6113E+00    0.1156E+04       110
  TIM IN SYS 1        0.2671E+02  0.8341E+01    0.3123E+00    0.1618E+02    0.5263E+02       215
  TIM IN SYS 2        0.1436E+03  0.1415E+03    0.9853E+00    0.7754E+02    0.9726E+03        39
  TIM IN SYS 3        0.1930E+03  0.3296E+03    0.1708E+01    0.4277E+02    0.1214E+04        71


                              **FILE STATISTICS**

  FILE                     AVERAGE       STANDARD      MAXIMUM    CURRENT     AVERAGE
  NUMBER  LABEL/TYPE       LENGTH        DEVIATION     LENGTH     LENGTH      WAITING TIME

     1    ADM  AWAIT       0.4143        1.1032           8         0         16.0393
     2    ORD  AWAIT       0.0000        0.0000           1         0          0.0000
     3         QUEUE       0.0192        0.1448           2         0          2.4959
     4         QUEUE       0.0000        0.0000           0         0          0.0000
     5         CALENDAR    4.4597        1.7648          12         4         13.3402


                          **REGULAR ACTIVITY STATISTICS**

  ACTIVITY         AVERAGE       STANDARD     MAXIMUM CURRENT    ENTITY
  INDEX/LABEL      UTILIZATION   DEVIATION    UTIL    UTIL       COUNT

     2               0.0000        0.0000        1       0          3


                          **SERVICE ACTIVITY STATISTICS**

  ACTIVITY   START NODE OR    SERVER     AVERAGE       STANDARD      CURRENT      AVERAGE      MAXIMUM IDLE    MAXIMUM BUSY    ENTITY
  INDEX      ACTIVITY LABEL   CAPACITY   UTILIZATION   DEVIATION     UTILIZATION  BLOCKAGE     TIME/SERVERS    TIME/SERVERS    COUNT

     1          QUEUE            1         0.1114        0.3146          0          0.0000       1143.8895         98.9658       111
     3          QUEUE            2         0.0757        0.2836          0          0.0000          2.0000          2.0000       111


                              **RESOURCE STATISTICS**

  RESOURCE   RESOURCE   CURRENT    AVERAGE       STANDARD      MAXIMUM       CURRENT
  NUMBER     LABEL      CAPACITY   UTILIZATION   DEVIATION     UTILIZATION   UTILIZATION

     1       ADMOF         0        0.5234        0.8178           2             0
     2       ORDLY         3        0.1496        0.4043           2             0


  RESOURCE   RESOURCE   CURRENT     AVERAGE       MINIMUM     MAXIMUM
  NUMBER     LABEL      AVAILABLE   AVAILABLE     AVAILABLE   AVAILABLE

     1       ADMOF         0         0.3099         -2           2
     2       ORDLY         3         2.8504          1           3
```

D9-3.7

9-4.   The solution to exercise 9-4 is given in D9-4.1. The problem concerns an assembly line where the arrivals to the line are paced, although the line itself is not. Movement of units from one server to the next can occur if the current operation has been finished and the next server is idle. Each server is allowed 15 time units from the time he receives an item to complete it. A network solution without user-written inserts is provided. The logic included in this model is complex and it is anticipated that different approaches may require user functions.

For I=1,2, and 3, ATRIB(I) is the time at station I, ATRIB(4) is the remaining processing time, and ATRIB(5) is the arrival time. The assignment of activity numbers causes SLAM II to print utilization statistics on the summary report for activities. For this example, this will provide utilization statistics on the stations and the total number of entities served by them. Note that the utilization of activity 1 must be added to the utilization of activity 2 to get the total utilization of the first server -- and similarly for the other 2 servers. By collecting statistics on XX(I) variables, the number of items routed offline from server I, an estimate of the number of entities diverted from each station is obtained. With the number served and the number diverted, an estimate of the fraction of entities diverted can be made.

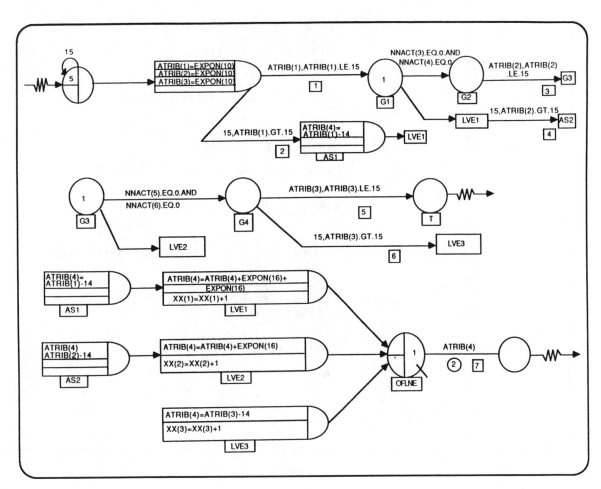

D9-4.1

In the statement model D9-4.2, a priority statement makes the secondary ranking of the event calendar low-value-first based on attribute 5, the arrival time of an entity. This secondary ranking causes events associated with entities already in the system to be processed before new entities. In this way, new entities that might begin processing at exactly the same time as an entity completing processing will not be counted as being in the activity concurrently. Although not specifying this secondary ranking for the event calendar does not change the average utilization, it would affect the maximum number of entities concurrently in a branch. A MONTR statement was used to obtain a trace of the events for the first 150 time units.

```
 1    GEN,ROLSTON,PROBLEM 9.4,9/19/80,1;
 2    LIMITS,1,5,100;
 3    PRIORITY/NCLNR,LVF(5);
 4    TIMST,XX(1),FROM1;
 5    TIMST,XX(2),FROM2;
 6    TIMST,XX(3),FROM3;
 7    NETWORK;
 8          CREATE,15,,5;
 9          ASSIGN,ATRIB(1)=EXPON(10),
10                 ATRIB(2)=EXPON(10),
11                 ATRIB(3)=EXPON(10);
12          ACT/1,ATRIB(1),ATRIB(1).LE.15,G1;
13          ACT/2,15,ATRIB(1).GT.15,AS1;
14    G1    GOON,1;
15          ACT,,NNACT(3).EQ.0.AND.NNACT(4).EQ.0,G2;
16          ACT,,,LVE1;
17    G2    GOON;
18          ACT/3,ATRIB(2),ATRIB(2).LE.15,G3;
19          ACT/4,1,ATRIB(2).GT.15,AS2;
20    G3    GOON,1;
21          ACT,,NNACT(5).EQ.0.AND.NNACT(6).EQ.0,G4;
22          ACT,,,LVE2;
23    G4    GOON;
24          ACT/5,ATRIB(3),ATRIB(3).LE.15,T;
25          ACT/6,15,ATRIB(3).GT.15,LVE3;
26    T     TERM;
27    AS1   ASSIGN,ATRIB(4)=ATRIB(1)-14;
28    LVE1  ASSIGN,ATRIB(4)=ATRIB(4)+EXPON(16)+EXPON(16),
29                 XX(1)=XX(1)+1;
30          ACT,,,OFLNE;
31    AS2   ASSIGN,ATRIB(4)=ATRIB(2)-14;
32    LVE2  ASSIGN,ATRIB(4)=ATRIB(4)+EXPON(16),
33                 XX(2)=XX(2)+1;
34          ACT,,,OFLNE;
35    LVE3  ASSIGN,ATRIB(4)=ATRIB(3)-14,
36                 XX(3)=XX(3)+1;
37    OFLNE QUEUE(1);
38          ACT(2)/7,ATRIB(4)
39          TERM;
40          END;
41    INIT,0,900;
42    MONTR,TRACE,0,150;
43    FIN;
```

D9-4.2

9-4,Embellishment(a). Network and statement models for this embellishment are given in D9-4.3 and D9-4.4. In this case, the line itself is paced as well as the arrivals, that is, the movement of items can only occur at multiples of 15 minutes.

The solution uses ACCUMULATE nodes to model the movement of items. Upon reaching the server, an entity will take one of the first two branches to represent the actual processing. In addition, an entity will be routed over a third branch which will delay the arrival to the ACCUMULATE node by 15 minutes. No branch will take longer than 15 minutes, so movement occurs every 15 minutes when the entity routed over the third branch reaches the ACCUMULATE node.

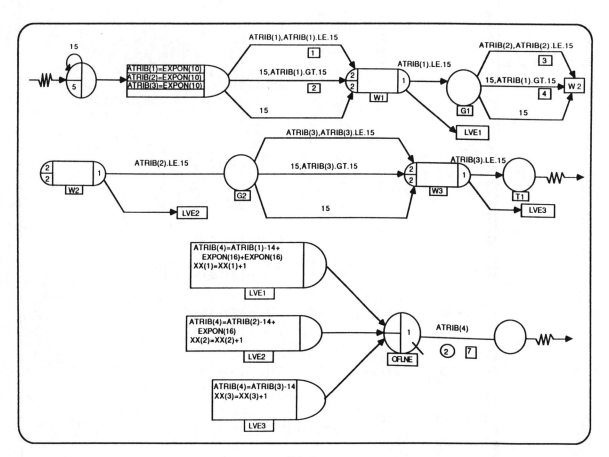

D9-4.3

```
 1    GEN,ROLSTON,PROBLEM 9.4A,7/21/80,1;
 2    LIMITS,1,5,50;
 3    PRIORITY/NCLNR,LVF(5);
 4    TIMST,XX(1),FROM1;
 5    TIMST,XX(2),FROM2;
 6    TIMST,XX(3),FROM3;
 7    NETWORK;
 8            CREATE,15,,5;
 9            ASSIGN,ATRIB(1)=EXPON(10),
10                ATRIB(2)=EXPON(10),
11                ATRIB(3)=EXPON(10);
12            ACT/1,ATRIB(1),ATRIB(1).LE.15,W1;
13            ACT/2,15,ATRIB(1).GT.15,W1;
14            ACT,15,,W1;
15    W1      ACCUMULATE,2,2,,1;
16            ACT,,ATRIB(1).LE.15,G1;
17            ACT,,,LVE1;
18    G1      GOON;
19            ACT/3,ATRIB(2),ATRIB(2).LE.15,W2;
20            ACT/4,15,ATRIB(2).GT.15,W2;
21            ACT,15,,W2;
22    W2      ACCUMULATE,2,2,,1;
23            ACT,,ATRIB(2).LE.15,G2;
24            ACT,,,LVE2;
25    G2      GOON;
26            ACT/5,ATRIB(3),ATRIB(3).LE.15,W3;
27            ACT/6,15,ATRIB(3).GT.15,W3;
28            ACT,15,,W3;
29    W3      ACCUMULATE,2,2,,1;
30            ACT,,ATRIB(3).LE.15,T1;
31            ACT,,,LVE3;
32    T1      TERM;
33    LVE1    ASSIGN,ATRIB(4)=ATRIB(1)-15,
34                ATRIB(4)=ATRIB(4)+1,
35                ATRIB(4)=ATRIB(4)+EXPON(16),
36                ATRIB(4)=ATRIB(4)+EXPON(16),
37                XX(1)=XX(1)+1;
38            ACT,,,OFLNE;
39    LVE2    ASSIGN,ATRIB(4)=ATRIB(2)-15,
40                ATRIB(4)=ATRIB(4)+1,
41                ATRIB(4)=ATRIB(4)+EXPON(16),
42                XX(2)=XX(2)+1;
43            ACT,,,OFLNE;
44    LVE3    ASSIGN,ATRIB(4)=ATRIB(3)-15,
45                ATRIB(4)=ATRIB(4)+1,
46                XX(3)=XX(3)+1;
47            ACT,,,OFLNE;
48    OFLNE   QUEUE(1);
49            ACT(2)/7,ATRIB(4);
50            TERM;
51            END;
52    INTLC,XX(1)=0,XX(2)=0,XX(3)=0;
53    INIT,0,900;
54    MONTR,TRACE,0,200;
55    FIN;
```

D9-4.4

9-4,Embellishment(b). In this case one entity is allowed to wait before each station on the assembly line. This change causes a new approach to be used as resources must now be included to allow for the waiting of entities. The network and statement models are shown in D9-4.5 and D9-4.6.

9-4,Embellishment(c). The network model for the solution to embellishment (c) is shown in D9-4.7. This model allows units to be recycled back to the on-line servers if they are idle. An additional attribute, ATRIB(6), is set equal to the server number to which the unit should be recycled. The offline servers are modeled as resources to allow the use of conditional branching after the resource is allocated. This solution illustrates the use of NNACT(I), NNQ(I), and ATRIB(I) as conditional branching variables.

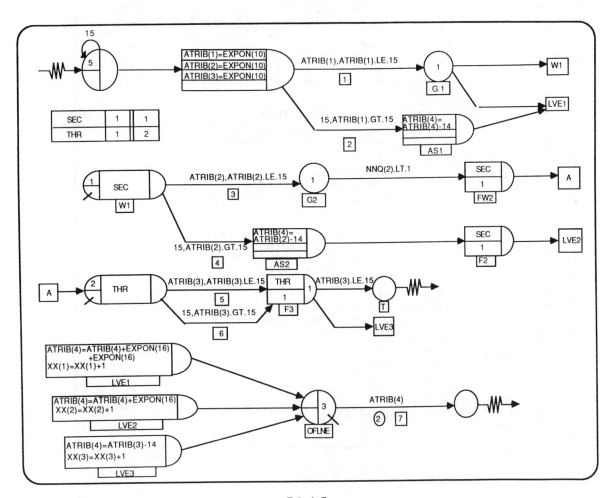

D9-4.5

```
 1    GEN,ROLSTON,PROBLEM 9.4B,7/21/80,1;
 2    LIMITS,3,6,50;
 3    PRIORITY/NCLNR,LVF(5);
 4    TIMST,XX(1),FROM 1;
 5    TIMST,XX(2),FROM 2;
 6    TIMST,XX(3),FROM 3;
 7    NETWORK;
 8              RESOURCE/SEC(1),2;
 9              RESOURCE/THR(1),3;
10              CREATE,15,,5;
11              ASSIGN,ATRIB(1)=EXPON(10),
12                     ATRIB(2)=EXPON(10),
13                     ATRIB(3)=EXPON(10),
14                     ATRIB(4)=TNOW+30,
15                     ATRIB(5)=TNOW+45,1;
16              ACT/1,ATRIB(1),ATRIB(1).LE.15,WAIT2;
17              ACT/2,15,ATRIB(1).GT.15,LVE1;
18    WAIT2     AWAIT(2),SEC/1;
19              ASSIGN,XX(4)=ATRIB(4)-TNOW,1;
20              ACT,ATRIB(2),ATRIB(2).LE.XX(4),FSEC;
21              ACT,XX(4),ATRIB(2).GT.XX(4),LVE2;
22    FSEC      FREE,SEC/1;
23              AWAIT(3),THR/1;
24              ASSIGN,XX(5)=ATRIB(5)-TNOW,1;
25              ACT,ATRIB(3),ATRIB(3).LE.XX(5),FTHR;
26              ACT,XX(5),ATRIB(3).GT.XX(5),LVE3;
27    FTHR      FREE,THR/1;
28              TERM;
29    LVE1      ASSIGN,ATRIB(6)=ATRIB(1)-15,
30                     ATRIB(6)=ATRIB(6)+1,
31                     ATRIB(6)=ATRIB(6)+EXPON(16)+EXPON(16),
32                     XX(1)=XX(1)+1;
33              ACT,,,OFLNE;
34    LVE2      ASSIGN,ATRIB(6)=ATRIB(2)-XX(4),
35                     ATRIB(6)=ATRIB(6)+1,
36                     ATRIB(6)=ATRIB(6)+EXPON(16),
37                     XX(2)=XX(2)+1;
38              FREE,SEC/1;
39              ACT,,,OFLNE;
40    LVE3      ASSIGN,ATRIB(6)=ATRIB(3)-XX(5),
41                     ATRIB(6)=ATRIB(6)+1,
42                     XX(3)=XX(3)+1;
43              FREE,THR/1;
44    OFLNE     QUEUE(1);
45              ACT(2)/3,ATRIB(6);
46              TERM;
47              END;
48    INIT,0,900;
49    INTLC,XX(1)=0,XX(2)=0,XX(3)=0;
50    FIN;
```

D9-4.6

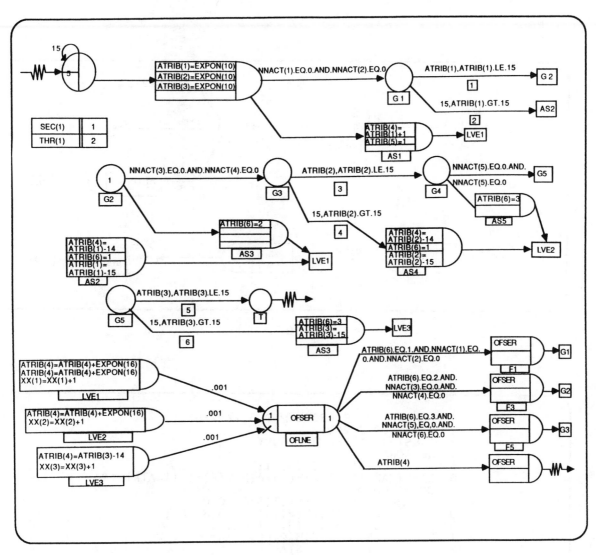

D9-4.7

9-5.   For an analytic and simulation discussion of this exercise see Pritsker, A.A.B., "Application of Multichannel Queueing Results to the Analysis of Conveyor Systems", The Journal of Industrial Engineering, Vol. XVII, No. 1, January 1966, pp. 14-21. The network and statement models for this example are shown in D9-5.1 and D9-5.2. This exercise illustrates the use of conditional branching based on the number of entities in an activity and an attribute based activity number. The simulation run is set to be run for 100 time units. A TIMST statement is used to collect statistics on the number busy which is represented by the XX(1) variable. The INTLC statement which sets XX(1) to 0 is not required as SLAM II does this automatically. However, it is a good practice to include the statement as it would be required if multiple runs are to be made. The utilization of a specific server is the utilization of the corresponding numbered activity.

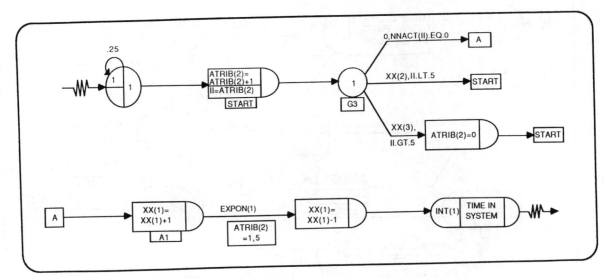

D9-5.1

```
 1   GEN,PRITSKER,PROBLEM 9.5,7/22/86,1;
 2   LIMITS,0,2,200;
 3   TIMST,XX(1),NUMBER BUSY;
 4   NETWORK;
 5           CREATE,.25,,1,,1;
 6   START ASSIGN,ATRIB(2)=ATRIB(2)+1,II=ATRIB(2);
 7   G3      GOON,1;
 8           ACTIVITY,,NNACT(II).EQ.0,A1;
 9           ACTIVITY,XX(2),II.LT.5,START;
10           ACTIVITY,XX(3),II.GE.5;
11           ASSIGN,ATRIB(2)=0;
12           ACTIVITY,,,START;
13   A1      ASSIGN,XX(1)=XX(1)+1;
14           ACTIVITY/ATRIB(2)=1,5,EXPON(1);
15           ASSIGN,XX(1)=XX(1)-1;
16           COLCT,INT(1),TIME IN SYSTEM;
17           TERMINATE;
18           END;
19   INIT,0,100;
20   INTLC,XX(1)=0,XX(2)=1,XX(3)=5;
21   FIN;
```

D9-5.2

9-5,Embellishment(a). Change the INTLC statement to set XX(2)=2 and XX(3)=10.

9-5,Embellishment(b). To allow infinite storage before the fifth server, make the fifth server a resource with a capacity of 1. Items routed to the fifth server are sent to an AWAIT node. A resource is used in order to increase XX(1) when an item is put into service and to decrease XX(1) when server 5 completes service. If statistics were not being calculated on the number of busy servers, a QUEUE node and service activity could be used to perform this function. A complete statement listing is shown in D9-5.3.

9-5,Embellishment(c). The network for the situation in which there is one space before each server is shown in D9-5.4. Each server is a resource and ATRIB(2) is defined as the resource number. Each of the files associated with the AWAIT node WORK has a capacity of 1. An item which arrives at a full workstation balks to the ASSIGN node labeled NEXT, where ATRIB(2) is indexed by 1. If an entity is recycled, it is routed back to the STRT node with a 5 minute delay. Activity 1 is used to represent the processing of all 5 servers. Statistics on the number of servers busy are obtained from the outputs for activity 1.

```
 1    GEN,PRITSKER,PROBLEM 9.5B,7/22/86,1;
 2    LIMITS,1,2,200;
 3    TIMST,XX(1),NUMBER BUSY;
 4    NETWORK;
 5           RESOURCE/FIVE(1),1;
 6           CREATE,.25,,1,,1;
 7    START  ASSIGN,ATRIB(2)=ATRIB(2)+1,II=ATRIB(2);
 8    G3     GOON,1;
 9           ACTIVITY,,NNACT(II).EQ.0,A1;
10           ACTIVITY,XX(2),II.LT.4,START;
11           ACTIVITY,XX(3),II.EQ.4;
12           AWAIT(1),FIVE/1;
13           ASSIGN,XX(1)=XX(1)+1;
14           ACTIVITY/5,EXPON(1);
15           ASSIGN,XX(1)=XX(1)-1;
16           FREE,FIVE/1;
17           ACTIVITY,,,CL1;
18    A1     ASSIGN,XX(1)=XX(1)+1;
19           ACTIVITY/ATRIB(2)=1,4,EXPON(1);
20           ASSIGN,XX(1)=XX(1)-1;
21    CL1    COLCT,INT(1),TIME IN SYSTEM;
22           TERMINATE;
23           END;
24    INIT,0,100;
25    INTLC,XX(1)=0,XX(2)=1,XX(3)=5;
26    FIN;
```

D9-5.3

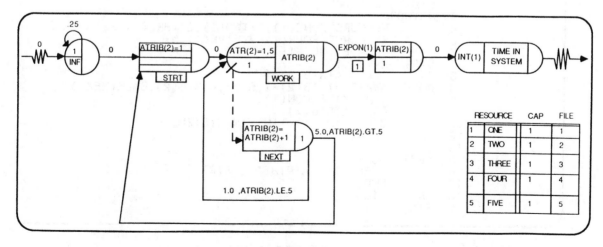

D9-5.4

9-5,Embellishment(d). Consideration should be given to revenues from parts produced, costs of in-process inventory, cost associated with additional service including machines, conveyors and people.

9-6.   The first approach will be to model a centralized control strategy. D9-6.1 shows a network and D9-6.2 the statements for a centralized control to route the arriving entities to a server based on a decision rule coded in the EVENT(I) subroutine (D9-6.3). The routing strategy is as follows: send the item to the closest server who is idle; if all are busy, send the item to the server farthest away with less than two in its queue; if all servers are busy and all queues are full, send the item around the conveyor. Branches for balking are included to test whether a server's queue is full when an item arrives. This is required because the decision to send an item to a server is based on the status and number in the queue at the time an item arrives to the system, without knowledge of how many items may be on the conveyor traveling to a station. When an item balks from a full queue, the cycle time back to the server assignment node STRT is 10-ATRIB(2) as the model assumes a reassignment is necessary.

The next solution (D9-6.4, D9-6.5 and D9-6.6) uses a similar network but in this case the items are routed to the servers in a cyclic manner using variable II. Branches for balking are included again for the reason discussed above.

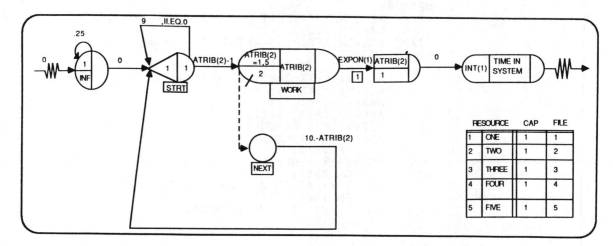

D9-6.1

```
 1    GEN,OREILLY,PROBLEM 9.6B,4/15/85,1;
 2    LIMITS,5,2,200;
 3    NETWORK;
 4          RESOURCE/ONE(1),1/TWO(1),2/THRE(1),3;
 5          RESOURCE/FOUR(1),4/FIVE(1),5;
 6          CREATE,.25,,1;
 7    STRT  EVENT,1,1;
 8          ACT,ATRIB(2)-1;
 9    WORK  AWAIT(ATRIB(2)=1,5/2),ATRIB(2),BALK(NEXT);
10          ACT/1,EXPON(1);
11          FREE,ATRIB(2);
12          COLCT,INT(1),TIME IN SYSTEM;
13          TERM;
14    ;
15    NEXT  GOON;
16          ACT,10.-ATRIB(2),,STRT;
17          END;
18    INIT,0,100;
19    FIN;
```

D9-6.2

```
      SUBROUTINE EVENT(I)
      COMMON/SCOM1/ATRIB(100),DD(100),DDL(100),DTNOW,II,MFA,MSTOP,NCLNR
     1,NCRDR,NPRNT,NNRUN,NNSET,NTAPE,SS(100),SSL(100),TNEXT,TNOW,XX(100)
      DO 10 J=1,5
      II=J
      IF(NRUSE(J).EQ.0) GO TO 90
10    CONTINUE
      DO 20 J=1,5
      K=6-J
      II=K
      IF(NNQ(K).LT.2) GO TO 90
20    CONTINUE
      II=0
90    ATRIB(2)=II
      RETURN
      END
```

D9-6.3

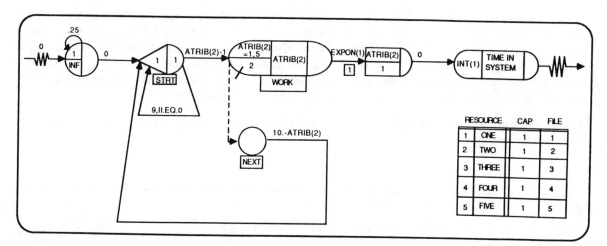

D9-6.4

```
 1    GEN,OREILLY,PROBLEM 9.6A,4/15/85,1;
 2    LIMITS,5,2,200;
 3    NETWORK;
 4            RESOURCE/ONE(1),1/TWO(1),2/THRE(1),3;
 5            RESOURCE/FOUR(1),4/FIVE(1),5;
 6            CREATE,.25,,1;
 7    STRT   EVENT,1,1;
 8              ACT,9,II.EQ.0,STRT;
 9              ACT,ATRIB(2)-1;
10    WORK   AWAIT(ATRIB(2)=1,5/2),ATRIB(2),BALK(NEXT);
11              ACT/1,EXPON(1);
12            FREE,ATRIB(2);
13            COLCT,INT(1),TIME IN SYSTEM;
14            TERM;
15          ;
16    NEXT   GOON;
17              ACT,10.-ATRIB(2),,STRT;
18            END;
19    INIT,0,100;
20    FIN;
```

D9-6.5

```
        SUBROUTINE INTLC
        COMMON/SCOM1/ATRIB(100),DD(100),DDL(100),DTNOW,II,MFA,MSTOP,NCLNR
       1,NCRDR,NPRNT,NNRUN,NNSET,NTAPE,SS(100),SSL(100),TNEXT,TNOW,XX(100)
        II=0
        RETURN
        END

        SUBROUTINE EVENT(I)
        COMMON/SCOM1/ATRIB(100),DD(100),DDL(100),DTNOW,II,MFA,MSTOP,NCLNR
       1,NCRDR,NPRNT,NNRUN,NNSET,NTAPE,SS(100),SSL(100),TNEXT,TNOW,XX(100)
        II=II+1
        IF(II.GT.5) II=1
        ATRIB(2)=II
        RETURN
        END
```

D9-6.6

The next solution (D9-6.7, D9-6.8 and D9-6.9) shows a decentralized decision rule. An item enters EVENT node STRT prior to arriving at any server. The decision as to whether the item is taken by the server or left on the conveyor is made at this point. If the server is idle at the time of arrival, the item is taken off the conveyor. If busy, it will be sent on if there is a server farther down the conveyor with a fewer number of items in service and in its queue. If the item reaches the fifth server, it is taken off the conveyor if the server is idle or if there is room in the queue. Otherwise, it is sent back to the first server.

A comparison of the routing strategies is shown in D9-6.10. There are many other decision rules that can be used to route items to servers.

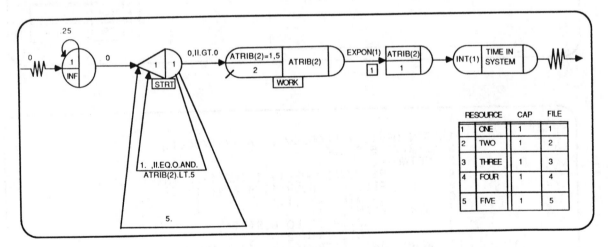

D9-6.7

```
    1   GEN,OREILLY,PROBLEM 9.6C,4/15/85,1;
    2   LIMITS,5,2,200;
    3   NETWORK;
    4           RESOURCE/ONE(1),1/TWO(1),2/THRE(1),3;
    5           RESOURCE/FOUR(1),4/FIVE(1),5;
    6           CREATE,.25,,1;
    7   STRT    EVENT,1,1;
    8             ACT,0,II.GT.0,WORK;
    9             ACT,1,II.EQ.0.AND.ATRIB(2).LT.5,STRT;
   10             ACT,5,,STRT;
   11   WORK    AWAIT(ATRIB(2)=1,5/2),ATRIB(2);
   12             ACT/1,EXPON(1);
   13           FREE,ATRIB(2);
   14           COLCT,INT(1),TIME IN SYSTEM;
   15           TERM;
   16           END;
   17   INIT,0,100;
   18   FIN;
```

D9-6.8

```
    SUBROUTINE EVENT(I)
    COMMON/SCOM1/ATRIB(100),DD(100),DDL(100),DTNOW,II,MFA,MSTOP,NCLNR
   1,NCRDR,NPRNT,NNRUN,NNSET,NTAPE,SS(100),SSL(100),TNEXT,TNOW,XX(100)
    II=ATRIB(2)+1.0
    ATRIB(2)=II
    IF (II.LT.5) THEN
      IF(NRUSE(II).EQ.0) RETURN
      K=II+1
      DO 10 J=K,5
        IF(NRUSE(II)+NNQ(II).GE.NRUSE(J)+NNQ(J)) GO TO 20
   10   CONTINUE
      RETURN
   20   II=0
    ELSE
      IF(NRUSE(II)+NNQ(II).LT.3) RETURN
      ATRIB(2)=1.0
      II=0
    ENDIF
    RETURN
    END
```

D9-6.9

		COMPARISON OF ROUTING STRATEGIES									
STRATEGY	TIME IN SYSTEM	AVERAGE QUEUE LENGTH					SERVER UTILIZATION				
		1	2	3	4	5	1	2	3	4	5
1	3.95	1.8	1.7	1.3	.73	.06	1.0	.98	.92	.71	.18
2	8.45	0	.67	.77	.63	.66	.88	.72	.68	.52	.52
3	4.38	.81	.58	.48	.53	.24	.86	.76	.76	.73	.63
4	3.04	.15	.09	.14	.13	.46	.91	.82	.78	.68	.64

D9-6.10

9-7.   The network portion of this model is shown in D9-7.1. All activities which may be started (all predecessor activities are completed) are placed in file 15 to await the opening of gate GFRA. The gate is opened at an OPEN node after a call to subroutine ENTER from an event routine. All the entities in file 15 are removed and evaluated against the condition to determine which shall be started.

The conditional branching to decide which activity should be started is done from a GOON node, instead of directly following the AWAIT node. This is necessary because of the way the entities are processed by SLAM II when the gate is opened. Modeling the opening of the gate in this manner ensures that the entities are evaluated against the branching condition in the same order they were ranked in the gate file. As a general rule, it is best to follow a GATE node with another node.

The XX(I) variables represent the amount of resource I available, and ATRIB(I) of the entity represents the amount of resource I required for each of the four resources. The first entity considered which meets the condition is routed over the branch and calls function USERF(IFN) (see D9-7.2). This function reduces the XX(I) variables by the corresponding ATRIB(I) and sets the activity time equal to ATRIB(6) of the entity. The first unit which does not meet the condition is routed to CLOSE node CL and closes the gate, so that all activities which are not started are placed back in the AWAIT file and wait for the gate to open again. The ranking of the entities in the AWAIT file determines the order in which they will be considered for resource allocation . The AWAIT file in this solution is ranked based on ATRIB(9) which is set to the activity time (ATRIB(6)).

When an activity finishes, subroutine EVENT(I) is called with I=1 from EVENT node E1. The resources used by the entity are made available and if activities are released, entities representing these activities are placed in the AWAIT file. Subroutine EVENT is given in D9-7.3.

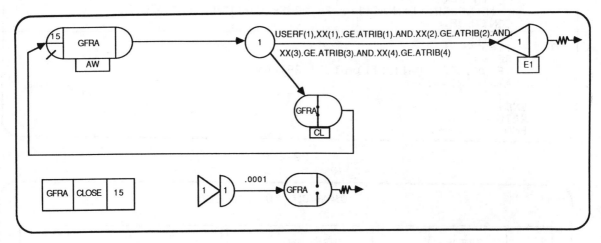

D9-7.1

```
        FUNCTION USERF(I)
        COMMON/SCOM1/ATRIB(100),DD(100),DDL(100),DTNOW,II,MFA,MSTOP,NCLNR
       1,NCRDR,NPRNT,NNRUN,NNSET,NTAPE,SS(100),SSL(100),TNEXT,TNOW,XX(100)
        DO 10 J=1,4
        XX(J)=XX(J)-ATRIB(J)
10      CONTINUE
        USERF=ATRIB(6)
        RETURN
        END
```

D9-7.2

```
        SUBROUTINE EVENT(I)
        COMMON/SCOM1/ATRIB(100),DD(100),DDL(100),DTNOW,II,MFA,MSTOP,NCLNR
       1,NCRDR,NPRNT,NNRUN,NNSET,NTAPE,SS(100),SSL(100),TNEXT,TNOW,XX(100)
        COMMON/UCOM1/NCA(14),NCN(14),TIME,KOUNT,REL(14),NUMC(22)
        INTEGER REL
C       FREE RESOURCES USED IN COMPLETING ACTIVITY
        DO 10 J=1,4
        XX(J)=XX(J)+ATRIB(J)
10      CONTINUE
C       DECREMENT NUMBER OF RELEASES FOR END NODE
        KI=ATRIB(7)
        REL(KI)=REL(KI)-1
C       IS NODE RELEASED?
        IF(REL(KI).GT.0) GOTO 20
C       YES,SAVE ACT. NO. AND START NODE FOR CRITICAL PATH CALCULATION
        NCA(KI)=ATRIB(5)
        NCN(KI)=ATRIB(8)
C       COLLECT STATS ON TIME NODE RELEASED
        TREL=TNOW-TIME
        CALL COLCT(TREL,KI)
C       IS THE NODE BEING RELEASED THE SINK NODE?
        IF(KI.EQ.10) GOTO 40
C       NO, PLACE EMANATING ACTIVITIES IN GATE FILE
        NIF=NNQ(KI)
        DO 30 N=1,NIF
        CALL COPY(N,KI,ATRIB)
        NACT=ATRIB(5)
        ATRIB(6)=ACTIM(NACT)
        ATRIB(9)=ATRIB(6)
        CALL FILEM(15,ATRIB)
30      CONTINUE
C       SCHEDULE OPENING OF GATE TO ALLOCATE RESOURCES TO ANOTHER ACTIVITY
20      CALL ENTER(1,ATRIB)
        RETURN
40      KOUNT=KOUNT+1
C       CALL ROUTINE TO REINITIALIZE VARIABLES FOR NEXT SIMULATION
C       AND TO CALCULATE CRITICAL PATH
        CALL REINT
C       IF NUMBER OF RUNS EQUALS 100, STOP
        IF(KOUNT.EQ.100) MSTOP=-1
        RETURN
        END
```

D9-7.3

The following is a list of definitions for the variables used in this exercise.

ATRIB(1)   Number of units of Resource 1 required
ATRIB(2)   Number of units of Resource 2 required
ATRIB(3)   Number of units of Resource 3 required
ATRIB(4)   Number of units of Resource 4 required
ATRIB(5)   Activity Number
ATRIB(6)   Activity Time
ATRIB(7)   End Node
ATRIB(8)   Start Node
ATRIB(9)   Priority

REL(J)   No. of Releases for Node (J)
XX(I)   No. of Units of Resource I Currently Available

NCA(I)   Critical Activity to Node I
NCN(I)   Start Node of Critical Activity
NUMC(I)   Number of Times Activity on Critical Path

File K is used to store attributes of activities emanating from Node K.

KOUNT   Number of Runs Completed
TIME   Time at which Node 1 Released

There are 14 files which maintain entities representing the 22 activities. In INTLC (D9-7.4), an entity representing an activity is placed in the file of its start node.

At the beginning of a run the attributes of the entities in the source node file are copied, the activity time is calculated using function ACTIM (D9-7.5), and the entities are placed in file 15 to await the opening of the gate. The gate is scheduled to open in .00001 time units, by putting an entity into the network at ENTER node 1. The .00001 time delay insures that all the entities are placed in the file before the gate is opened.

Each time an activity is completed, EVENT 1 is called and the resources used to perform the activity freed. The variable REL(J) is decremented by one where J is the end node of the activity completed. When REL(J) reaches zero, the node is realized. The activity number and start node of the activity causing the realization are saved to calculate the critical path. Statistics are collected on the time to realize the node, relative to the time the source node was released. If the node is not the sink node, copies of the entities in the realized node's file are placed in the AWAIT file in the same manner as was done in subroutine INTLC. Then the gate is scheduled to open so that the freed resources can be allocated. If the sink node is realized, the number of simulation runs is incremented, and subroutine REINT (D9-7.6) is called to reinitialize the REL(J) and TIME variables, to calculate the critical path, and to place the activity entities for the source node in the gate file. The call to CLOSX(1) is made after the call to FILEM but since the ranking for events is LIFO, the closing of the gate will occur first. The gate is scheduled to open in .00001 time units by a call to ENTER(1, ATRIB) for the same reasons discussed previously.

Subroutine OTPUT (D9-7.7) is called at the end of the simulation and is used to calculate and print out the criticality indices. The input statements for this exercise are given in D9-7.8. A printout of the criticality indices is given in D9-7.9 and a portion of the summary report is shown in D9-7.10.

This example illustrates the use of gates and requires more complex logic than previous exercises. A likely place for errors would be in the opening and closing of the gate in relation to the filing of entities in the AWAIT file. Knowledge of the order in which events are placed on the event calendar and executed is required. This problem also illustrates the use of the USERF(IFN) function to change the state of the system as well as returning a value to the network.

```
      SUBROUTINE INTLC
      COMMON/SCOM1/ATRIB(100),DD(100),DDL(100),DTNOW,II,MFA,MSTOP,NCLNR
     1,NCRDR,NPRNT,NNRUN,NNSET,NTAPE,SS(100),SSL(100),TNEXT,TNOW,XX(100)
      COMMON/UCOM1/NCA(14),NCN(14),TIME,KOUNT,REL(14),NUMC(22)
C     INITIALIZE NUMBER OF RELEASES FOR EACH NODE
      INTEGER REL
      REL(1)=0
      REL(2)=1
      REL(3)=1
      REL(4)=2
      REL(5)=3
      REL(6)=1
      REL(7)=2
      REL(8)=2
      REL(9)=1
      REL(10)=3
      REL(11)=1
      REL(12)=1
      REL(13)=1
      REL(14)=3
      DO 10 I=1,22
10    NUMC(I)=0
      TIME=0.0
      KOUNT=0
C     PLACE ACTIVITIES EMANATING FROM START NODE IN GATE FILE
      NIF=NNQ(1)
      DO 30 N=1,NIF
      CALL COPY(N,1,ATRIB)
      NACT=ATRIB(5)
      ATRIB(6)=ACTIM(NACT)
      ATRIB(9)=ATRIB(6)
30    CALL FILEM(15,ATRIB)
C     SCHEDULE OPENING OF GATE
      CALL ENTER(1,ATRIB)
      RETURN
      END
```

D9-7.4

```
      FUNCTION ACTIM(NACT)
      GOTO (1,2,3,4,5,4,7,8,2,4,5,12,13,3,15,2,4,12,19,20,21,4),NACT
1     ACTIM=RLOGN(7.,2.,1)
      GOTO 100
2     ACTIM=3.
      GOTO 100
3     ACTIM=10.
      GOTO 100
4     ACTIM=5.
      GOTO 100
5     ACTIM=0.0
      GOTO 100
7     ACTIM=UNFRM(2.66,3.33,1)
      GOTO 100
8     ACTIM=GAMA(.75,10.66,1)
      GOTO 100
12    ACTIM=2.0
      GOTO 100
13    ACTIM=6.
      GOTO 100
15    ACTIM=EXPON(10.,1)
      GOTO 100
19    ACTIM=RNORM(5.,4.,1)
      GOTO 100
20    ACTIM=1.
      GOTO 100
21    ACTIM=15.
100   IF(ACTIM.LT.0.0) ACTIM=0.0
      RETURN
      END
```

D9-7.5

```
      SUBROUTINE REINT
      COMMON/SCOM1/ATRIB(100),DD(100),DDL(100),DTNOW,II,MFA,MSTOP,NCLNR
     1,NCRDR,NPRNT,NNRUN,NNSET,NTAPE,SS(100),SSL(100),TNEXT,TNOW,XX(100)
      COMMON/UCOM1/NCA(14),NCN(14),TIME,KOUNT,REL(14),NUMC(22)
C     INITIALIZE NUMBER OF RELEASES FOR EACH NODE
      INTEGER REL
      REL(1)=0
      REL(2)=1
      REL(3)=1
      REL(4)=2
      REL(5)=3
      REL(6)=1
      REL(7)=2
      REL(8)=2
      REL(9)=1
      REL(10)=3
      REL(11)=1
      REL(12)=1
      REL(13)=1
      REL(14)=3
C     FIND CRITICAL PATH, INCREMENT NUMC FOR ACTIVITIES ON CRITICAL PATH
      LCN=10
20    LCA=NCA(LCN)
      NUMC(LCA)=NUMC(LCA)+1
      LCN=NCN(LCN)
      IF(LCN.EQ.1) GOTO 10
      GOTO 20
C     SET BEGINNING TIME OF SIMULATION TO TNOW
10    TIME=TNOW
C     PLACE ACTIVITIES EMANATING FROM START NODE IN GATE FILE
      NIF=NNQ(1)
      DO 30 N=1,NIF
      CALL COPY(N,1,ATRIB)
      NACT=ATRIB(5)
      ATRIB(6)=ACTIM(NACT)
      ATRIB(9)=ATRIB(6)
30    CALL FILEM(15,ATRIB)
      CALL CLOSX(1)
C     SCHEDULE OPENING OF GATE
      CALL ENTER(1,ATRIB)
      RETURN
      END
```

D9-7.6

```
      SUBROUTINE OTPUT
      COMMON/SCOM1/ATRIB(100),DD(100),DDL(100),DTNOW,II,MFA,MSTOP,NCLNR
     1,NCRDR,NPRNT,NNRUN,NNSET,NTAPE,SS(100),SSL(100),TNEXT,TNOW,XX(100)
      COMMON/UCOM1/NCA(14),NCN(14),TIME,KOUNT,REL(14),NUMC(22)
      DIMENSION XNUMC(22)
      DO 10 J=1,22
      XNUMC(J)=FLOAT(NUMC(J))/FLOAT(KOUNT)
      WRITE(NPRNT,20) J,XNUMC(J)
10    CONTINUE
20    FORMAT(/,35X,9HACT. NO. ,I2,3X,12HCRIT. INDEX ,F5.3)
      RETURN
      END
```

D9-7.7

```
 1   GEN,ROLSTON,PROBLEM 9.7,8/13/80,1;
 2   LIMITS,15,9,100;
 3   PRIORITY/15,LVF(9);
 4   TIMST,XX(1),SYSTEMS ANALYST;
 5   TIMST,XX(2),MARKETING;
 6   TIMST,XX(3),MAINTENANCE;
 7   TIMST,XX(4),ENGINEERING;
 8   STAT,1,NODE 1;
 9   STAT,2,NODE 2,20/8/2;
10   STAT,3,NODE 3;
11   STAT,4,NODE 4,20/10/2;
12   STAT,5,NODE 5,20/31/1;
13   STAT,6,NODE 6,20/16/2;
14   STAT,7,NODE 7,20/21/1;
15   STAT,8,NODE 8,20/36/1;
16   STAT,9,NODE 9,20/42/1;
17   STAT,11,NODE 11;
18   STAT,12,NODE 12,20/7/1;
19   STAT,13,NODE 13,20/10/2;
20   STAT,14,NODE 14,20/29/1;
21   STAT,10,PROJ COMPLETE,20/54/1;
22   NETWORK;
23        GATE/GFRA,CLOSE,15;
24   AW   AWAIT(15),GFRA;
25        GOON,1;
26        ACT,USERF(1),XX(1).GE.ATRIB(1).AND.XX(2).GE.ATRIB(2).AND.
27             XX(3).GE.ATRIB(3).AND.XX(4).GE.ATRIB(4),E1;
28        ACT,,,CL;
29   E1   EVENT,1;
30        TERM;
31   CL   CLOSE,GFRA;
32        ACT,,,AW;
33        ENTER,1,1;
34        ACT,0.0001;
35        OPEN,GFRA;
36        TERM;
37        END;
38   INTLC,XX(1)=3,XX(2)=2,XX(3)=3,XX(4)=3;
39   ENTRY/1,2,0,1,3,1,0,2,1,0/1,2,0,1,1,2,0,3,1,0;
40   ENTRY/1,0,2,0,0,3,0,7,1,0/1,1,1,1,1,4,0,11,1,0;
41   ENTRY/2,0,0,0,0,5,0,4,2,0/2,2,0,3,2,6,0,6,2,0;
42   ENTRY/3,1,0,2,3,7,0,4,3,0;
43   ENTRY/4,3,0,0,1,8,0,5,4,0/4,1,0,0,0,9,0,5,4,0;
44   ENTRY/5,2,0,1,3,10,0,8,5,0;
45   ENTRY/6,0,0,0,0,11,0,5,6,0/6,2,1,2,2,12,0,8,6,0;
46   ENTRY/7,0,1,1,1,17,0,14,7,0;
47   ENTRY/8,1,1,0,0,13,0,9,8,0;
48   ENTRY/9,0,0,0,0,14,0,10,9,0;
49   ENTRY/11,1,1,1,0,15,0,7,11,0/11,1,1,2,0,16,0,12,11,0;
50   ENTRY/12,0,1,0,3,18,0,13,12,0/12,1,0,0,2,19,0,14,12,0;
51   ENTRY/12,1,1,0,0,20,0,14,12,0;
52   ENTRY/13,0,1,2,0,21,0,10,13,0;
53   ENTRY/14,0,1,0,0,22,0,10,14,0;
54   FIN;
```

D9-7.8

```
                    **INTERMEDIATE RESULTS**

          ACT. NO.   1    CRIT. INDEX 0.700

          ACT. NO.   2    CRIT. INDEX 0.190

          ACT. NO.   3    CRIT. INDEX 0.000

          ACT. NO.   4    CRIT. INDEX 0.110

          ACT. NO.   5    CRIT. INDEX 0.390

          ACT. NO.   6    CRIT. INDEX 0.310

          ACT. NO.   7    CRIT. INDEX 0.190

          ACT. NO.   8    CRIT. INDEX 0.580

          ACT. NO.   9    CRIT. INDEX 0.000

          ACT. NO.  10    CRIT. INDEX 0.830

          ACT. NO.  11    CRIT. INDEX 0.250

          ACT. NO.  12    CRIT. INDEX 0.060

          ACT. NO.  13    CRIT. INDEX 0.890

          ACT. NO.  14    CRIT. INDEX 0.890

          ACT. NO.  15    CRIT. INDEX 0.020

          ACT. NO.  16    CRIT. INDEX 0.090

          ACT. NO.  17    CRIT. INDEX 0.020

          ACT. NO.  18    CRIT. INDEX 0.040

          ACT. NO.  19    CRIT. INDEX 0.050

          ACT. NO.  20    CRIT. INDEX 0.000

          ACT. NO.  21    CRIT. INDEX 0.040

          ACT. NO.  22    CRIT. INDEX 0.070
```

D9-7.9

```
                         S L A M   I I   S U M M A R Y   R E P O R T

               SIMULATION PROJECT PROBLEM 9.7          BY ROLSTON

               DATE  8/13/1980                         RUN NUMBER   1 OF   1


               CURRENT TIME    0.6467E+04
               STATISTICAL ARRAYS CLEARED AT TIME  0.0000E+00


                        **STATISTICS FOR VARIABLES BASED ON OBSERVATION**

                       MEAN       STANDARD     COEFF. OF    MINIMUM      MAXIMUM      NUMBER OF
                       VALUE      DEVIATION    VARIATION    VALUE        VALUE        OBSERVATIONS

       NODE 1                      NO VALUES RECORDED
       NODE 2        0.1566E+02   0.4628E+01   0.2956E+00   0.8352E+01   0.3878E+02     100
       NODE 3        0.3000E+01   0.5502E-04   0.1834E-04   0.3000E+01   0.3000E+01     100
       NODE 4        0.1690E+02   0.4051E+01   0.2397E+00   0.1127E+02   0.3878E+02     100
       NODE 5        0.4204E+02   0.1096E+02   0.2608E+00   0.2360E+02   0.8162E+02     100
       NODE 6        0.3095E+02   0.1298E+02   0.4195E+00   0.1627E+02   0.5776E+02     100
       NODE 7        0.2936E+02   0.1220E+02   0.4154E+00   0.1559E+02   0.7465E+02     100
       NODE 8        0.4818E+02   0.1057E+02   0.2195E+00   0.2936E+02   0.8662E+02     100
       NODE 9        0.5418E+02   0.1057E+02   0.1952E+00   0.3536E+02   0.9262E+02     100
       PROJ COMPLETE 0.6467E+02   0.1026E+02   0.1586E+00   0.4636E+02   0.1026E+03     100
       NODE 11       0.5000E+01   0.5512E-04   0.1102E-04   0.5000E+01   0.5000E+01     100
       NODE 12       0.1398E+02   0.6694E+01   0.4789E+00   0.8000E+01   0.3267E+02     100
       NODE 13       0.2588E+02   0.6231E+01   0.2408E+00   0.1035E+02   0.4261E+02     100
       NODE 14       0.4026E+02   0.1071E+02   0.2659E+00   0.2240E+02   0.7965E+02     100


                        **STATISTICS FOR TIME-PERSISTENT VARIABLES**

                       MEAN       STANDARD     MINIMUM      MAXIMUM      TIME         CURRENT
                       VALUE      DEVIATION    VALUE        VALUE        INTERVAL     VALUE

       SYSTEMS ANALYST 0.1377E+01  0.1033E+01  0.0000E+00   0.3000E+01   0.6467E+04   0.3000E+01
       MARKETING       0.8333E+00  0.7802E+00  0.0000E+00   0.2000E+01   0.6467E+04   0.2000E+01
       MAINTENANCE     0.1494E+01  0.1177E+01  0.0000E+00   0.3000E+01   0.6467E+04   0.3000E+01
       ENGINEERING     0.1489E+01  0.1214E+01  0.0000E+00   0.3000E+01   0.6467E+04   0.3000E+01
```

D9-7.10

9-8.   This exercise requires file searching which is not covered until Chapters 11 and 12. The solution presented employs function NFIND, subroutine COPY, the use of a pointer to find and copy an entity's attributes and the direct changing of an attribute of an entity in a file. The exercise should not be assigned unless Chapter 12 capabilities are first discussed. The network model is shown in D9-8.1. EVENT and ENTER nodes are included in the model to accommodate the complex logic associated with the ending of CPU burst (time slice). The discussion of the model follows the network given in D9-8.1.

The arrival of jobs is modeled using a CREATE node. Upon arrival, the following attributes are assigned:

ATRIB(1)   Mark time, also used as job ID
ATRIB(2)   Total CPU time required for job
ATRIB(3)   CPU time remaining until job ends
ATRIB(4)   Disk assigned for I/O
ATRIB(5)   I/O rate
ATRIB(6)   Main memory required for job

A simple sequential assignment of disks is used. XX(3) is the disk number to be assigned to the next arriving job. It is initalized to 1 and incremented by 1 each time a job arrives. When it reaches 4, it is reset to 1.

The central memory, CPUs, channel, and disks are all modeled as resources. At AWAIT node AMEM, the job is allocated the proper amount of the memory resource as specified by ATRIB(6). Next, the job awaits a CPU. ASSIGN node ACPT sets ATRIB(7), the CPU burst time. The following activity represents the CPU burst.

At the end of this activity, event 1 is called as the job entity arrives to an EVENT node. At the end of a CPU burst, the job will be ready for another CPU burst or will have to wait until a pending I/O operation is completed. In the former case, the job is routed to ENTER node 2. In the latter case, it is filed in file 3. These operations are performed in subroutine ENDSLC to be described later.

The second portion of the network represents the I/O activity. Each disk is modeled as a resource. The entities representing an I/O request are inserted into the network through ENTER nodes by calls to subroutine ENTER in event 1. They then await the proper disk resource. The activity following the AWAIT node represents the seek time on the disk. After the appropriate data has been found, the channel resource is needed. When it has been allocated, the activity representing data transfer occurs. Its duration is given by ATRIB(3) which was previously assigned. After data tranfer, the channel is freed and event 2 is called by routing the entity to EVENT node 2. At the end of the I/O operation, the job entity is either routed to ENTER node 1 to request CPU space or to ENTER node 10 which models the end of processing for the job. The coding of the end of the I/O operation is contained in event 2, ENDIO, to be discussed later.

When the job is finished, an entity is placed in the network through ENTER node 10. The memory resources are freed, then statistics are collected on both the time in system and the waiting time. This completes the description of the network (D9-8.1).

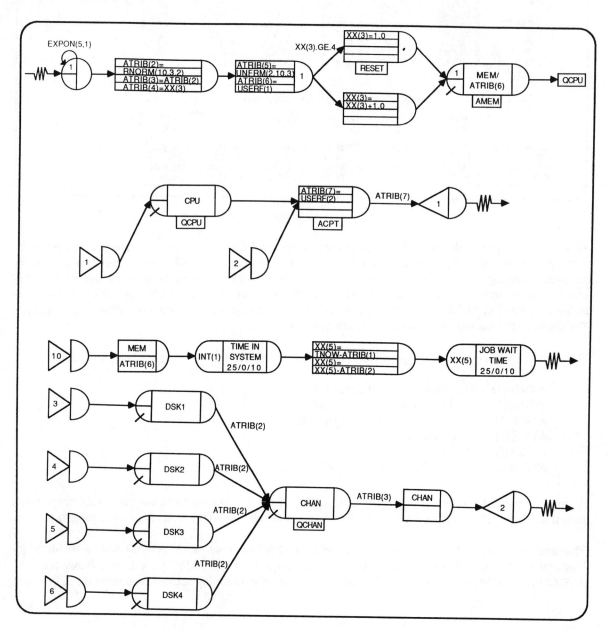

D9-8.1

```
66          ASSIGN,XX(3)=XX(3)+1.0;              ADVANCE DISK ASSIGNMENT
67          ACT,,,AMEM;                          BYPASS RESET
68   RESET  ASSIGN,XX(3)=1.0;
69   AMEM   AWAIT(1),MEM/ATRIB(6);               ALLOCATE MEMORY
70          ACT,,,QCPU;                          GO QUEUE FOR CPU
71   ;
72          ENTER,1;                             RESTART CPU, I/O DONE
73   ;                                           (FROM EVENT 3)
74   QCPU   AWAIT,CPU;                           OBTAIN CPU
75   ACPT   ASSIGN,ATRIB(7)=USERF(2);            CPU BURST TIME
76          ACT,ATRIB(7);                        CPU BURST
77          EVENT,1;                             END CPU BURST;
78          TERM;
79   ;
80          ENTER,2;
81          ACT,,,ACPT;
82   ;
83   ;==========================================================
84   ;
85   ;    I/O REQUEST ENTITY
86   ;
87   ;    ATRIB()          CONTENTS
88   ;    =======          ================================
89   ;       1             JOB ID
90   ;       2             SEEK TIME ON ASSIGNED DISK
91   ;       3             TRANSFER TIME BETWEEN CHAN AND DISK
92   ;       4             DISK ASSIGNED FOR THIS I/O OPERATION
93   ;       5             PRIORITY (WHEN QUEUED FOR SPECIFIC
94   ;                     DISK OR FOR CHANNEL)
95   ;
96   ;==========================================================
97   ;
98          ENTER,3;                             WAIT FOR RESOURCE
99          AWAIT,DSK1;                          ALLOCATE DISK
100         ACT,ATRIB(2),,QCHAN;                 SEEK,THEN QUEUE FOR CHAN
101  ;
102         ENTER,4;
103         AWAIT,DSK2;
104         ACT,ATRIB(2),,QCHAN;
105  ;
106         ENTER,5;
107         AWAIT,DSK3;
108         ACT,ATRIB(2),,QCHAN;
109  ;
110         ENTER,6;
111         AWAIT,DSK4;
112         ACT,ATRIB(2);
113  ;
114  QCHAN  AWAIT,CHAN;                          WAIT FOR CHANNEL
115         ACT,ATRIB(3);                        TRANSFER TIME
116         FREE,CHAN;                           FREE CHANNEL RESOURCE
117         EVENT,2;                             END OF I/O OPERATION
118         TERM;
119  ;
120  ;==========================================================
121  ;
122  ;    END OF JOB PROCESSING
123  ;
124         ENTER,10;                            END OF JOB
125         FREE,MEM/ATRIB(6);                   FREE RESOURCE
126         COLCT,INT(1),TIME IN SYSTEM,25/0/10;
127         ASSIGN,XX(5)=TNOW-ATRIB(1),
128               XX(5)=XX(5)-ATRIB(2);          WAIT TIME
129         COLCT,XX(5),JOB WAIT TIME,25/0/10;
130         TERM;
131         END;
132  FIN;
```

The control and network statements for this exercise are given in D9- 8.2. The MAIN program and subroutine
EVENT are presented in D9-8.3 and D9-8.4 respectively.

```
1   GEN,ROLSTON,PROBLEM 9.8,10/29/80,1;
2   LIMITS,8,7,100;
3   INTLC,XX(3)=1.0;
4   ;
5   ;=================================================
6   ;    LOW VALUE FIRST IN ALL QUEUES AS PER JOB PRIORITY
7   ;
8   PRI/1,LVF(6);
9   PRI/2,LVF(6);
10  PRI/4,LVF(6);
11  PRI/5,LVF(6);
12  PRI/6,LVF(6);
13  PRI/7,LVF(6);
14  PRI/8,LVF(6);
15  INIT,0,1200;
16  NETWORK;
17  ;=================================================
18  ;
19  ;    FILE USAGE
20  ;
21  ;    FILE              ENTITY/RESOURCE        STATE
22  ;    ====              ===============        ==========================
23  ;     1                    JOB                WAITING FOR MEMORY
24  ;     2                    JOB                WAITING FOR CPU
25  ;     3                    JOB                WAITING FOR I/O COMPLETION
26  ;     4                    IOR                WAITING FOR CHANNEL
27  ;     5                    IOR                WAITING FOR DISK # 1
28  ;     6                    IOR                WAITING FOR DISK # 2
29  ;     7                    IOR                WAITING FOR DISK # 3
30  ;     8                    IOR                WAITING FOR DISK # 4
31  ;
32  ;=================================================
33       RES/CPU(2),2;
34       RES/MEM(131),1;
35       RES/DSK1,5;
36       RES/DSK2,6;
37       RES/DSK3,7;
38       RES/DSK4,8;
39       RES/CHAN,4;
40  ;
41  ;=================================================
42  ;    BEGIN NETWORK PROCESSING
43  ;
44  ;    JOB ENTITY
45  ;
46  ;    ATRIB()           CONTENTS
47  ;    =======           ================================
48  ;       1              MARK ATRIB, ALSO USED AS JOB ID
49  ;       2              TOTAL CPU TIME RQD FOR JOB
50  ;       3              CPU TIME REMAINING UNTIL JOB END
51  ;       4              DISK ASSIGNED FOR I/O ACTIVITY
52  ;       5              I/O RATE
53  ;       6              MAIN MEMORY REQUIRED FOR JOB
54  ;       7              NORMAL FILES : CURRENT CPU SLICE
55  ;                      IN FILE 3 (I/O WAIT)  : NUM OF I/O OUTSTANDING
56  ;
57  ;=================================================
58       CREATE,EXPON(5.,1),,1;                    ARRIVING JOBS
59       ASSIGN,ATRIB(2)=RNORM(10.,3.,2),
60              ATRIB(3)=ATRIB(2),
61              ATRIB(4)=XX(3);
62       ASSIGN,ATRIB(5)=UNFRM(2.0,10.0,3),
63              ATRIB(6)=USERF(1),1;
64       ACT,,XX(3).GE.4.0,RESET;                  CYCLE THRU 4 DISKS
65       ACT;                                      ELSE CONTINUE
```

```
      COMMON/SCOM1/ATRIB(100),DD(100),DDL(100),DTNOW,II,MFA,MSTOP,NCLNR
     1,NCRDR,NPRNT,NNRUN,NNSET,NTAPE,SS(100),SSL(100),TNEXT,TNOW,XX(100)
      DIMENSION NSET(5000)
      COMMON QSET(5000)
      EQUIVALENCE(NSET(1),QSET(1))
      NNSET=5000
      NCRDR=5
      NPRNT=6
      NTAPE=7
      CALL SLAM
      STOP
      END
```

D9-8.3

```
      SUBROUTINE EVENT(I)
      COMMON/SCOM1/ATRIB(100),DD(100),DDL(100),DTNOW,II,MFA,MSTOP,NCLNR
     1,NCRDR,NPRNT,NNRUN,NNSET,NTAPE,SS(100),SSL(100),TNEXT,TNOW,XX(100)
      GOTO (1,2),I
C
C===============================================
C     END OF CPU BURST (TIME SLICE) EVENT
C
1     CALL ENDSLC
      RETURN
C
C===============================================
C     END OF I/O OPERATION EVENT
C
2     CALL ENDIO
      RETURN
      END
```

D9-8.4

Event 1, end of CPU burst, is modeled in subroutine ENDSLC, shown in D9-8.5. The first statement subtracts the CPU burst time just completed from the total CPU time required for the job. Then the attributes of an I/O request entity are set and the entity is inserted into the next portion of the network which models I/O activity. The attributes of this entity are as specified in the comments given in the subroutine listing (D9-8.5). If the job already has an I/O request pending, then its processing must be suspended and the CPU relinquished. This is done in the following manner. Whenever an I/O request entity is created, an entity is placed in file 3. The attributes of this entity are the same as the job itself, except that ATRIB(7) represents the number of I/O requests pending. When a CPU burst is completed, this file is checked to see if the job has an I/O request pending. If an entity is not found in this file, then one is placed in it with ATRIB(7) equal to 1. If the job is not finished, the entity is inserted back into the network through ENTER node 2. If an entity was found in file 3, the number of I/O requests pending is incremented by 1, and the CPU resource is freed. The job will become eligible for a CPU again when one of the I/O requests is completed. This is modeled in end of I/O event, event 2. When the CPU time required has been satisfied, the completion of the last I/O request is also the end of job processing.

```
      SUBROUTINE ENDSLC
      COMMON/SCOM1/ATRIB(100),DD(100),DDL(100),DTNOW,II,MFA,MSTOP,NCLNR
     1,NCRDR,NPRNT,NNRUN,NNSET,NTAPE,SS(100),SSL(100),TNEXT,TNOW,XX(100)
      DIMENSION AIOR(9),DISCARD(9)
C
C==========================================================
C     CREDIT CPU TO JOB'S REQD CPU TIME
C
      ATRIB(3)=ATRIB(3)-ATRIB(7)
C
C==========================================================
C     BUILD I/O REQUEST
C
C     CELL          DESCRIPTION
C     ====          ===========
C      1            JOB ID (TIME STAMP)
C      2            SEEK TIME
C      3            TRANSFER TIME
C      4            DISK ASSIGNED
C      5            PRIORITY
C
      AIOR(1)=ATRIB(1)
      AIOR(2)=UNFRM(0.0,.075,6)
      AIOR(3)=.001*(2.5+UNFRM(0.0,25.0,7))
      AIOR(4)=ATRIB(4)
      AIOR(5)=ATRIB(6)
      CALL ENTER(2+IFIX(ATRIB(4)),AIOR)
C
C==========================================================
C     IF JOB HAS I/O PENDING THEN SUSPEND CPU PROCESSING
C     AND ENTER JOB IN A FILE WAITING FOR I/O
C
C     IF AN ENTITY EXISTS IN FILE 3 THAN AN I/O IS
C     STILL OUTSTANDING
C
      NRANK=NFIND(1,3,1,0,ATRIB(1),0.0)
      IF(NRANK.NE.0) GOTO 10
      ATRIB(7)=1.0
      GOTO 20
10    CALL RMOVE(NRANK,3,DISCARD)
      ATRIB(7)=DISCARD(7)+1.0
20    CALL FILEM(3,ATRIB)
      IF(ATRIB(3).LE.0.0) GOTO 30
      IF(ATRIB(7).LE.1.0) GOTO 40
      CALL FREE(1,1)
      RETURN
40    CALL ENTER(2,ATRIB)
      RETURN
30    CALL FREE(1,1)
      RETURN
      END
```

D9-8.5

Event code 2 corresponds to subroutine ENDIO, shown in D9-8.6. The disk resource is freed first. Then the pointer to the entity in file 3 representing the job is found. ATRIB(7) is checked to determine the number of pending I/O requests. If only one is pending, then the number of I/O requests is decremented by one, and a job entity is entered into the network at ENTER node 1. If this is the last I/O operation for the job, an entity is entered into the network at ENTER node 10.

In this exercise, function USERF(IFN) is employed to calculate memory requirements (IFN=1) and to calculate the duration of the next CPU burst (IFN=2). Function USERF is shown in D9-8.7. Note that the memory requirements are computed as an integer in the range 20 to 59 with each integer being equally likely.

A summary report for this exercise is given in D9-8.8.

```
      SUBROUTINE ENDIO
      COMMON/SCOM1/ATRIB(100),DD(100),DDL(100),DTNOW,II,MFA,MSTOP,NCLNR
     1,NCRDR,NPRNT,NNRUN,NNSET,NTAPE,SS(100),SSL(100),TNEXT,TNOW,XX(100)
      DIMENSION NSET(1)
      COMMON QSET(1)
      EQUIVALENCE(NSET(1),QSET(1))
C
C===================================================================
C     FREE DISK RESOURCE
C
      CALL FREE(IFIX(ATRIB(4))+2,1)
C
C===================================================================
C     IF I/O IS LAST I/O OUTSTANDING THEN
C     REMOVE I/O WAITING ENTITY FROM FILE 3
C
C     IF THE COMPLETION OF THIS I/O OPERATION
C     BRINGS THE I/O WAITING COUNT BELOW THE
C     POINT WHERE PROCESSING WAS HALTED
C     THEN SCHEDULE THE NEXT CPU BURST
C
      I=NFIND(-MMFE(3),3,1,0,ATRIB(1),0.0)
      IF(IFIX(QSET(I+7)).EQ.1) GOTO 10
      QSET(I+7)=QSET(I+7)-1.0
      CALL COPY(-I,3,ATRIB)
      IF(ATRIB(3).LE.0.0) RETURN
      CALL ENTER(1,ATRIB)
      RETURN
10    CALL RMOVE(-I,3,ATRIB)
      IF(ATRIB(3).GT.0.0) RETURN
      CALL ENTER(10,ATRIB)
      RETURN
      END
```

D9-8.6

```
      FUNCTION USERF(I)
      COMMON/SCOM1/ATRIB(100),DD(100),DDL(100),DTNOW,II,MFA,MSTOP,NCLNR
     1,NCRDR,NPRNT,NNRUN,NNSET,NTAPE,SS(100),SSL(100),TNEXT,TNOW,XX(100)
      GOTO (1,2),I
C
C===============================================
C      CALCULATE MEMORY REQUIREMENTS
C
1     USERF=IFIX(UNFRM(20.,60.,4))
      RETURN
C
C===============================================
C     CALCULATE DURATION OF TIME SLICE
C     IF GREATER THAN THE AMOUNT REMAINING
C     THEN USE AMOUNT REMAINING TO COMPLETE
C     THE JOB
C
2      USERF=EXPON(1.0/ATRIB(5),5)
      IF(USERF.GT.ATRIB(3)) USERF=ATRIB(3)
      RETURN
      END
```

D9-8.7

```
                    S L A M   I I   S U M M A R Y   R E P O R T


            SIMULATION PROJECT PROBLEM 9.8            BY ROLSTON

            DATE 10/29/1980                           RUN NUMBER   1 OF   1



            CURRENT TIME   0.1200E+04
            STATISTICAL ARRAYS CLEARED AT TIME  0.0000E+00



                    **STATISTICS FOR VARIABLES BASED ON OBSERVATION**

                    MEAN        STANDARD     COEFF. OF    MINIMUM      MAXIMUM     NUMBER OF
                    VALUE       DEVIATION    VARIATION    VALUE        VALUE       OBSERVATIONS

    TIME IN SYSTEM  0.5831E+02  0.1067E+03   0.1830E+01   0.4531E+01   0.6813E+03      232
    JOB WAIT TIME   0.4833E+02  0.1067E+03   0.2208E+01   0.7822E-01   0.6693E+03      232



                    **FILE STATISTICS**

    FILE                     AVERAGE      STANDARD     MAXIMUM    CURRENT     AVERAGE
    NUMBER  LABEL/TYPE       LENGTH       DEVIATION    LENGTH     LENGTH      WAITING TIME

      1     AMEM AWAIT       19.2968      10.1646        34         31        87.3817
      2     QCPU AWAIT        0.7865       0.6284         2          0         0.2600
      3                       0.6465       0.7325         4          0         0.0571
      4     QCHA AWAIT        0.0110       0.1066         2          0         0.0010
      5          AWAIT        0.0359       0.2056         3          0         0.0116
      6          AWAIT        0.0348       0.2119         3          0         0.0127
      7          AWAIT        0.0358       0.2120         3          0         0.0134
      8          AWAIT        0.0416       0.2371         3          0         0.0148
      9          CALENDAR     3.5260       0.6853         9          3         0.0281



                    **RESOURCE STATISTICS**

    RESOURCE  RESOURCE  CURRENT   AVERAGE      STANDARD    MAXIMUM      CURRENT
    NUMBER    LABEL     CAPACITY  UTILIZATION  DEVIATION   UTILIZATION  UTILIZATION

      1       CPU         2        1.9310      0.2719          2            2
      2       MEM       131      108.4734     19.7038        131           98
      3       DSK1        1        0.1653      0.3715          1            0
      4       DSK2        1        0.1470      0.3541          1            0
      5       DSK3        1        0.1432      0.3503          1            0
      6       DSK4        1        0.1505      0.3575          1            0
      7       CHAN        1        0.1702      0.3758          1            0
```

D9–8.8

9-9.   This exercise is a continuation of Exercise 6-9 and the solution to Exercise 6-9 should be read to provide the basic solution to the problem. Included in this exercise is the modeling of a cart to transport parts from arrival to processing and then the finished part to packaging. Also included in this example is maintenance operations and the subcontracting of part processing if the queue is too large when maintenance is to be performed. The SLAM II network model is shown in D9-9.1 and the corresponding statement model in D9-9.2.

The cart is modeled as a resource with a capacity of 1 and parts waiting for the cart are maintained in file 2. At BATCH node BAT1, four part entities are gathered to form a batch and the batched entity is sent to the AWAIT node to request the cart. In Chapter 16, the QBATCH node is introduced which allows for the combining of this batching and resource waiting capability. An activity with a duration of 5 time units is used to model the cart travel to the machine after which the cart is freed and the batched entity is routed to an UNBATCH node where the individual entities of the batch are released by specifying that attribute 5 contains a pointer to the entities in the batch.

At AWAIT node CART, one unit of the cart is also requested to transport the finished product to the packaging area. The two different part types are assembled at the SELECT node PACK prior to the packaging activity.

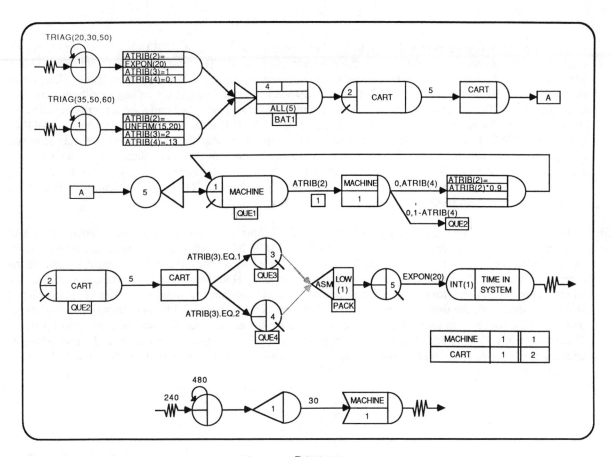

D9-9.1

```
 1   GEN,OREILLY,PROBLEM 9.9,6/26/1986,1;
 2   LIMITS,5,5,200;
 3   NETWORK;
 4        RESOURCE/MACHINE,1;                        DEFINE MACHINE
 5        RESOURCE/CART(1),2;                        DEFINE CART
 6        CREATE,TRIAG(20,30,50),,1;                 ARRIVAL PART TYPE 1
 7        ASSIGN,ATRIB(2)=EXPON(20),ATRIB(3)=1,
 8             ATRIB(4)=0.10;                        ASSIGN CHARACTERISTICS
 9        ACTIVITY,,,BAT1;                           ROUTE TO BATCH
10        CREATE,TRIAG(35,50,60),,1;                 ARRIVAL PART TYPE 2
11        ASSIGN,ATRIB(2)=UNFRM(15,20),ATRIB(3)=2,
12             ATRIB(4)=0.13;                        ASSIGN CHARACTERISTICS
13 BAT1  BATCH,,4,,ALL(5);                           BATCH PARTS FOR CART
14       AWAIT(2),CART;                              WAIT FOR CART
15       ACTIVITY,5;                                 CART TRAVEL TO MACHINE
16       FREE,CART;                                  RELEASE CART
17       UNBATCH,5;                                  UNLOAD PARTS
18 QUE1  AWAIT(1),MACHINE;                           WAIT FOR MACHINE
19       ACTIVITY/1,ATRIB(2);                        PERFORM MACHINING
20       FREE,MACHINE,1;                             RELEASE MACHINE
21       ACTIVITY,,1-ATRIB(4),QUE2;                  PASS INSPECTION
22       ACTIVITY,,ATRIB(4);                         FAIL INSPECTION
23       ASSIGN,ATRIB(2)=ATRIB(2)*0.9;               ASSIGN NEW MACHINING TIME
24       ACTIVITY,,,QUE1;                            ROUTE BACK TO MACHINE
25 QUE2  AWAIT(2),CART;                              WAIT FOR CART
26       ACTIVITY,5;                                 CART TRAVEL TO PACKAGING
27       FREE,CART;                                  RELEASE CART
28       ACTIVITY,,ATRIB(3).EQ.1,QUE3;               ROUTE PART TYPE 1 TO WAIT
29       ACTIVITY,,ATRIB(3).EQ.2,QUE4;               ROUTE PART TYPE 2 TO WAIT
30 QUE3  QUEUE(3),,,,PACK;                           WAIT TO PACKAGE
31 QUE4  QUEUE(4),,,,PACK;                           WAIT TO PACKAGE
32 PACK  SELECT,ASM/LOW(1),,,QUE3,QUE4;              COMBINE PART TYPES
33       ACTIVITY;                                   CONTINUE
34       QUEUE(5);                                   WAIT FOR PACKAGING
35       ACTIVITY,EXPON(20);                         PACKAGE PARTS
36       COLCT,INT(1),TIME IN SYSTEM;                COLLECT TIME IN SYSTEM
37       TERM;                                       DEPART FROM SYSTEM
38   ;
39       CREATE,480,240;                             ARRIVAL SCHEDULED MAINTENANCE
40       EVENT,1;                                    REMOVE PARTS FOR SUBCONTRACTING
41       ACTIVITY,30;                                PERFORM MAINTENANCE
42       ALTER,MACHINE,1;                            RELEASE MACHINE
43       TERM;                                       DEPART FROM SYSTEM
44       END;
45 INIT,0,2400;
46 FIN;
```

D9-9.2

A disjoint network is used to model the scheduled maintenance activities. A CREATE node is used to put an entity into the network representing a scheduled maintenance event. This entity is routed with event code 1 where parts will be removed for subcontracting if the number of parts in file 1 is greater than 3. The use of subroutine ALTER in subroutine EVENT shown in D9-9.3 is used to indicate the calling of subroutines from subroutine EVENT. Subroutine RMOVE is used to remove entities from file 1 and they are removed from the end of the queue by having the first argument of subroutine RMOVE to be NNQ(1). In subroutine RMOVE, the value of NNQ(1) is updated to be the number in the queue so that the last entity in file 1 is removed through the use of this statement. No additional processing is done for the entity since subcontracting is not modeled in this exercise. Statistics could be collected on the time in the system for entities that are routed to subcontracting or time between subcontracting could also be collected. The SLAM II Summary Report for this exercise is shown in D9-9.4.

9-9,Embellishment. The modeling changes required for this embellishment are to user functions to replace the travel time for the cart.   In the statement model given in D9-9.2, USERF(1) replaces the value 5 in line number 15 and USERF(2) replaces the value 5 at line 26. In function USERF(I), the value XX(1) is used to indicate the current cart location. The total distance the cart must travel is equal to the time to travel to the desired location which can be ascertained from the user function number and the distance required to transfer the part which is also determined from the user function number. The total distance is divided by the cart speed of 100 to get the travel time which is set into USERF which is returned to the network as the duration of the travel activity. The value of XX(1) is updated to be the cart location. XX(1) is used to facilitate additional embellishments which would require knowledge of the current cart's location. The summary report for this embellishment is givein in D9-9.5.

```
            SUBROUTINE EVENT(I)
            DIMENSION A(10)
            CALL ALTER(1,-1)
            NUMBER=NNQ(1)-3
            IF (NUMBER.LE.0) RETURN
            DO 10 I=1,NUMBER
            CALL RMOVE(NNQ(1),1,A)
        10  CONTINUE
            RETURN
            END
```

D9-9.3

```
                    S L A M   I I   S U M M A R Y   R E P O R T

        SIMULATION PROJECT PROBLEM 9.9              BY OREILLY

        DATE  6/26/1986                             RUN NUMBER    1 OF    1


        CURRENT TIME   0.2400E+04
        STATISTICAL ARRAYS CLEARED AT TIME  0.0000E+00
```

STATISTICS FOR VARIABLES BASED ON OBSERVATION

	MEAN VALUE	STANDARD DEVIATION	COEFF. OF VARIATION	MINIMUM VALUE	MAXIMUM VALUE	NUMBER OF OBSERVATIONS
TIME IN SYSTEM	0.5053E+03	0.2036E+03	0.4029E+00	0.1052E+03	0.8646E+03	47

FILE STATISTICS

FILE NUMBER	LABEL/TYPE	AVERAGE LENGTH	STANDARD DEVIATION	MAXIMUM LENGTH	CURRENT LENGTH	AVERAGE WAITING TIME
1	QUE1 AWAIT	2.7955	1.9698	8	3	51.2159
2	AWAIT	0.0252	0.1618	2	0	0.4146
3	QUE3 QUEUE	10.4175	6.5417	21	20	367.6769
4	QUE4 QUEUE	0.0042	0.0646	1	0	0.2095
5	QUEUE	0.0477	0.2131	1	0	2.3841
6	CALENDAR	4.5802	0.6848	10	5	7.1149

REGULAR ACTIVITY STATISTICS

ACTIVITY INDEX/LABEL	AVERAGE UTILIZATION	STANDARD DEVIATION	MAXIMUM UTIL	CURRENT UTIL	ENTITY COUNT
1 PERFORM MACH	0.8860	0.3178	1	1	127

SERVICE ACTIVITY STATISTICS

ACTIVITY INDEX	START NODE OR ACTIVITY LABEL	SERVER CAPACITY	AVERAGE UTILIZATION	STANDARD DEVIATION	CURRENT UTILIZATION	AVERAGE BLOCKAGE	MAXIMUM IDLE TIME/SERVERS	MAXIMUM BUSY TIME/SERVERS	ENTITY COUNT
0	PACK SELECT	1	0.0000	0.0000	0	0.0000	150.5511	0.0000	
0	QUEUE	1	0.3275	0.4693	1	0.0000	126.1769	112.5549	

RESOURCE STATISTICS

RESOURCE NUMBER	RESOURCE LABEL	CURRENT CAPACITY	AVERAGE UTILIZATION	STANDARD DEVIATION	MAXIMUM UTILIZATION	CURRENT UTILIZATION
1	MACHINE	1	0.8860	0.3178	1	1
2	CART	1	0.3042	0.4601	1	0

RESOURCE NUMBER	RESOURCE LABEL	CURRENT AVAILABLE	AVERAGE AVAILABLE	MINIMUM AVAILABLE	MAXIMUM AVAILABLE
1	MACHINE	0	0.0515	-1	1
2	CART	1	0.6958	0	1

D9-9.4

```
                    S L A M   I I   S U M M A R Y   R E P O R T

            SIMULATION PROJECT PROBLEM 9.9A            BY OREILLY

            DATE  6/26/1986                            RUN NUMBER   1 OF   1


            CURRENT TIME   0.2400E+04
            STATISTICAL ARRAYS CLEARED AT TIME  0.0000E+00


                    **STATISTICS FOR VARIABLES BASED ON OBSERVATION**

              MEAN        STANDARD      COEFF. OF     MINIMUM      MAXIMUM      NUMBER OF
              VALUE       DEVIATION     VARIATION     VALUE        VALUE        OBSERVATIONS

TIME IN SYSTEM  0.5198E+03  0.2340E+03   0.4502E+00   0.9676E+02   0.8775E+03       43


                    **FILE STATISTICS**

FILE                AVERAGE      STANDARD      MAXIMUM    CURRENT    AVERAGE
NUMBER  LABEL/TYPE  LENGTH       DEVIATION     LENGTH     LENGTH     WAITING TIME

  1     QUE1 AWAIT   4.2313       2.4113         10          9        74.6693
  2          AWAIT   0.0023       0.0479          1          0         0.0389
  3     QUE3 QUEUE  11.1741       7.6697         24         23       400.2679
  4     QUE4 QUEUE   0.0044       0.0663          1          0         0.2410
  5          QUEUE   0.0399       0.1962          2          0         2.1765
  6          CALENDAR 4.3291      0.5608          9          5         6.8670


                    **REGULAR ACTIVITY STATISTICS**

ACTIVITY          AVERAGE      STANDARD      MAXIMUM CURRENT    ENTITY
INDEX/LABEL       UTILIZATION  DEVIATION     UTIL    UTIL       COUNT

  1 PERFORM MACH    0.9129       0.2820         1       1        123


                    **SERVICE ACTIVITY STATISTICS**

ACTIVITY  START NODE OR   SERVER     AVERAGE      STANDARD    CURRENT      AVERAGE    MAXIMUM IDLE    MAXIMUM BUSY    ENTITY
INDEX     ACTIVITY LABEL  CAPACITY   UTILIZATION  DEVIATION   UTILIZATION  BLOCKAGE   TIME/SERVERS    TIME/SERVERS    COUNT

  0       PACK SELECT        1       0.0000       0.0000          0        0.0000      135.5117         0.0000
  0            QUEUE         1       0.2916       0.4545          1        0.0000      126.7269       117.6581


                    **RESOURCE STATISTICS**

RESOURCE  RESOURCE  CURRENT    AVERAGE      STANDARD      MAXIMUM       CURRENT
NUMBER    LABEL     CAPACITY   UTILIZATION  DEVIATION     UTILIZATION   UTILIZATION

   1      MACHINE      1       0.9129       0.2820           1             1
   2      CART         1       0.0621       0.2413           1             0


RESOURCE  RESOURCE  CURRENT    AVERAGE      MINIMUM     MAXIMUM
NUMBER    LABEL     AVAILABLE  AVAILABLE    AVAILABLE   AVAILABLE

   1      MACHINE      0       0.0246         -1           1
   2      CART         1       0.9379          0           1
```

CHAPTER 10

Network Modeling With Continuous Variables

10-1.   Inclusion of an exponentially decreasing input rate to the storage tank can be accomplished by setting

RATIN=300.*(1.-EXP(-SS(1)/750.))

Alternatively, a differential equation can be used which is set equal to the above when there is input to the storage tank. The latter modeling approach would appear to answer the exercise more precisely. In either case, the introduction of an exponentially decreasing input rate vastly affects the results of the simulation. With the parameters included in the exercise, it takes over 7.7 days to unload a tanker which results in unacceptable performance. The 7.7 days is obtained by solving the differential equation:

$$\frac{ds}{dt} = -300(1-\exp(-s/750))$$

which results in the following equation for t given that s(0)=150

$$t = -(s/300 + 2.5 \quad n(1-\exp(-s/750))) - 3.769.$$

A faster unloading rate is required to accommodate the tanker fleet.

10-2.   Many different models could be built for the statement of the problem in this exercise. A basic form for solution is shown below:

DD(1)=-(A*SS(1)+B*SS(2)+C*XX(4))

DD(2)=-DD(1)-RATOUT

where XX(4) is used to define the viscosity of the oil in the tanker being unloaded and A, B, and C are coefficients to make the flow rate a function of the level of crude in the tanker, the level of crude in the storage tank and the viscosity of the crude. If A, B, and C are constants they could be assigned in subroutine INTLC and included in a user COMMON block. Alternatively, A, B, and C could be made function statements in subroutine STATE to allow them to be dependent on other aspects of the model. The viscosity of the crude could be assigned in the network at the ASSIGN node where SS(1) is set.

10-3.   The easiest way to modify the original problem for a three shift operation is to set XX(1) = 1 in an INTLC statement. The shift startup/shut down subprocess as shown in Figure 10-5 should be deleted from the network statements.

The model with the three off-shore unloading docks is shown in D10-3.1. The network statements for this modification are shown in D10-3.2 and the revised subroutine STATE in D10-3.3. A comparison of the results shows that the three dock operation is more efficient with a significant reduction in waiting time.

The use of three docks on a one shift basis does not decrease the average round-trip time. The reason for this lack of decrease is that the operation of the unloading dock for only one shift a day causes tankers to spend at least two separate portions of a day at the unloading dock because it takes at least a half a day to unload a tanker. Thus, the proposed redesign enforces at least a two-thirds of a day wait in the dock for each tanker. For some tankers that arrive near the end of a working shift, an additional two-thirds of a day is spent in the dock.

The output shows that tankers on the average wait .4463 days to gain access to one of the unloading docks. They spend .75 days being unloaded. Subtracting these values and the average three days for loading and nine days for traveling yields an expected time spent in the dock of 1.93 days. Thus, unloading over a three day period should be anticipated.

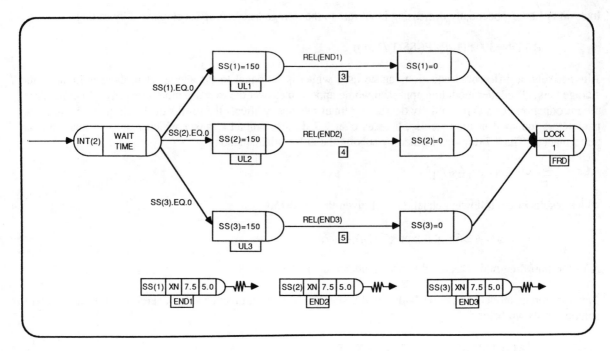

D10-3.1

```
 1    GEN,ROLSTON,PROBLEM 10.3,8/18/80,1;
 2    LIMITS,1,2,100;
 3    TIMST,XX(3),REFN INPUT AVAIL;
 4    CONT,0,4,.0025,.25,.25;
 5    RECORD,TNOW,DAYS,0,P,.25,200,250;
 6    VAR,SS(1),1,TANKER 1 LEVEL,0,300;
 7    VAR,SS(2),2,TANKER 2 LEVEL,0,300;
 8    VAR,SS(3),3,TANKER 3 LEVEL,0,300;
 9    VAR,SS(4),S,STORAGE LEVEL,0,2000;
10    TIMST,SS(4),STORAGE LEVEL;
11    INTLC,SS(4)=1000,XX(2)=1,XX(3)=1,XX(1)=0;
12    NETWORK;
13    ;
14    ;TANKER FLOW SUBPROCESS
15    ;----------------------
16          RESOURCE/DOCK(3),1;
17          CREATE,.5,0,,15;
18    VLDZ  ASSIGN,ATRIB(1)=TNOW;
19          ACT/1,UNFRM(2.9,3.1);
20          GOON;
21          ACT/2,RNORM(5.,1.5);
22          ASSIGN,ATRIB(2)=TNOW;
23          AWAIT,DOCK;
24          COLCT,INT(2),WAITING TIME,,1;
25          ACT,,SS(1).EQ.0,UL1;
26          ACT,,SS(2).EQ.0,UL2;
27          ACT,,SS(3).EQ.0,UL3;
28    UL1   ASSIGN,SS(1)=150;
29          ACT/3,REL(END1);
30          ASSIGN,SS(1)=0;
31          ACT,,,FRD;
32    UL2   ASSIGN,SS(2)=150;
33          ACT/4,REL(END2);
34          ASSIGN,SS(2)=0;
35          ACT,,,FRD;
36    UL3   ASSIGN,SS(3)=150;
37          ACT/5,REL(END3);
38          ASSIGN,SS(3)=0;
39          ACT,,,FRD;
40    FRD   FREE,DOCK;
41          ACT/6,RNORM(4.,1.);
42          COLCT,INT(1),TRIP TIME;
43          ACT,,,VLDZ;
44    ;
45    ;SHIFT START UP/SHUT DOWN SUBPROCESS
46    ;-----------------------------------
47          CREATE,1,,,25;
48          ASSIGN,XX(1)=1;
49          ACT,0.3333;
50          ASSIGN,XX(1)=0;
51          TERM;
52    ;
53    ;STATE EVENT SUBPROCESSES
54    ;------------------------
55    END1  DETECT,SS(1),XN,7.5,5;
56          TERM;
57    END2  DETECT,SS(2),XN,7.5,5;
58          TERM;
59    END3  DETECT,SS(3),XN,7.5,5;
60          TERM;
61          DETECT,SS(4),XN,5,5;
62          ASSIGN,XX(3)=0;
63          TERM;
64          DETECT,SS(4),XP,50,5;
65          ASSIGN,XX(3)=1;
66          TERM;
67          DETECT,SS(4),XP,2000,5;
68          ASSIGN,XX(2)=0;
69          TERM;
70          DETECT,SS(4),XN,1600,5;
71          ASSIGN,XX(2)=1;
72          TERM;
73          END;
74    INIT,0,365;
75    MONTR,CLEAR,65;
76    MONTR,SUMRY,165;
77    MONTR,CLEAR,165;
78    FIN;
```

```
                     S L A M   I I   S U M M A R Y   R E P O R T

        SIMULATION PROJECT PROBLEM 10.3          BY ROLSTON

        DATE  8/18/1980                          RUN NUMBER   1 OF   1


        CURRENT TIME   0.3650E+03
        STATISTICAL ARRAYS CLEARED AT TIME  0.1650E+03


                  **STATISTICS FOR VARIABLES BASED ON OBSERVATION**

                 MEAN        STANDARD     COEFF. OF    MINIMUM      MAXIMUM      NUMBER OF
                 VALUE       DEVIATION    VARIATION    VALUE        VALUE        OBSERVATIONS

WAITING TIME    0.4271E+00   0.6314E+00   0.1479E+01   0.0000E+00   0.3063E+01      209
TRIP TIME       0.1438E+02   0.1872E+01   0.1302E+00   0.9582E+01   0.1947E+02      208


                  **STATISTICS FOR TIME-PERSISTENT VARIABLES**

                 MEAN        STANDARD     MINIMUM      MAXIMUM      TIME         CURRENT
                 VALUE       DEVIATION    VALUE        VALUE        INTERVAL     VALUE

REFN INPUT AVAIL 0.9977E+00  0.4743E-01   0.0000E+00   0.1000E+01   0.2000E+03   0.1000E+01
STORAGE LEVEL    0.3620E+03  0.1185E+03   0.4963E+01   0.6757E+03   0.2000E+03   0.2244E+03


                  **FILE STATISTICS**

FILE                 AVERAGE      STANDARD     MAXIMUM   CURRENT   AVERAGE
NUMBER  LABEL/TYPE   LENGTH       DEVIATION    LENGTH    LENGTH    WAITING TIME

  1     AWAIT        0.4463       0.7963          4        0       0.4271
  2     CALENDAR    13.5845       1.5007         19       15       0.9985


                  **REGULAR ACTIVITY STATISTICS**

ACTIVITY        AVERAGE      STANDARD     MAXIMUM CURRENT  ENTITY
INDEX/LABEL     UTILIZATION  DEVIATION    UTIL    UTIL     COUNT

   1            3.1274       1.2623         7       3       209
   2            5.0379       1.5035        10       7       209
   3            0.8073       0.3944         1       0        74
   4            0.7987       0.4009         1       0        72
   5            0.6964       0.4598         1       1        64
   6            4.0860       1.3325         8       4       208


                  **RESOURCE STATISTICS**

RESOURCE  RESOURCE  CURRENT   AVERAGE       STANDARD     MAXIMUM       CURRENT
NUMBER    LABEL     CAPACITY  UTILIZATION   DEVIATION    UTILIZATION   UTILIZATION

   1      DOCK         3      2.3025        0.8358          3             1


RESOURCE  RESOURCE  CURRENT    AVERAGE      MINIMUM      MAXIMUM
NUMBER    LABEL     AVAILABLE  AVAILABLE    AVAILABLE    AVAILABLE

   1      DOCK         2       0.6975          0            3


                  **STATE AND DERIVATIVE VARIABLES**

                 (I)        SS(I)          DD(I)
                  1       0.0000E+00     0.0000E+00
                  2       0.0000E+00     0.0000E+00
                  3       0.1667E+02     0.0000E+00
                  4       0.2244E+03     0.0000E+00
```

10-4.   The network model for the tanker problem when an elaborated version of the refinery is included does not change significantly as shown in the statement model given in D10-4.1. State variables for the initial storage, intermediate storage, and final storage are now required along with DETECT nodes for determining when state events occur. In this solution to the problem, filters will be scheduled to be changed every 30 days. XX(7) is used to reference the time of change of the filter. Lines 73 to 77 in D10-4.1 are used to model the filter replacement process.

Subroutine STATE for this exercise is shown in D10-4.2. The procedures for setting the input and output rates are similar to those presented in Example 10-1. The output rate from the filter is made a function of TNOW-XX(7) as shown in the statement for SRATOUT. Another change is the cyclic demand for the output which was modeled for this exercise through the use of a SIN function in the computation of DAND. Other than these changes, the coding for subroutine STATE is direct. The outputs obtained from this exercise are shown in D10-4.3.

10-5.   To make the modifications to the tanker problem, define the following:

XX(4) = number of tankers remaining to be retired

XX(5) = unloading speed factor (1 for regular tankers; and 1.5 for the super tanker)

ATRIB(3) = identity of tanker (less than 70 is a regular tanker)

The network modifications for Embellishments a, b, and c are shown in D10-5.1 and described below.

10-5,Embellishment(a).   The super tanker is introduced at a CREATE node and only one such entity is created. The number of tankers remaining to be retired is set at 3 at an ASSIGN node. Branching based on ATRIB(3) is done where required to distinguish the regular tankers from the super tanker.

10-5,Embellishment(b).   When a regular tanker ends a trip, conditional branching based on XX(4) is done and, if XX(4) is not equal to 0, the regular tanker is retired and not sent back to Valdez (VLDZ).

10-5,Embellishment(c).   The super tanker is given priority over the other tankers waiting for unloading by having file 1 ranked on high-value-first based on ATRIB(3) using the following statement:

PRIORITY/1,HVF(3);

10-5,Embellishment(d). This embellishment requires different equations for representing the unloading of tankers based on tanker type. Define XX(5) as 0 for regular tankers and as 1 for the super tanker. In subroutine STATE, a test on the type of tanker being unloaded is made to specify the appropriate differential equations. This is shown in D10-5.2.

10-6.   To include the maintenance man in the model involves defining a resource type called MAINTENANCE which has a capacity of 1. An additional file is necessary to have reactors await the availability of maintenance. This requires that the LIMIT statements in line 2 be changed to indicate that two files are in the model. At line 64 of Figure 10-13, an AWAIT node is placed before activity 3 and a FREE node is placed after the activity 3 to model the requirement for a maintenance man before the C_AND_R activity can be performed.

```
 1    GEN,ROLSTON,PROBLEM 10.4,8/28/80,1;
 2    LIMITS,1,2,100;
 3    TIMST,XX(3),PROC UTIL;
 4    TIMST,XX(6),FILT UTIL;
 5    TIMST,XX(5),DEMAND MET;
 6    CONT,0,4,.0025,.25,.25;
 7    RECORD,TNOW,DAYS,0,P,.25,300;
 8    VAR,SS(1),T,TANKER LEVEL,0,300;
 9    VAR,SS(2),I,INITIAL STOR,0,2000;
10    VAR,SS(3),M,INTERMED STOR,0,2000;
11    VAR,SS(4),F,FINAL STOR,0,2000;
12    TIMST,SS(2),INIT STOR;
13    TIMST,SS(3),MED STOR;
14    TIMST,SS(4),FINAL STOR;
15    INTLC,SS(2)=1000,SS(3)=1200,SS(4)=800;
16    INTLC,XX(2)=1,XX(3)=1,XX(4)=1,XX(5)=1,XX(6)=1,XX(7)=0;
17    NETWORK;
18          RESOURCE/DOCK,1;
19          CREATE,.5,0,,15;
20    VLDZ  ASSIGN,ATRIB(1)=TNOW;
21          ACT/1,UNFRM(2.9,3.1)
22          GOON;
23          ACT/2,RNORM(5.,1.5);
24          ASSIGN,ATRIB(2)=TNOW;
25          AWAIT,DOCK;
26          COLCT,INT(2),WAITING TIME;
27          ASSIGN,SS(1)=150;
28          ACT/3,REL(ENDU);
29          ASSIGN,SS(1)=0;
30          FREE,DOCK;
31          ACT/4,RNORM(4.,1.);
32          COLCT,INT(1),TRIP TIME;
33          ACT,,,VLDZ;
34    ENDU  DETECT,SS(1),XN,7.5,5;
35          TERM;
36          DETECT,SS(2),XN,5,5;
37          ASSIGN,XX(3)=0
38          TERM;
39          DETECT,SS(2),XP,50,5;
40          ASSIGN,XX(3)=1;
41          TERM;
42          DETECT,SS(2),XP,2000,5;
43          ASSIGN,XX(2)=0;
44          TERM;
45          DETECT,SS(2),XN,1600,5;
46          ASSIGN,XX(2)=1;
47          TERM;
48          DETECT,SS(3),XN,5,5;
49          ASSIGN,XX(4)=0;
50          TERM;
51          DETECT,SS(3),XP,50,5;
52          ASSIGN,XX(4)=1;
53          TERM;
54          DETECT,SS(3),XP,2000,5;
55          ASSIGN,XX(3)=0;
56          TERM;
57          DETECT,SS(3),XN,1600,5;
58          ASSIGN,XX(3)=1;
59          TERM;
60          DETECT,SS(4),XN,5,5;
61          ASSIGN,XX(5)=0;
62          TERM;
63          DETECT,SS(4),XP,50,5;
64          ASSIGN,XX(5)=1;
65          TERM;
66          DETECT,SS(4),XP,2000,5;
67          ASSIGN,XX(4)=0;
68          TERM;
69          DETECT,SS(4),XN,1600,5;
70          ASSIGN,XX(4)=1;
71          TERM;
72
73          CREATE,,30;
74    AS1   ASSIGN,XX(6)=0;
75          ACT,EXPON(.5,1);
76          ASSIGN,XX(6)=1,XX(7)=TNOW;
77          ACT,30,,AS1;
78          END;
79    INIT,0,365;
80    MONTR,CLEAR,65;
81    MONTR,SUMRY,165;
82    MONTR,CLEAR,165;
83    FIN;
```

```
SUBROUTINE STATE
COMMON/SCOM1/ATRIB(100),DD(100),DDL(100),DTNOW,II,MFA,MSTOP,NCLNR
1,NCRDR,NPRNT,NNRUN,NNSET,NTAPE,SS(100),SSL(100),TNEXT,TNOW,XX(100)
DATA PI/3.14159/
FRATIN=300
IF(XX(2)*SS(1).EQ.0.0) FRATIN=0
FRATOUT=150.*XX(3)
SRATIN=FRATOUT
SRATOUT=(200.-100./30.*(TNOW-XX(7)))*XX(4)*XX(6)
TRATIN=SRATOUT
DAND=150.+50.*SIN(PI*TNOW/45.)
DMAND=DAND*XX(5)
TRATOUT=DMAND
SS(1)=SSL(1)-DTNOW*FRATIN
SS(2)=SSL(2)+DTNOW*(FRATIN-FRATOUT)
SS(3)=SSL(3)+DTNOW*(SRATIN-SRATOUT)
SS(4)=SSL(4)+DTNOW*(TRATIN-TRATOUT)
RETURN
END
```

D10-4.2

S L A M   I I   S U M M A R Y   R E P O R T

SIMULATION PROJECT PROBLEM 10.4   BY ROLSTON

DATE 8/28/1980   RUN NUMBER   1 OF   1

CURRENT TIME   0.3650E+03
STATISTICAL ARRAYS CLEARED AT TIME   0.1650E+03

STATISTICS FOR VARIABLES BASED ON OBSERVATION

	MEAN VALUE	STANDARD DEVIATION	COEFF. OF VARIATION	MINIMUM VALUE	MAXIMUM VALUE	NUMBER OF OBSERVATIONS
WAITING TIME	0.1576E+01	0.1237E+01	0.7849E+00	0.0000E+00	0.4720E+01	208
TRIP TIME	0.1448E+02	0.2271E+01	0.1569E+00	0.8996E+01	0.2069E+02	208

STATISTICS FOR TIME-PERSISTENT VARIABLES

	MEAN VALUE	STANDARD DEVIATION	MINIMUM VALUE	MAXIMUM VALUE	TIME INTERVAL	CURRENT VALUE
PROC UTIL	0.9895E+00	0.1018E+00	0.0000E+00	0.1000E+01	0.2000E+03	0.1000E+01
FILT UTIL	0.9892E+00	0.1031E+00	0.0000E+00	0.1000E+01	0.2000E+03	0.1000E+01
DEMAND MET	0.9854E+00	0.1197E+00	0.0000E+00	0.1000E+01	0.2000E+03	0.1000E+01
INIT STOR	0.1808E+04	0.1130E+03	0.1596E+04	0.2005E+04	0.2000E+03	0.1611E+04
MED STOR	0.1582E+04	0.1712E+03	0.1263E+04	0.2000E+04	0.2000E+03	0.1672E+04
FINAL STOR	0.8488E+03	0.5097E+03	0.3737E+01	0.1607E+04	0.2000E+03	0.1288E+04

FILE STATISTICS

FILE NUMBER	LABEL/TYPE	AVERAGE LENGTH	STANDARD DEVIATION	MAXIMUM LENGTH	CURRENT LENGTH	AVERAGE WAITING TIME
1	AWAIT	1.5888	1.2520	6	0	1.5277
2	CALENDAR	13.4640	1.3361	17	15	1.3628

REGULAR ACTIVITY STATISTICS

ACTIVITY INDEX/LABEL	AVERAGE UTILIZATION	STANDARD DEVIATION	MAXIMUM UTIL	CURRENT UTIL	ENTITY COUNT
1	3.1107	1.4902	7	3	207
2	5.2108	1.4487	9	8	204
3	0.9472	0.2236	1	1	208
4	4.1424	1.4330	7	3	208

RESOURCE STATISTICS

RESOURCE NUMBER	RESOURCE LABEL	CURRENT CAPACITY	AVERAGE UTILIZATION	STANDARD DEVIATION	MAXIMUM UTILIZATION	CURRENT UTILIZATION
1	DOCK	1	0.9472	0.2236	1	1

RESOURCE NUMBER	RESOURCE LABEL	CURRENT AVAILABLE	AVERAGE AVAILABLE	MINIMUM AVAILABLE	MAXIMUM AVAILABLE
1	DOCK	0	0.0528	0	1

STATE AND DERIVATIVE VARIABLES

(I)	SS(I)	DD(I)
1	0.8109E+02	0.0000E+00
2	0.1611E+04	0.0000E+00
3	0.1672E+04	0.0000E+00
4	0.1288E+04	0.0000E+00

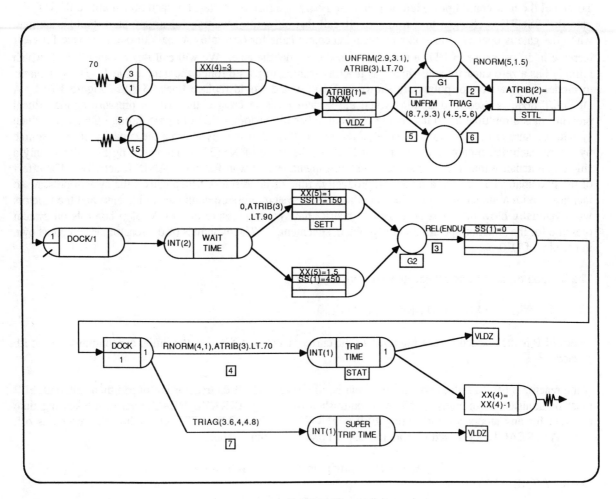

D10-5.1

```
       SUBROUTINE STATE
       COMMON/SCOM1/ATRIB(100),DD(100),DDL(100),DTNOW,II,MFA,MSTOP,NCLNR
      1,NCRDR,NPRNT,NNRUN,NNSET,NTAPE,SS(100),SSL(100),TNEXT,TNOW,XX(100)
       RATIN=300
       IF(XX(1)*XX(2)*SS(1).EQ.0.) RATIN=0
       RATOUT=150.*XX(3)
       IF(XX(5)*RATIN.EQ.0.) GOTO 10
       DD(1)=-SS(1)
       DD(2)=SS(1)-RATOUT
       RETURN
   10  DD(1)=-RATIN
       DD(2)=RATIN-RATOUT
       RETURN
       END
```

D10-5.2

To model the new control procedure which allows only one reactor to start for each 50 psi above PCRIT, the gate PRESSURE is changed to a resource. Recall that all entities waiting for a gate pass through the gate when the gate is opened. Since this problem statement calls for one entity to pass through the gate for each increase in 50 psi above PCRIT, a resource is used to model the gate. We will call this resource GATE which initially has a zero capacity and which has reactors waiting to pass through the gate in file 1. The statements to model the new control policy are shown in D10-6.1 and would replace lines 75-80 of Figure 10-13. A DETECT node is used to determine when the system pressure crosses the critical pressure or the critical pressure plus a multiple of 50 depending on the number of reactors waiting to pass through the gate. When system pressure crosses the value XX(29), a reactor is allowed to start by altering the capacity of the resource by 1. At each detection of the crossing of XX(29), the value of XX(29) is increased by 50. When only a single maintenance man is available, reactors must queue up and wait for the C_AND_R activity. The effect of the two changes is to increase the non-productive time of the reactors which causes the system pressure to increase. With a model run under these conditions, the system pressure continues to increase and the reactors will eventually blow up. It is necessary to restrict the system pressure and to develop controls on system pressure for the situation specified in the problem statement. The dynamics of the reactor system are shown in the plot of D10-6.2.

10-7.   The solution of the differential equation for concentration is

$$SS(I) = XX(I) * EXP(-RK(I)*PEFF*ELP)$$

where ELP is the elapsed processing time for the current batch and XX(I) is the initial concentration of the Ith reactor.

If the reactor is off, ELP does not advance, that is, $SS(I) = SSL(I)$. If the reactor is stopped then restarted, ELP must be adjusted accordingly. This is accomplished by setting ARRAY(2,REACTOR) to the starting time adjusted for any stoppages and setting ELP = TNOW - ARRAY(2,REACTOR). While the reactor is off, ARRAY(2,REACTOR) is used to store the elapsed processing time. Thus,

$$ARRAY(2,REACTOR) = \begin{cases} \text{Equivalent starting time if REACTOR is on} \\ \text{Elapsed processing time if REACTOR is off} \end{cases}$$

The revised subroutine STATE is given in D10-7.1 and the statement model in D10-7.2.

10-8.   This exercise is a modification of Exercise 6-9 to include tool breakdown based on processing time rather than elapsed time. (Other embellishments for this model are given in Exercise 9-9). The basic model is the same as given in the solution to Exercise 6-9 with the duration of activity 1 prescribed as the processing time which is defined as ATRIB(2) of the entity multiplied by XX(1) where XX(1) is the adjustment to processing time based on the number of parts waiting in file 1. To change the probability of passing inspection based on the speed on which the part is processed, the variable XX(2) is defined to be the probability of not passing inspection. Thus XX(1) is set to 0.9 and XX(2) to 0.25 when NNQ(1) crosses the value 4 in the positive direction. These values are reset to their normal values of 1 and 0.15 when the number in the queue crosses the threshold value of 1 in the negative direction. The statement model for this exercise is shown in D10-8.1. To determine the time of machine tool breakdown, a DETECT node is used to monitor SS(1) which is the cumulative processing time on the machine tool. When this value crosses the threshold 480 with a tolerance of 30, a breakdown occurs. SS(1) is defined in subroutine STATE shown in D10-8.2 where the rate of change of SS(1) is dependent on the number of entities in activity 1. In this way, SS(1) is used to accumulate the processing time of the machine tool as represented by the time activity 1 has a part in it. When SS(1) crosses 480 in the positive direction, the amount of processing time since the last failure is reset to 0, the machine is preempted, the repair is performed and following the repair time, the machine is freed. This portion of the model is shown in lines 22-27 of D10-8.1. The summary report for Exercise 10-8 is shown in D10-8.3.

```
 1   GEN,PRITSKER,CHEM REACTOR,1/9/86;
 2   LIMITS,2,2,50;
 3   TIMST,XX(5),NO. BUSY REACT;
 4   TIMST,SS(5),SURGE PRESSURE;
 5   CONTINUOUS,5,0,.001,.1,.1;
 6   ARRAY(1,4)/0,0,0,0;                          REACTOR STATUS
 7   EQUIVALENCE/ATRIB(1),REACTOR/EXPON(1),DISCHARGE;
 8   EQUIVALENCE/XX(7),PNOM/XX(8),PCRIT/RNORM(1,.5),C_AND_R;
 9   ;
10   ;   XX(I) = INITIAL CONCENTRATION FOR REACTOR I, I=1,4
11   ;   XX(5) = NUMBER OF BUSY REACTORS
12   ;   XX(6) = LAST REACTOR STARTED, XX(7) = NOMINAL PRESSURE
13   ;   XX(8) = CRITICAL PRESSURE, XX(9) = FCOMP, XX(10)=RTV
14   ;   XX(21:24) = REACTION CONSTANTS 1-4
15   ;   XX(25:28) = VOLUME OF REACTORS 1-4
16   ;
17   INTLC,SS(1)=0.1,      SS(2)=0.4,     SS(3)=0.2,     SS(4)=0.5, SS(5)=500;
18   INTLC,XX(1)=0.1,      XX(2)=0.4,     XX(3)=0.2,     XX(4)=0.5;
19   INTLC,XX(7)=150,      XX(8)=100,     XX(9)=4.352,   XX(10)=118.03;
20   INTLC,XX(21)=0.03466, XX(22)=0.00866,XX(23)=0.01155,XX(24)=0.00770;
21   INTLC,XX(25)=10,      XX(26)=15,     XX(27)=20,     XX(28)=25;
22   INTLC,XX(29)=150;
23   ;
24   RECORD,TNOW,TIME,0,P,.1,0,20;
25      VAR,SS(1),1,CONC 1,0,.5;
26      VAR,SS(2),2,CONC 2,0,.5;
27      VAR,SS(3),3,CONC 3,0,.5;
28      VAR,SS(4),4,CONC 4,0,.5;
29      VAR,SS(5),P,PRESSURE,0,1000;
30      VAR,PNOM,!,PNOM,0,1000;
31      VAR,PCRIT,!,PCRIT,0,1000;
32   INIT,0,300;
33   ;
34   NETWORK;
35         RESOURCE,MAINTENANCE(1),2;
36         RESOURCE,GATE(0),1;
37         CREATE,.5,,,4;                   CREATE INITIAL BATCHES.
38            ACT/10;
39         ASSIGN,REACTOR=NNCNT(10);        ASSIGN REACTOR NUMBER.
40   START GOON,1;                          BEGIN A BATCH.
41            ACT,,SS(5).LT.PNOM,LOWP;      IF PRESSURE.LT.PNOM, GO TO LOWP
42            ACT;                          ELSE
43         ASSIGN,ARRAY(1,REACTOR)=1,
44               XX(5)=XX(5)+1,
45               XX(6)=REACTOR;            TURN ON REACTOR.
46         TERM;
47   LOWP  ASSIGN,XX(29)=PCRIT+50;
48         AWAIT(1),GATE;                  WAIT HERE IF PRESSURE IS TOO LOW
49         ACT,,,START;                    TO START A BATCH.
50   ;
51         DETECT,SS(1),XN,.01,.001;       STOP REACTOR 1: BATCH COMPLETE
52         ASSIGN,REACTOR=1;
53            ACT,,,ENDR;
54         DETECT,SS(2),XN,.04,.001;       STOP REACTOR 2: BATCH COMPLETE
55         ASSIGN,REACTOR=2;
56            ACT,,,ENDR;
57         DETECT,SS(3),XN,.02,.001;       STOP REACTOR 3: BATCH COMPLETE
58         ASSIGN,REACTOR=3;
59            ACT,,,ENDR;
60         DETECT,SS(4),XN,.05,.001;       STOP REACTOR 4: BATCH COMPLETE
61         ASSIGN,REACTOR=4;
62   ENDR  ASSIGN,ARRAY(1,REACTOR)=0,
63               XX(5)=XX(5)-1,
64               II=REACTOR, SS(II)=XX(II),ATRIB(2)=TNOW;
65            ACT/2,DISCHARGE;             DISCHARGE
66         GOON;
67         AWAIT(2),MAINTENANCE;
68            ACT/3,C_AND_R;               CLEAN&RECHRG
69         FREE,MAINTENANCE;
70         COLCT,INT(2),DOWNTIME,10/0/.5;
71            ACT,,,START;
72   ;
73         DETECT,SS(5),XN,PCRIT,1;        STOP LAST REACTOR STARTED IF
74            ACT,,XX(5).GT.0;             SURGE TANK PRESSURE FALLS
75   ;                                     BELOW CRITICAL PRESSURE.
76         ASSIGN,REACTOR=XX(6),
77               ARRAY(1,REACTOR)=0,XX(5)=XX(5)-1;
78         ACT/4,,,LOWP;                   INTERRUPT
79   ;
80         DETECT,SS(5),XP,XX(29),1;       ALLOW REACTORS TO PROCEED WHEN
81         ASSIGN,XX(29)=XX(29)+50;        SURGE TANK PRESSURE RETURNS TO
82         ALTER,GATE,1;                   NOMINAL PRESSURE.
83         TERM;
84         END;
85   FIN;
```

```
                              **PLOT NUMBER  1**
                               RUN  NUMBER  1

                                SCALES OF PLOT
1=CONC 1      0.0000E+00        0.1250E+00        0.2500E+00        0.3750E+00        0.5000E+00
2=CONC 2      0.0000E+00        0.1250E+00        0.2500E+00        0.3750E+00        0.5000E+00
3=CONC 3      0.0000E+00        0.1250E+00        0.2500E+00        0.3750E+00        0.5000E+00
4=CONC 4      0.0000E+00        0.1250E+00        0.2500E+00        0.3750E+00        0.5000E+00
P=PRESSURE    0.0000E+00        0.2500E+03        0.5000E+03        0.7500E+03        0.1000E+04
!=PNOM        0.0000E+00        0.2500E+03        0.5000E+03        0.7500E+03        0.1000E+04
!=PCRIT       0.0000E+00        0.2500E+03        0.5000E+03        0.7500E+03        0.1000E+04

            0   5  10  15  20  25  30  35  40  45  50  55  60  65  70  75  80  85  90  95  100 DUPLICATES

TIME

0.0000E+00  +     !     1       +           3       P                  +    2            4
0.1000E+00  +     !   1!        +           3         + P              +    2            4
0.2000E+00  +     !   1         +           3           P              +    2            4 1!
0.3000E+00  +    1 !   !        +           3       +       P          +    2            4
0.4000E+00  +   1  !   !        +           3       +         P        +    2            4
0.5000E+00  +  1   !   !        +           3       +           P      +    2            4
0.6000E+00  + 1    !   !        +           3       +             P  2 +    2            4
0.7000E+00  + 1    !   !   1    +           3       +             P 2 2 +   2            4
0.8000E+00  +      !   !   1    +           3       +           2 P    +    2            4
0.9000E+00  +      !   !   1    +           3       +         2   P    +    2            4
0.1000E+01  +      !   !   1    +           3       + 2     P          +    2            4
0.1100E+01  +      !   !   1    +         3       2 +     P            +    2            4
0.1200E+01  +      !   !   1    +   3           2   +   P              +    2            4
0.1300E+01  +      !   !   1    + 3         2       + P                +    2            4
0.1400E+01  +      !   !   1    3         2         P                  +    2            4
0.1500E+01  +      !   !   13   +       2         P                    +    2            4
0.1600E+01  +      !   !   1    +     2       P     +                  +  4 +    13
0.1700E+01  +      !   ! 31     + 2       P         +              4   +    +
0.1800E+01  +      !   13  1    P2        +         +          4       +    +
0.1900E+01  +      ! 33 P 1   22 +         +       4 +4                +    + 1P 3!
0.2000E+01  +    !P33 !    122  +         +           44              +    + !P
0.2100E+01  +    !3P !       +         +           4              +    + 12
0.2200E+01  +    3  P! 2 1   +         +           4              +    + 3!
0.2300E+01  +    3! P P 22 1 +         +          44              +    + P!
0.2400E+01  +   3 PP   !2    +         +        4                 +    + P! !2
0.2500E+01  +   3  !   2!    1 +       +        4                 +    + 2P !P
0.2600E+01  +   3  ! P2 !    1 +       +       4                  +    +
0.2700E+01  +   3  PP2   1   +         +    4                     +    + P!
0.2800E+01  +   3  !2 P!    1 +        +   4                      +    + !P
0.2900E+01  + 3   2 P !    1 +     4   +                          +    + 2!
0.3000E+01  + 3   2!P !    1 +     3     4 +                      +    +
0.3100E+01  + 3    2 !P !  1 +       3   4 +                      +    +
0.3200E+01  + 3    2 P  !  1 +      344    +                      +  2 + P!
0.3300E+01  +      !P  !   1 +    4 3      +                      +  2 +
0.3400E+01  +      !P !    1 +   4   3     +                      +  2 +
0.3500E+01  +      ! P !   1 +  4     3    +                      +  2 + P!
0.3600E+01  +      !  P    1 + 4      3    +                      +  2 +
0.3700E+01  +      !  !P   1 + 4      3    +                      +  2 +
0.3800E+01  +      !  !  P1  44       3    +                      +  2 + 1P
0.3900E+01  +      !  ! 1 P  4+       3    +                      +  2 +
0.4000E+01  +      ! ! 1    P 4       3    +                      +  2 +
0.4100E+01  +   1 !    !    P4        3    +                      +  2 +
0.4200E+01  +   1  !   !   4 P        3    +                      +  2 +
0.4300E+01  +  1   !   !   4   P +    3    +                      +  2 +
0.4400E+01  + 1    !    !44     PP    3    +                      +  2 + 4!
0.4500E+01  + 1    !   4 !    + PPP   3    +                      +  2 +
0.4600E+01  +      ! 4! 1    +     P  3    +                      +  2 +
0.4700E+01  +      ! 4 !  1  +      PP 3   +                      +  2 + 3P
0.4800E+01  +      !4 !   1  +       P3    +                      +  2 +
0.4900E+01  +      !4 !   1  +      3 P    +                      +  2 + 4 !4
0.5000E+01  +      !4 !   1  +      3  P PP +                     +  2 +
0.5100E+01  +      ! !    1  +      3    P P                      +  2 + 4
0.5200E+01  +      ! !    1  +      3      P P                    +  2 + 4
0.5300E+01  +      ! !    1  +      3        P P                  +  2 + 4
0.5400E+01  +      ! !    1  +      3          P P                +  2 + 4
0.5500E+01  +      ! !    1  +      3            P P              +  2 + 4
0.5600E+01  +      ! !    1  +      3 3            PP  2          +  4
0.5700E+01  +      ! !    1  +    3               +P    2        + 4
0.5800E+01  +      ! !    1  +   3                +P    2        + 4
0.5900E+01  +      ! !    1  + 3                  + P 2          + 4
0.6000E+01  +      ! !    1  33+                  + P2           4 2P
0.6100E+01  +      ! !      1 3 +                 + 2P           4
0.6200E+01  +      ! !    31   +                  + 2  P         4
0.6300E+01  +      ! ! 33 1    +                  + 2   PP       4
0.6400E+01  +      ! 3 !  1    +                  + 2      PP P  4 3!
0.6500E+01  +      ! 3! 1      +                  + 2         P  4
```

D10-6.2(1)

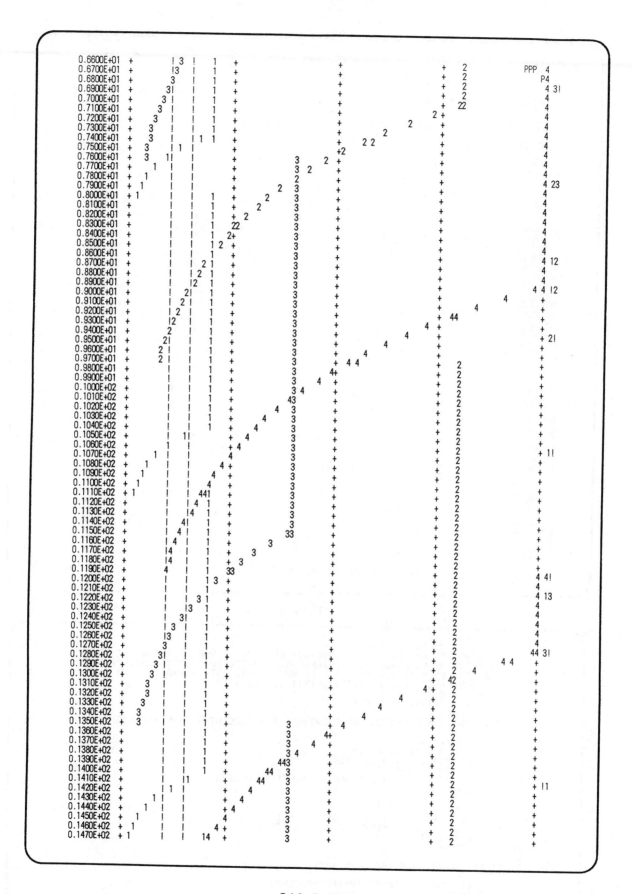

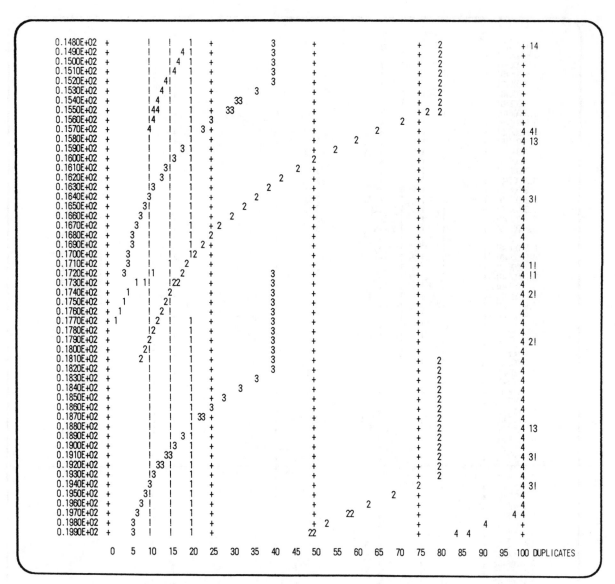

D10-6.2(3)

```
            SUBROUTINE STATE
            DIMENSION RK(4),F(4),V(4)
            COMMON/SCOM1/ ATRIB(100),DD(100),DDL(100),DTNOW,II,MFA,MSTOP,NCLNR
           1,NCRDR,NPRNT,NNRUN,NNSET,NTAPE,SS(100),SSL(100),TNEXT,TNOW,XX(100)
            EQUIVALENCE (PCRIT,XX(8)),  (FCOMP,XX(9)),  (RTV,XX(10))
            EQUIVALENCE (RK(1),XX(21)), (V(1),XX(25))
            SUMF = 0
            PEFF = AMIN1(SS(5),PCRIT)
      C
      C *** DEFINE THE DIFFERENTIAL EQUATIONS FOR CONCENTRATE IN REACTORS 1 - 4
      C
            DO 10 I=1,4
               STATUS = GETARY(1,I)
               IF (STATUS.GT.0.0) THEN
                  EPT=GETARY(2,I)
                  SS(I)=XX(I)*EXP(-RK(I)*PEFF*(TNOW-EPT))
               ELSE
                  SS(I)=SSL(I)
               ENDIF
               F(I)=RK(I)*SS(I)*STATUS*PEFF*V(I)
      C**** SET TOTAL MOLAL FLOW FOR ALL REACTORS
               SUMF = SUMF+F(I)
      10    CONTINUE
      C
      C *** SET SURGE TANK PRESSURE RATE
      C
            SS(5)=SSL(5)+DTNOW*RTV*(FCOMP-SUMF)
            RETURN
            END
```

D10-7.1

```
 1   GEN,PRITSKER,CHEM REACTOR,1/9/86;
 2   LIMITS,1,2,50;
 3   TIMST,XX(5),NO. BUSY REACT;
 4   TIMST,SS(5),SURGE PRESSURE;
 5   CONTINUOUS,5,0,.001,.1,.1;
 6   ARRAY(1,4)/0,0,0,0;                              REACTOR STATUS
 7   ARRAY(2,4)/0,0,0,0;                              TIME LAST START
 8   EQUIVALENCE/ATRIB(1),REACTOR/EXPON(1),DISCHARGE;
 9   EQUIVALENCE/XX(7),PNOM/XX(8),PCRIT/RNORM(1,.5),C_AND_R;
10   ;
11   ;   XX(I) = INITIAL CONCENTRATION FOR REACTOR I, I=1,4
12   ;   XX(5) = NUMBER OF BUSY REACTORS
13   ;   XX(6) = LAST REACTOR STARTED, XX(7) = NOMINAL PRESSURE
14   ;   XX(8) = CRITICAL PRESSURE, XX(9) = FCOMP, XX(10)=RTV
15   ;   XX(21:24) = REACTION CONSTANTS 1-4
16   ;   XX(25:28) = VOLUME OF REACTORS 1-4
17   ;
18   INTLC,SS(1)=0.1,        SS(2)=0.4,        SS(3)=0.2,        SS(4)=0.5, SS(5)=500;
19   INTLC,XX(1)=0.1,        XX(2)=0.4,        XX(3)=0.2,        XX(4)=0.5;
20   INTLC,XX(7)=150,        XX(8)=100,        XX(9)=4.352,      XX(10)=118.03;
21   INTLC,XX(21)=0.03466,   XX(22)=0.00866,   XX(23)=0.01155,   XX(24)=0.00770;
22   INTLC,XX(25)=10,        XX(26)=15,        XX(27)=20,        XX(28)=25;
23
24   RECORD,TNOW,TIME,0,P,.1,0,20;
25      VAR,SS(1),1,CONC 1,0,.5;
26      VAR,SS(2),2,CONC 2,0,.5;
27      VAR,SS(3),3,CONC 3,0,.5;
28      VAR,SS(4),4,CONC 4,0,.5;
29      VAR,SS(5),P,PRESSURE,0,1000;
30      VAR,PNOM,!,PNOM,0,1000;
31      VAR,PCRIT,!,PCRIT,0,1000;
32   INIT,0,300;
33   ;
34   NETWORK;
35        GATE/PRESSURE,CLOSED,1;
36        CREATE,.5,,,4;                       CREATE INITIAL BATCHES.
37           ACT/10;
38        ASSIGN,REACTOR=NNCNT(10);            ASSIGN REACTOR NUMBER.
39   START GOON,1;                             BEGIN A BATCH.
40        ACT,,SS(5).LT.PNOM,LOWP;             IF PRESSURE.LT.PNOM, GO TO LOWP
41        ACT;                                 ELSE
42        ASSIGN,ARRAY(1,REACTOR)=1,
43              XX(5)=XX(5)+1,
44              XX(6)=REACTOR,
45              ARRAY(2,REACTOR)=TNOW-ARRAY(2,REACTOR);
46   ;                                         SET EQUIVALENT START TIME
47        TERM;
48   LOWP  AWAIT(1),PRESSURE;                  WAIT HERE IF PRESSURE IS TOO LOW
49        ACT,,,START;                         TO START A BATCH.
50   ;
51        DETECT,SS(1),XN,.01,.001;            STOP REACTOR 1: BATCH COMPLETE
52        ASSIGN,REACTOR=1;
53           ACT,,,ENDR;
54        DETECT,SS(2),XN,.04,.001;            STOP REACTOR 2: BATCH COMPLETE
55        ASSIGN,REACTOR=2;
56           ACT,,,ENDR;
57        DETECT,SS(3),XN,.02,.001;            STOP REACTOR 3: BATCH COMPLETE
58        ASSIGN,REACTOR=3;
59           ACT,,,ENDR;
60        DETECT,SS(4),XN,.05,.001;            STOP REACTOR 4: BATCH COMPLETE
61        ASSIGN,REACTOR=4;
62   ENDR  ASSIGN,ARRAY(1,REACTOR)=0,
63              ARRAY(2,REACTOR)=0,
64              XX(5)=XX(5)-1,
65              II=REACTOR, SS(II)=XX(II),ATRIB(2)=TNOW;
66        ACT/2,DISCHARGE;                     DISCHARGE
67        GOON;
68           ACT/3,C_AND_R;                    CLEAN&RECHRG
69        COLCT,INT(2),DOWNTIME,10/0/.5;
70           ACT,,,START;
71   ;
72        DETECT,SS(5),XN,PCRIT,1;             STOP LAST REACTOR STARTED IF
73           ACT,,XX(5).GT.0;                  SURGE TANK PRESSURE FALLS
74   ;                                         BELOW CRITICAL PRESSURE.
75        ASSIGN,REACTOR=XX(6),
76              ARRAY(2,REACTOR)=TNOW-ARRAY(2,REACTOR),
77              ARRAY(1,REACTOR)=0,XX(5)=XX(5)-1;
78        ACT/4,,,LOWP;                        INTERRUPT
79   ;
80        DETECT,SS(5),XP,PNOM,1;              ALLOW REACTORS TO PROCEED WHEN
81        OPEN,PRESSURE;                       SURGE TANK PRESSURE RETURNS TO
82        CLOSE,PRESSURE;                      NOMINAL PRESSURE.
83        TERM;
84        END;
85   FIN;
```

```
1    GEN,VASEK,PROBLEM 10.8,6/26/86,1;
2    LIMITS,2,2,100;
3    INTLC,XX(1)=1.,XX(2)=.15;
4    CONTINUOUS,1,0,10.,60.;
5    NETWORK;
6            RESOURCE/MACHINE(1),1;                DEFINE MACHINE
7            CREATE,TRIAG(20,30,50),,1;            ARRIVAL OF PART TYPE 1
8            ASSIGN,ATRIB(2)=EXPON(20);            ASSIGN PROCESS TIME
9            ACTIVITY,,,QUE1;                      ROUTE TO MACHINE
10           CREATE,TRIAG(35,50,60),,1;            ARRIVAL OF PART TYPE 2
11           ASSIGN,ATRIB(2)=UNFRM(15,20);         ASSIGN PROCESS TIME
12   QUE1    AWAIT(1),MACHINE;                     WAIT FOR MACHINE TO BE AVAILABLE
13           ACTIVITY/1,ATRIB(2)*XX(1);            PROCESS PART BY MACHINE
14           FREE,MACHINE,1;                       RELEASE MACHINE
15           ACTIVITY,,1-XX(2),PASS;               ROUTE TO PASS INSPECTION
16           ACTIVITY,,XX(2);                      FAIL INSPECTION
17           ASSIGN,ATRIB(2)=ATRIB(2)*.9;          REASSIGN PROCESS TIME
18           ACTIVITY,,,QUE1;                      ROUTE TO MACHINE
19   PASS    COLCT,INT(1),TIME IN SYSTEM;          COLLECT STAT ON TIME IN SYS
20           TERMINATE;                            DEPART FROM SYSTEM
21   ;
22           DETECT,SS(1),XP,480.,30.;             MACHINE TOOL BREAKDOWN
23           ASSIGN,SS(1)=0.;                      RESET BREAKDOWN TIME
24           PREEMPT(2),MACHINE,QUE1;              SEIZE MACHINE
25           ACTIVITY/2,RNORM(30,8);               REPAIR THE MACHINE TOOL
26           FREE,MACHINE;                         RELEASE THE MACHINE
27           TERMINATE;                            DEPART FROM SYSTEM
28   ;
29           DETECT,NNQ(1),XP,4.;                  NUMBER WAITING ON MACHINE
30           ASSIGN,XX(1)=.9,XX(2)=.25;            ADJUST VARIABLES
31           TERMINATE;                            DEPART FROM SYSTEM
32   ;
33           DETECT,NNQ(1),XN,1.;                  NUMBER WAITING ON MACHINE
34           ASSIGN,XX(1)=1.,XX(2)=.15;            READJUST VARIABLES
35           TERMINATE;                            DEPART FROM SYSTEM
36           END;
37   INIT,0,2400;
38   FIN;
```

D10-8.1

```
SUBROUTINE STATE
COMMON/SCOM1/ ATRIB(100),DD(100),DDL(100),DTNOW,II,MFA,MSTOP,NCLNR
1,NCRDR,NPRNT,NNRUN,NNSET,NTAPE,SS(100),SSL(100),TNEXT,TNOW,XX(100)
STATUS=NNACT(1)
DD(1)=STATUS
RETURN
END
```

D10-8.2

```
                    S L A M   I I   S U M M A R Y   R E P O R T

          SIMULATION PROJECT PROBLEM 10.8          BY VASEK

          DATE  6/26/1986                          RUN NUMBER  1 OF   1


          CURRENT TIME   0.2400E+04
          STATISTICAL ARRAYS CLEARED AT TIME  0.0000E+00


                  **STATISTICS FOR VARIABLES BASED ON OBSERVATION**

                  MEAN        STANDARD    COEFF. OF   MINIMUM     MAXIMUM     NUMBER OF
                  VALUE       DEVIATION   VARIATION   VALUE       VALUE       OBSERVATIONS

   TIME IN SYSTEM  0.1262E+03  0.1160E+03  0.9191E+00  0.1370E+01  0.5686E+03      103


                          **FILE STATISTICS**

   FILE               AVERAGE     STANDARD    MAXIMUM   CURRENT   AVERAGE
   NUMBER  LABEL/TYPE  LENGTH      DEVIATION   LENGTH    LENGTH    WAITING TIME

      1   QUE1 AWAIT   5.9316      4.7908         17       17     90.1005
      2        PREEMPT 0.0000      0.0000          0        0      0.0000
      3        CALENDAR 2.9616     0.1921          4        3      8.6366


                       **REGULAR ACTIVITY STATISTICS**

   ACTIVITY        AVERAGE      STANDARD    MAXIMUM CURRENT   ENTITY
   INDEX/LABEL     UTILIZATION  DEVIATION   UTIL    UTIL      COUNT

      1 PROCESS PART   0.9129      0.2820        1       1       136
      2 REPAIR THE M   0.0488      0.2153        1       0         4


                          **RESOURCE STATISTICS**

   RESOURCE  RESOURCE  CURRENT   AVERAGE       STANDARD    MAXIMUM       CURRENT
   NUMBER    LABEL     CAPACITY  UTILIZATION   DEVIATION   UTILIZATION   UTILIZATION

      1      MACHINE     1        0.9616        0.1921         1             1


   RESOURCE  RESOURCE  CURRENT    AVERAGE      MINIMUM     MAXIMUM
   NUMBER    LABEL     AVAILABLE  AVAILABLE    AVAILABLE   AVAILABLE

      1      MACHINE     0        0.0384         0            1


                     **STATE AND DERIVATIVE VARIABLES**

                  (I)       SS(I)        DD(I)
                   1      0.2395E+03   0.1000E+01
```

D10–8.3

CHAPTER 11

Discrete Event Modeling and Simulation

11-1. There are two events for this exercise, the arrival (ARVL) of items and the end-of-service (ENDSV) for items. The subroutines for modeling these events are shown in D11-1.1 and D11-1.2.

File 1 is used to maintain entities in the queue of the server. The following variable definitions are used in this problem:

ATRIB(1)   arrival time of item
ATRIB(2)   type of item
BUSY   server status
TQUE   time in queue

```
      SUBROUTINE ARVL
      COMMON/SCOM1/ATRIB(100),DD(100),DDL(100),DTNOW,II,MFA,MSTOP,NCLNR
     1,NCRDR,NPRNT,NNRUN,NNSET,NTAPE,SS(100),SSL(100),TNEXT,TNOW,XX(100)
      EQUIVALENCE (XX(1),BUSY)
      IF(ATRIB(2).EQ.2.) GOTO 10
C     SCHEDULE NEXT ARRIVAL OF TYPE 1
      CALL SCHDL(1,EXPON(4.,1),ATRIB)
      GOTO 20
C     SCHEDULE NEXT ARRIVAL OF TYPE 2
10    CALL SCHDL(1,EXPON(2.,1),ATRIB)
C     SAVE ARRIVAL TIME
20    ATRIB(1)=TNOW
      CALL FILEM(1,ATRIB)
      IF(BUSY.EQ.0. .AND. NNQ(1).EQ.1) CALL SCHDL(2,0,ATRIB)
      RETURN
      END
```

D11-1.1

```
      SUBROUTINE ENDSV
      COMMON/SCOM1/ATRIB(100),DD(100),DDL(100),DTNOW,II,MFA,MSTOP,NCLNR
     1,NCRDR,NPRNT,NNRUN,NNSET,NTAPE,SS(100),SSL(100),TNEXT,TNOW,XX(100)
      EQUIVALENCE (XX(1),BUSY)
      IF (BUSY.EQ.0.) GOTO 30
C     CALCULATE TIME IN SYSTEM AND COLLECT STATISTICS
      TSYS=TNOW-ATRIB(1)
      IF(ATRIB(2).EQ.2.) GOTO 10
      CALL COLCT(TSYS,3)
      GOTO 20
10    CALL COLCT(TSYS,4)
20    IF(NNQ(1).GT.0) GOTO 30
C     NO JOBS WAITING, SET SERVER TO IDLE
      BUSY=0
      RETURN
C     JOBS WAITING, REMOVE FIRST FROM QUEUE
C     CALCULATE TIME IN QUEUE AND COLLECT STATISTICS
30    BUSY=1
      CALL RMOVE(1,1,ATRIB)
      TQUE=TNOW-ATRIB(1)
      IF(ATRIB(2).EQ.2.) GOTO 40
      CALL COLCT(TQUE,1)
      GOTO 50
40    CALL COLCT(TQUE,2)
C     SCHEDULE END OF SERVICE
50    CALL SCHDL(2,EXPON(1.,2),ATRIB)
      RETURN
      END
```

140

D11-1.2

The comments in the listings provide documentation for the events. Initial arrival events for each type of item are established in subroutine INTLC shown in D11-1.3. The server is made idle initially by setting BUSY to 0. Subroutine EVENT shown in D11-1.4 is used to decode the event code. The input statements for this problem are shown in D11-1.5. The STAT statements indicate that four variables based on observation will be collected representing the queue time for each item and the system time for each item. The priority of file 1 is established as low value first based on item type (ATRIB(2)). Thus, in file 1, item type 1 is given priority over item type 2.

The summary report for this problem is shown in D11-1.6. The average waiting time in file 1 is estimated for the items that waited in File 1. Thus, if an item arrives and the server is idle, the item is never inserted into File 1 and its waiting time is not included in the computation of the average waiting time. Note, this is different than in a network model where an item always flows through a QUEUE or an AWAIT node before reaching a server. It is also different from the QUEUE TIME obtained from observations of the time spent in the queue which includes zero waiting times. If it is desired to compute the average waiting time including the zero waiting time for those items that do not wait then the item should be put into the file and removed after the test on the server status is made.

```
        SUBROUTINE INTLC
        COMMON/SCOM1/ATRIB(100),DD(100),DDL(100),DTNOW,II,MFA,MSTOP,NCLNR
       1,NCRDR,NPRNT,NNRUN,NNSET,NTAPE,SS(100),SSL(100),TNEXT,TNOW,XX(100)
        EQUIVALENCE (XX(1),BUSY)
        BUSY=0
C       SCHEDULE FIRST ARRIVAL OF TYPE 1
        ATRIB(2)=1
        CALL SCHDL(1,0.0,ATRIB)
C       SCHEDULE FIRST ARRIVAL OF TYPE 2
        ATRIB(2)=2
        CALL SCHDL(1,0.0,ATRIB)
        RETURN
        END
```

D11-1.3

```
        SUBROUTINE EVENT(I)
        GOTO (1,2),I
1       CALL ARVL
        RETURN
2       CALL ENDSV
        RETURN
        END
```

D11-1.4

```
 1    GEN,ROLSTON,PROBLEM 11.1, 8/12/86,1;
 2    LIMITS,1,2,150;
 3    PRIORITY/1,LVF(2);
 4    STAT,1,QUEUE TIME 1;
 5    STAT,2,QUEUE TIME 2;
 6    STAT,3,SYSTEM TIME 1;
 7    STAT,4,SYSTEM TIME 2;
 8    TIMST,XX(1),SERVER BUSY TIME;
 9    INIT,0,300;
10    FIN;
```

D11-1.5

SLAM II SUMMARY REPORT

SIMULATION PROJECT PROBLEM 11.1   BY ROLSTON

DATE 8/12/1986   RUN NUMBER   1 OF   1

CURRENT TIME   0.3000E+03
STATISTICAL ARRAYS CLEARED AT TIME 0.0000E+00

STATISTICS FOR VARIABLES BASED ON OBSERVATION

	MEAN VALUE	STANDARD DEVIATION	COEFF. OF VARIATION	MINIMUM VALUE	MAXIMUM VALUE	NUMBER OF OBSERVATIONS
QUEUE TIME 1	0.1479E+01	0.1463E+01	0.9894E+00	0.0000E+00	0.6218E+01	85
QUEUE TIME 2	0.1086E+02	0.8348E+01	0.7686E+00	0.0000E+00	0.2716E+02	167
SYSTEM TIME 1	0.2757E+01	0.1588E+01	0.5760E+00	0.5919E+00	0.7293E+01	84
SYSTEM TIME 2	0.1189E+02	0.8435E+01	0.7094E+00	0.7721E-02	0.2817E+02	167

STATISTICS FOR TIME-PERSISTENT VARIABLES

	MEAN VALUE	STANDARD DEVIATION	MINIMUM VALUE	MAXIMUM VALUE	TIME INTERVAL	CURRENT VALUE
SERVER BUSY TIME	0.9257E+00	0.2622E+00	0.0000E+00	0.1000E+01	0.3000E+03	0.1000E+01

FILE STATISTICS

FILE NUMBER	LABEL/TYPE	AVERAGE LENGTH	STANDARD DEVIATION	MAXIMUM LENGTH	CURRENT LENGTH	AVERAGE WAITING TIME
1		6.4655	5.2130	22	0	7.6970
2	CALENDAR	2.9257	0.2622	3	3	1.6624

D11-1.6

11-1,Embellishment. To include the interruption of low priority items when a high priority item arrives, a variable, TYPE, is used to define the item type being served. When the item type being served is a low priority item then service on it must be stopped when a high priority item arrives. To accomplish this a search of the event calendar, NCLNR, is made and the event is canceled by removing it from the event file. The service time remaining is then calculated and a histogram of remaining times is obtained through a call to subroutine COLCT. The interrupted item is then placed in the file of interrupted jobs. In the end-of-service event, additional logic is required to determine the next queue to examine in order to determine which item should be processed. The revised ARVL and ENDSV subroutines are shown in D11- 1.7 and D11-1.8.

```
        SUBROUTINE ARVL
        COMMON/SCOM1/ATRIB(100),DD(100),DDL(100),DTNOW,II,MFA,MSTOP,NCLNR
       1,NCRDR,NPRNT,NNRUN,NNSET,NTAPE,SS(100),SSL(100),TNEXT,TNOW,XX(100)
        COMMON/UCOM1/TYPE
        DIMENSION A(5)
        EQUIVALENCE (XX(1),BUSY)
        IF(ATRIB(2).EQ.2.) GOTO 10
C       SCHEDULE NEXT ARRIVAL OF TYPE 1 JOB
        CALL SCHDL(1,EXPON(4.,1),ATRIB)
        GOTO 20
C       SCHEDULE NEXT ARRIVAL OF TYPE 2 JOB
10      CALL SCHDL(1,EXPON(2.,1),ATRIB)
C       SAVE ARRIVAL TIME
20      ATRIB(1)=TNOW
        IF(BUSY.EQ.0.) GOTO 30
        IF(ATRIB(2).LT.2.) GOTO 35
C       IF SERVER IS BUSY, AND JOB IS TYPE 2
C       PLACE IN FILE 2 TO WAIT
        CALL FILEM(2,ATRIB)
        RETURN
C       IF JOB IS TYPE 1, FIND JOB TYPE OF JOB CURRENTLY
C       IN SERVICE ACTIVITY
35      IF(TYPE.GT.1.) GOTO 36
C       JOB IN SERVICE IS TYPE 1
C       PLACE ARRIVING JOB IN FILE 1 TO WAIT
        CALL FILEM(1,ATRIB)
        RETURN
C       JOB IN SERVICE IS TYPE 2
C       PREEMPT JOB - REMOVE END OF SERVICE EVENT FROM CALENDAR
C       PLACE JOB IN FILE 3, AFTER SAVING REMAINING PROCESSING
C          TIME IN ATTRIBUTE 3
36      NRANK=NFIND(1,NCLNR,4,0,2.0,0.0)
        CALL RMOVE(NRANK,NCLNR,A)
        A(3)=A(5)-TNOW
        CALL COLCT(A(3),5)
        CALL FILEM(3,A)
C       SET SERVER TO BUSY, SCHEDULE END OF SERVICE
30      BUSY=1
        TYPE=ATRIB(2)
        CALL SCHDL(2,EXPON(1.,2),ATRIB)
        RETURN
        END
```

D11-1.7

```
        SUBROUTINE ENDSV
        COMMON/SCOM1/ATRIB(100),DD(100),DDL(100),DTNOW,II,MFA,MSTOP,NCLNR
       1,NCRDR,NPRNT,NNRUN,NNSET,NTAPE,SS(100),SSL(100),TNEXT,TNOW,XX(100)
        COMMON/UCOM1/TYPE
        EQUIVALENCE (XX(1),BUSY)
C       CALCULATE TIME IN SYSTEM AND COLLECT STATISTICS
        TSYS=TNOW-ATRIB(1)
        IF(ATRIB(2).EQ.2.) GOTO 10
        CALL COLCT(TSYS,3)
        GOTO 20
10      CALL COLCT(TSYS,4)
20      IF(NNQ(1).GT.0) GOTO 30
        IF(NNQ(3).GT.0) GOTO 32
        IF(NNQ(2).GT.0) GOTO 34
C       NO JOBS WAITING, SET SERVER STATUS TO IDLE
        BUSY=0
        RETURN
C       REMOVE PREEMPTED JOB FROM WAIT FILE AND
C       SCHEDULE END OF SERVICE
32      CALL RMOVE(1,3,ATRIB)
        TYPE=ATRIB(2)
        CALL SCHDL(2,ATRIB(3),ATRIB)
        RETURN
34      CALL RMOVE(1,2,ATRIB)
        GOTO 36
30      CALL RMOVE(1,1,ATRIB)
C       REMOVE FIRST JOB FROM QUEUE
C       CALCULATE TIME IN QUEUE AND COLLECT STATISTICS
36      TQUE=TNOW-ATRIB(1)
        IF(ATRIB(2).EQ.2.) GOTO 40
        CALL COLCT(TQUE,1)
        GOTO 50
40      CALL COLCT(TQUE,2)
C       SCHEDULE END OF SERVICE
50      TYPE=ATRIB(2)
        CALL SCHDL(2,EXPON(1.,2),ATRIB)
        RETURN
        END
```

D11-1.8

The histogram for the remaining service time is shown in D11-1.9. Since only 82 values were obtained, it is not obvious that an exponential distribution obtains. Note that the average is close to 1.00 as expected.

This embellishment can be done without searching the event file for the end-of-service event by using concepts presented in Chapter 12. Whenever an item is placed into service, the variable NRANKP can be set to MFA prior to scheduling the end-of-service event. MFA is the pointer to the location where the event will be stored on the event calendar. When an end-of-service event is to be cancelled the statement

CALL RMOVE(-NRANKP,NCLNR,A)

will cancel the end-of-service event. In Chapter 12, the use of a negative first argument for RMOVE is explained.

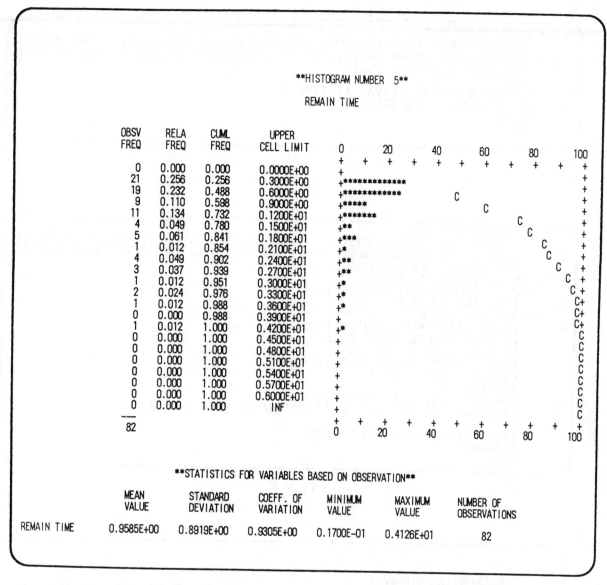

```
                                    **HISTOGRAM NUMBER  5**

                                         REMAIN TIME

         OBSV     RELA     CUML      UPPER
         FREQ     FREQ     FREQ    CELL LIMIT    0        20       40       60       80      100
                                                +    +    +    +    +    +    +    +    +    +
            0    0.000    0.000   0.0000E+00     +                                            +
           21    0.256    0.256   0.3000E+00     +*************                               +
           19    0.232    0.488   0.6000E+00     +************                                +
            9    0.110    0.598   0.9000E+00     +*****                      C                +
           11    0.134    0.732   0.1200E+01     +*******                         C           +
            4    0.049    0.780   0.1500E+01     +**                                   C      +
            5    0.061    0.841   0.1800E+01     +***                                    C    +
            1    0.012    0.854   0.2100E+01     +*                                       C   +
            4    0.049    0.902   0.2400E+01     +**                                       C  +
            3    0.037    0.939   0.2700E+01     +**                                        C +
            1    0.012    0.951   0.3000E+01     +*                                          C+
            2    0.024    0.976   0.3300E+01     +*                                          C+
            1    0.012    0.988   0.3600E+01     +*                                          C+
            0    0.000    0.988   0.3900E+01     +                                           C+
            1    0.012    1.000   0.4200E+01     +*                                          C+
            0    0.000    1.000   0.4500E+01     +                                           C
            0    0.000    1.000   0.4800E+01     +                                           C
            0    0.000    1.000   0.5100E+01     +                                           C
            0    0.000    1.000   0.5400E+01     +                                           C
            0    0.000    1.000   0.5700E+01     +                                           C
            0    0.000    1.000   0.6000E+01     +                                           C
            0    0.000    1.000      INF         +                                           C
          ---                                    +    +    +    +    +    +    +    +    +    +
           82                                    0        20       40       60       80      100
```

STATISTICS FOR VARIABLES BASED ON OBSERVATION

	MEAN VALUE	STANDARD DEVIATION	COEFF. OF VARIATION	MINIMUM VALUE	MAXIMUM VALUE	NUMBER OF OBSERVATIONS
REMAIN TIME	0.9585E+00	0.8919E+00	0.9305E+00	0.1700E-01	0.4126E+01	82

D11-1.9

11-2. This is a single server queueing situation and the listing for subroutine INTLC, ARVL and ENDSV are shown in D11-2.1 through D11-2.3. The input statements for the exercise are shown in D11-2.4.

From the input statements, it is seen that three runs are requested and that the seed value is to be reinitialized on each run. Random number stream 1 is used for generating interarrival times and random number stream 2 is used to generate service times. In this way, the same sequence of interarrival times and service times will be used and a direct comparison between the three priority rules can be made. This is an illustration of the use of common streams, for reducing the variance between alternatives. When using common streams it should be recognized that the results of the comparison between procedures is being made for a single situation, that is, the results can only be stated in terms of the interarrival and service time values generated for the runs.

A table of pertinent values from the summary reports for the three runs is shown in D11-2.5.

```
      SUBROUTINE INTLC
      COMMON/SCOM1/ATRIB(100),DD(100),DDL(100),DTNOW,II,MFA,MSTOP,NCLNR
     1,NCRDR,NPRNT,NNRUN,NNSET,NTAPE,SS(100),SSL(100),TNEXT,TNOW,XX(100)
      EQUIVALENCE (XX(1),BUSY)
      BUSY=0
C     SCHEDULE FIRST ARRIVAL AT TIME 0
      CALL SCHDL(1,0.0,ATRIB)
      RETURN
      END
```

D11-2.1

```
      SUBROUTINE ARVL
      COMMON/SCOM1/ATRIB(100),DD(100),DDL(100),DTNOW,II,MFA,MSTOP,NCLNR
     1,NCRDR,NPRNT,NNRUN,NNSET,NTAPE,SS(100),SSL(100),TNEXT,TNOW,XX(100)
      EQUIVALENCE (XX(1),BUSY)
C     SCHEDULE NEXT ARRIVAL
      CALL SCHDL(1,EXPON(1.25,1),ATRIB)
C     SAVE ARRIVAL TIME
      ATRIB(1)=TNOW
C     PLACE ESTIMATED SERVICE TIME IN ATRIB(2)
      ATRIB(2)=UNFRM(.75,1.25,2)
      IF(BUSY.EQ.0.) GOTO 10
C     SERVER BUSY, PLACE IN WAIT FILE
      CALL FILEM(1,ATRIB)
      RETURN
C     SERVER IDLE, SET STATUS TO BUSY
C     SCHEDULE END OF SERVICE AT ESTIMATED SERVICE TIME
10    BUSY=1
      CALL SCHDL(2,ATRIB(2),ATRIB)
      RETURN
      END
```

D11-2.2

```
      SUBROUTINE ENDSV
      COMMON/SCOM1/ATRIB(100),DD(100),DDL(100),DTNOW,II,MFA,MSTOP,NCLNR
     1,NCRDR,NPRNT,NNRUN,NNSET,NTAPE,SS(100),SSL(100),TNEXT,TNOW,XX(100)
      EQUIVALENCE (XX(1),BUSY)
C     CALCULATE TIME IN SYSTEM AND COLLECT STATISTICS
      TSYS=TNOW-ATRIB(1)
      CALL COLCT(TSYS,1)
      IF(NNQ(1).GT.0) GOTO 10
C     NO JOBS IN QUEUE, SET SERVER TO IDLE
      BUSY=0
      RETURN
C     REMOVE FIRST JOB FORM QUEUE AND SCHEDULE END OF SERVICE
10    CALL RMOVE(1,1,ATRIB)
      CALL SCHDL(2,ATRIB(2),ATRIB)
      RETURN
      END
```

D11-2.3

```
 1    GEN,ROLSTON,PROBLEM 11.2,7/23/80,3;
 2    LIMITS,1,2,100;
 3    PRIORITY/1,LVF(2);
 4    STAT,1,TIME IN SYSTEM;
 5    TIMST,XX(1),SERVER BUSY TIME;
 6    SEEDS,1274321477(1)/YES,2135124613(2)/YES;
 7    INIT,0,500;
 8    SIMULATE;
 9    PRIORITY/1,HVF(2);
10    SIMULATE;
11    PRIORITY/1,FIFO;
12    FIN;
```

D11-2.4

RUN NO	QUEUE RULE	TIME IN SYSTEM			NUMBER IN QUEUE			AVERAGE WAITING TIME
		MEAN	STD. DEV.	MAX	MEAN	STD. DEV.	MAX.	
1	Shortest Proc. Time	3.45	6.65	92.96	2.09	2.24	11	2.80
2	Longest Proc. Time	4.20	7.56	69.76	2.73	2.97	14	3.65
3	FIFO	3.76	2.56	12.56	2.35	2.54	12	3.16

D11-2.5

On each run, there were 427 observations on items being processed and no items were in the queue at the end of the run. The number of arrivals in 500 time units is Poisson distributed with a mean of 400 (also a variance of 400). Thus, the number of observations is higher than expected but within 2 standard deviations. From D11-2.5, it is seen that the estimated mean time in the system is lower when using the shortest processing time rule. This agrees with theoretical results. Note, however, the reduced standard deviation and maximum estimates for the time in the system when using a FIFO rule.

11-2,Embellishment. Before scheduling an end-of-service event, the estimated service time should be augmented by a sample from a normal distribution using RNORM(0.0,.2,3). SLAM II makes a test to insure that the sum of attribute 2 and the normal sample is greater than or equal to 0. If a negative value results, then the value is set to 0. This truncation could change the mean of the service time distribution.

11-3.   The listings of the subroutines required for the discrete event model of the two work stations in series are given in D11-3.1 through D11-3.5. The listing of the input statements is shown in D11-3.6. Comment statements are inserted throughout the subroutines to facilitate the description of the code.

```
SUBROUTINE INTLC
COMMON/SCOM1/ATRIB(100),DD(100),DDL(100),DTNOW,II,MFA,MSTOP,NCLNR
1,NCRDR,NPRNT,NNRUN,NNSET,NTAPE,SS(100),SSL(100),TNEXT,TNOW,XX(100)
COMMON/UCOM1/K
EQUIVALENCE (XX(1),BUSF)
EQUIVALENCE (XX(2),BUSS)
K=0
BUSF=0
BUSS=0
CALL SCHDL(1,0.0,ATRIB)
XX(3)=0
RETURN
END
```

D11-3.1

```
SUBROUTINE EVENT(I)
GOTO (1,2,3),I
1    CALL ARVL
RETURN
2    CALL ENDF
RETURN
3    CALL ENDS
RETURN
END
```

D11-3.2

```
      SUBROUTINE ARVL
      COMMON/SCOM1/ATRIB(100),DD(100),DDL(100),DTNOW,II,MFA,MSTOP,NCLNR
     1,NCRDR,NPRNT,NNRUN,NNSET,NTAPE,SS(100),SSL(100),TNEXT,TNOW,XX(100)
      COMMON/UCOM1/K
      EQUIVALENCE (XX(1),BUSF)
      EQUIVALENCE (XX(2),BUSS)
C     SCHEDULE NEXT ARRIVAL
      CALL SCHDL(1,EXPON(.4,1),ATRIB)
C     SAVE ARRIVAL TIME
      ATRIB(1)=TNOW
      IF(BUSF.EQ.1.) GOTO 10
C     FIRST SERVER IDLE, SET TO BUSY AND
C     SCHEDULE END OF SERVICE
      BUSF=1
      CALL SCHDL(2,EXPON(.25,2),ATRIB)
      RETURN
C     FIRST SERVER BUSY, CHECK NO. IN QUEUE
10    IF(NNQ(1).GT.3) GOTO 20
C     QUEUE NOT FULL, PLACE ARRIVING JOB IN QUEUE
      CALL FILEM(1,ATRIB)
      RETURN
C     QUEUE FULL, JOB BALKS
C     COLLECT STATISTICS ON TIME BETWEEN BALKS
20    K=K+1
      IF(K.GT.1) GOTO 30
      TLAST=TNOW
      RETURN
30    TBLK=TNOW-TLAST
      TLAST=TNOW
      CALL COLCT(TBLK,1)
      RETURN
      END
```

D11-3.3

```
      SUBROUTINE ENDF
      COMMON/SCOM1/ATRIB(100),DD(100),DDL(100),DTNOW,II,MFA,MSTOP,NCLNR
     1,NCRDR,NPRNT,NNRUN,NNSET,NTAPE,SS(100),SSL(100),TNEXT,TNOW,XX(100)
      EQUIVALENCE (XX(1),BUSF)
      EQUIVALENCE (XX(2),BUSS)
      IF(NNQ(2).GT.1) GOTO 10
      IF(NNQ(2).EQ.0) GOTO 20
C     SERVER 2 BUSY, QUEUE NOT FULL
5     CALL FILEM(2,ATRIB)
      GOTO 30
20    IF(BUSS.EQ.1) GOTO 5
C     SERVER 2 IDLE, SET TO BUSY, SCHEDULE END OF SERVICE
      BUSS=1
      CALL SCHDL(3,EXPON(.5,3),ATRIB)
30    IF(NNQ(1).NE.0) GOTO 40
C     NO JOBS WAITING FOR FIRST SERVER, SET TO IDLE
      BUSF=0
      RETURN
C     REMOVE FIRST JOB FROM QUEUE, SCHEDULE END OF SERVICE
40    CALL RMOVE(1,1,ATRIB)
      CALL SCHDL(2,EXPON(.25,2),ATRIB)
      RETURN
C     QUEUE 2 FULL, SET BLOCKING INDICATOR AND
C     PLACE BLOCKED JOB IN FILE 3
10    XX(3)=1
      CALL FILEM(3,ATRIB)
      RETURN
      END
```

D11-3.4

```
      SUBROUTINE ENDS
      COMMON/SCOM1/ATRIB(100),DD(100),DDL(100),DTNOW,II,MFA,MSTOP,NCLNR
     1,NCRDR,NPRNT,NNRUN,NNSET,NTAPE,SS(100),SSL(100),TNEXT,TNOW,XX(100)
      EQUIVALENCE (XX(1),BUSF)
      EQUIVALENCE (XX(2),BUSS)
C     CALCULATE TIME IN SYSTEM AND COLLECT STATISTICS
      TSYS=TNOW-ATRIB(1)
      CALL COLCT(TSYS,2)
      IF(NNQ(2).NE.0) GOTO 10
C     NO JOBS WAITING FOR SECOND SERVER, SET TO IDLE
      BUSS=0
      RETURN
10    IF(NNQ(2).GT.1) GOTO 20
C     ONLY ONE JOB WAITING FOR SECOND SERVER
C      REMOVE IT FROM FILE AND SCHEDULE END OF SERVICE
      CALL RMOVE(1,2,ATRIB)
      CALL SCHDL(3,EXPON(.5,3),ATRIB)
      RETURN
C     TWO JOBS WAITING FOR SECOND SERVER
C      REMOVE FIRST FROM WAIT FILE AND SCHEDULE END OF SERVICE
C      CHECK TO SEE IF SERVER ONE BLOCKED
20    CALL RMOVE(1,2,ATRIB)
      CALL SCHDL(3,EXPON(.5,3),ATRIB)
      IF(NNQ(3).NE.0) GOTO 30
      RETURN
C     SERVER ONE WAS BLOCKED, REMOVE BLOCKED JOB FROM FILE,
C     PLACE IN FILE 2, SET BLOCKING INDICATOR TO 0
C     CHECK FOR JOBS WAITING FOR FIRST SERVER
30    CALL RMOVE(1,3,ATRIB)
      CALL FILEM(2,ATRIB)
      XX(3)=0
      IF(NNQ(1).GT.0) GOTO 40
C     NO JOBS WAITING, SET SERVER ONE TO IDLE
      BUSF=0
      RETURN
C     JOBS WAITING, REMOVE FIRST AND SCHEDULE END OF SERVICE
40    CALL RMOVE(1,1,ATRIB)
      CALL SCHDL(2,EXPON(.25,2),ATRIB)
      RETURN
      END
```

D11-3.5

```
1    GEN,ROLSTON,PROBLEM 11.3,7/23/80,1;
2    LIMITS,3,3,50;
3    STAT,1,TIME BET BALKS;
4    STAT,2,TIME IN SYSTEM,20/0/.25;
5    TIMST,XX(1),SERV 1 BUSY;
6    TIMST,XX(2),SERV 2 BUSY;
7    TIMST,XX(3),SERV 1 BLOCK;
8    INIT,0,300;
9    FIN;
```

D11-3.6

11-3,Embellishment(a). The only changes required for this embellishment are to store the end-of-service times in the variable TIME(I). By checking TNOW against TIME(I), I=1,2, it can be determined if the new arrival should return or balk. The way the problem statement reads, a check must be made against an end-of-service for either station although a better problem statement probably would consider only an end-of-service for the first station. For the first station, a check must also be made to determine if the station is blocked in which case the second station's end-of-service time must be checked. When station 1 is blocked, TIME(1) can be set to a negative value to indicate this condition. The listings of subroutines and input statements are given in D11-3.7 through D11- 3.12.

11-3,Embellishment(b). There are two ways to model a situation in which the arrivals are from a finite population. The most direct procedure is to generate an arrival time for each of the 100 units equal to the failure time for the unit. All 100 arrival times are then placed in a separate file, say file 4, ranked on low-value-first of the arrival time. Each time an arrival occurs, the first entry in file 4 is removed and scheduled on the event calendar as an arrival. As each machine is repaired, its next arrival time is scheduled and it is placed in file 4 if the arrival time is after the next scheduled arrival that is on the event calendar. If its arrival time is before the next scheduled arrival time, the next scheduled arrival time is removed from the event calendar and placed back in file 4 and the machine with the earlier arrival time is placed on the event calendar.

Another change required is that machines which are subcontracted must be scheduled to return to operation in order that their next arrival time to the maintenance facility can be included. The time for repair when a unit is subcontracted is specified as triangularly distributed with a mode of 4, a minimum of 3, and a maximum of 6. We assume the repair time is the entire time the unit is down.

```
      SUBROUTINE INTLC
      COMMON/SCOM1/ATRIB(100),DD(100),DDL(100),DTNOW,II,MFA,MSTOP,NCLNR
     1,NCRDR,NPRNT,NNRUN,NNSET,NTAPE,SS(100),SSL(100),TNEXT,TNOW,XX(100)
      COMMON/UCOM1/K
      EQUIVALENCE (XX(1),BUSF)
      EQUIVALENCE (XX(2),BUSS)
      K=0
      BUSF=0
      BUSS=0
      ATRIB(2)=1
      CALL SCHDL(1,0.0,ATRIB)
      XX(3)=0
      RETURN
      END
```

D11-3.7

```
      SUBROUTINE EVENT(I)
      GOTO (1,2,3),I
    1 CALL ARVL
      RETURN
    2 CALL ENDF
      RETURN
    3 CALL ENDS
      RETURN
      END
```

D11-3.8

```
        SUBROUTINE ARVL
        COMMON/SCOM1/ATRIB(100),DD(100),DDL(100),DTNOW,II,MFA,MSTOP,NCLNR
       1,NCRDR,NPRNT,NNRUN,NNSET,NTAPE,SS(100),SSL(100),TNEXT,TNOW,XX(100)
        COMMON/UCOM1/K
        COMMON/UCOM2/TIME(2)
        EQUIVALENCE (XX(1),BUSF)
        EQUIVALENCE (XX(2),BUSS)
        IF(ATRIB(2).EQ.2.) GOTO 5
C       NEW JOB, SCHEDULE ARRIVAL OF NEXT JOB
        CALL SCHDL(1,EXPON(.4,1),ATRIB)
C       SAVE ARRIVAL TIME
        ATRIB(1)=TNOW
5       IF(BUSF.EQ.1.) GOTO 10
C       SERVER 1 IDLE, SET TO BUSY AND
C       SCHEDULE END OF SERVICE
        BUSF=1
        DT=EXPON(.25,2)
        TIME(1)=DT+TNOW
        CALL SCHDL(2,DT,ATRIB)
        RETURN
10      IF(NNQ(1).GT.3) GOTO 20
C       SERVER 1 BUSY, QUEUE 1 NOT FULL
C       PLACE JOB IN FILE 1 TO WAIT
        CALL FILEM(1,ATRIB)
        RETURN
20      IF(NNQ(3).GT.0) GOTO 22
C       SERVER 1 NOT BLOCKED, WILL HIS END OF SERVICE
C       OCCUR WITHIN .1 TIME UNIT?
        IF(TIME(1)-TNOW.GT..1) GOTO 25
C       END OF SERVICE WITHIN .1 TIME UNIT, SET ATRIB(2)=2 TO
C       SPECIFY RETURNING JOB, SCHEDULE ARRIVAL IN .2 TIME UNITS
        ATRIB(2)=2
        CALL SCHDL(1,.2,ATRIB)
        RETURN
25      IF(BUSS.EQ.0) GOTO 27
C       SERVER 2 BUSY, WILL HIS END OF SERVICE OCCUR WITHIN
C       .1 TIME UNITS?
22      IF(TIME(2)-TNOW.GT..1) GOTO 27
C       END OF SERVICE FOR SERVER 2 WITHIN .1 TIME UNITS
C       SET ATRIB(2)=2 TO SPECIFY RETURNING JOB, SCHEDULE ARRIVAL
        ATRIB(2)=2
        CALL SCHDL(1,.2,ATRIB)
26      RETURN
C       JOB BALKS, COLLECT STATISTICS ON TIME BETWEEN BALKS
27      K=K+1
        IF(K.GT.1) GOTO 30
        TLAST=TNOW
        RETURN
30      TBLK=TNOW-TLAST
        TLAST=TNOW
        CALL COLCT(TBLK,1)
        RETURN
        END
```

D11-3.9

```
      SUBROUTINE ENDF
      COMMON/SCOM1/ATRIB(100),DD(100),DDL(100),DTNOW,II,MFA,MSTOP,NCLNR
     1,NCRDR,NPRNT,NNRUN,NNSET,NTAPE,SS(100),SSL(100),TNEXT,TNOW,XX(100)
      COMMON/UCOM2/TIME(2)
      EQUIVALENCE (XX(1),BUSF)
      EQUIVALENCE (XX(2),BUSS)
      IF(NNQ(2).GT.1) GOTO 10
      IF(NNQ(2).EQ.0) GOTO 20
C     SERVER 2 BUSY, QUEUE NOT FULL
5     CALL FILEM(2,ATRIB)
      GOTO 30
20    IF(BUSS.EQ.1) GOTO 5
C     SERVER 2 IDLE, SET TO BUSY, SCHEDULE END OF SERVICE
      BUSS=1
      DT=EXPON(.5,3)
      TIME(2)=DT+TNOW
      CALL SCHDL(3,DT,ATRIB)
30    IF(NNQ(1).NE.0) GOTO 40
C     NO JOBS WAITING FOR FIRST SERVER, SET TO IDLE
      BUSF=0
      RETURN
C     REMOVE FIRST JOB FROM QUEUE, SCHEDULE END OF SERVICE
40    CALL RMOVE(1,1,ATRIB)
      DT=EXPON(.25,2)
      TIME(1)=DT+TNOW
      CALL SCHDL(2,DT,ATRIB)
      RETURN
C     QUEUE 2 FULL, SET BLOCKING INDICATOR AND
C     PLACE BLOCKED JOB IN FILE 3
10    XX(3)=1
      CALL FILEM(3,ATRIB)
      RETURN
      END
```

D11-3.10

```
          SUBROUTINE ENDS
          COMMON/SCOM1/ATRIB(100),DD(100),DDL(100),DTNOW,II,MFA,MSTOP,NCLNR
         1,NCRDR,NPRNT,NNRUN,NNSET,NTAPE,SS(100),SSL(100),TNEXT,TNOW,XX(100)
          COMMON/UCOM2/TIME(2)
          EQUIVALENCE (XX(1),BUSF)
          EQUIVALENCE (XX(2),BUSS)
C         CALCULATE TIME IN SYSTEM AND COLLECT STATISTICS
          TSYS=TNOW-ATRIB(1)
          CALL COLCT(TSYS,2)
          IF(NNQ(2).NE.0) GOTO 10
C         NO JOBS WAITING FOR SECOND SERVER, SET TO IDLE
          BUSS=0
          RETURN
10        IF(NNQ(2).GT.1) GOTO 20
C         ONLY ONE JOB WAITING FOR SECOND SERVER
C          REMOVE IT FROM FILE AND SCHEDULE END OF SERVICE
          CALL RMOVE(1,2,ATRIB)
          DT=EXPON(.5,3)
          TIME(2)=DT+TNOW
          CALL SCHDL(3,DT,ATRIB)
          RETURN
C         TWO JOBS WAITING FOR SECOND SERVER
C          REMOVE FIRST FROM WAIT FILE AND SCHEDULE END OF SERVICE
C          CHECK TO SEE IF SERVER ONE BLOCKED
20        CALL RMOVE(1,2,ATRIB)
          DT=EXPON(.5,3)
          TIME(2)=DT+TNOW
          CALL SCHDL(3,DT,ATRIB)
          IF(NNQ(3).NE.0) GOTO 30
          RETURN
C         SERVER ONE WAS BLOCKED, REMOVE BLOCKED JOB FROM FILE,
C         PLACE IN FILE 2, SET BLOCKING INDICATOR TO 0
C         CHECK FOR JOBS WAITING FOR FIRST SERVER
30        CALL RMOVE(1,3,ATRIB)
          CALL FILEM(2,ATRIB)
          XX(3)=0
          IF(NNQ(1).GT.0) GOTO 40
C         NO JOBS WAITING, SET SERVER ONE TO IDLE
          BUSF=0
          RETURN
C         JOBS WAITING, REMOVE FIRST AND SCHEDULE END OF SERVICE
40        CALL RMOVE(1,1,ATRIB)
          DT=EXPON(.25,2)
          TIME(1)=DT+TNOW
          CALL SCHDL(2,DT,ATRIB)
          RETURN
          END
```

D11-3.11

```
     1    GEN,ROLSTON,PROBLEM 11.3A,7/23/80,1;
     2    LIMITS,3,3,100;
     3    STAT,1,TIME BET BALKS;
     4    STAT,2,TIME IN SYSTEM,20/0/.25;
     5    TIMST,XX(1),SERV 1 BUSY;
     6    TIMST,XX(2),SERV 2 BUSY;
     7    TIMST,XX(3),SERV 1 BLOCK;
     8    INIT,0,300;
     9    MONTR,TRACE,0,25;
    10    FIN;
```

D11-3.12

An alternative approach is to keep track of the number of working machines and to schedule the next failure time to occur with an exponential distribution having a mean of 40 hours divided by the number of working machines. This new time for failure of the next machine must replace the arrival time that is currently on the event calendar. This second approach makes use of the information that the minimum of a set of exponentially distributed random variables with mean time, $1/\mu$, is exponentially distributed with a mean equal to $1/\mu$ divided by the number of variables over which the minimization is performed. It also makes use of the forgetfulness property of the exponential distribution. Since the first approach doesn't require the assumption of an exponential failure distribution, only the coding for that approach is shown in D11-3.13 through D11-3.16.

```
      SUBROUTINE INTLC
      COMMON/SCOM1/ATRIB(100),DD(100),DDL(100),DTNOW,II,MFA,MSTOP,NCLNR
     1,NCRDR,NPRNT,NNRUN,NNSET,NTAPE,SS(100),SSL(100),TNEXT,TNOW,XX(100)
      COMMON/UCOM1/K,TIME
      EQUIVALENCE (XX(1),BUSF)
      EQUIVALENCE (XX(2),BUSS)
      K=0
      BUSF=0
      BUSS=0
C     PLACE MACHINES AND THEIR TIMES TIL BREAKDOWN IN FILE
C     RANKED LOWEST TIME FIRST
      DO 10 I=1,100
      FAIL=EXPON(40.,1)
      ATRIB(2)=FAIL
      CALL FILEM(4,ATRIB)
10    CONTINUE
C     REMOVE FIRST MACHINE FROM FILE, SCHEDULE ITS ARRIVAL TO
C      THE SHOP AT BREAKDOWN TIME
      CALL RMOVE(1,4,ATRIB)
      CALL SCHDL(1,ATRIB(2),ATRIB)
      XX(3)=0
      RETURN
      END
```

D11-3.13

```
      SUBROUTINE CHKCAL
      COMMON/SCOM1/ATRIB(100),DD(100),DDL(100),DTNOW,II,MFA,MSTOP,NCLNR
     1,NCRDR,NPRNT,NNRUN,NNSET,NTAPE,SS(100),SSL(100),TNEXT,TNOW,XX(100)
      COMMON/UCOM1/K,TIME
      DIMENSION A(4)
C     IF NO ARRIVAL SCHEDULED, SCHEDULE FAILURE OF
C     SUBCONTRACTED UNIT
      IF(TIME.EQ.0.0) GOTO 10
C     CHECK TIME TO FAIL AGAINST THAT OF THE SUBCONTRACTED UNIT
      IF(ATRIB(2)+TNOW.GT.TIME) GOTO 20
C     THE SUBCONTRACTED UNIT WILL FAIL FIRST
C     PLACE IT ON THE EVENT CALENDAR
C     PLACE THE ONE CURRENTLY ON THE CALENDAR IN FILE 4
      NRANK=NFIND(1,NCLNR,3,0,1.0,0.0)
      CALL RMOVE(NRANK,NCLNR,A)
      A(2)=A(4)+TNOW
      CALL FILEM(4,A)
10    CALL SCHDL(1,ATRIB(2),ATRIB)
      RETURN
C     THE ONE ON THE EVENT CALENDAR FAILS FIRST
C     PLACE SUBCONRTRACTED UNIT IN FILE 4
20    ATRIB(2)=ATRIB(2)+TNOW
      CALL FILEM(4,ATRIB)
      RETURN
      END
```

D11-3.14

```
      SUBROUTINE ARVL
      COMMON/SCOM1/ATRIB(100),DD(100),DDL(100),DTNOW,II,MFA,MSTOP,NCLNR
     1,NCRDR,NPRNT,NNRUN,NNSET,NTAPE,SS(100),SSL(100),TNEXT,TNOW,XX(100)
      COMMON/UCOM1/K,TIME
      EQUIVALENCE (XX(1),BUSF)
      EQUIVALENCE (XX(2),BUSS)
      TIME=0.
      IF(NNQ(4).EQ.0) GOTO 5
C     SCHEDULE NEXT BREAKDOWN
      CALL RMOVE(1,4,ATRIB)
      TIME=ATRIB(2)
      DT=ATRIB(2)-TNOW
      CALL SCHDL(1,DT,ATRIB)
C     SAVE ARRIVAL TIME
5     ATRIB(1)=TNOW
      IF(BUSF.EQ.1.) GOTO 10
C     SERVER 1 IDLE, SET TO BUSY AND
C     SCHEDULE END OF SERVICE
      BUSF=1
      CALL SCHDL(2,EXPON(.25,2),ATRIB)
      RETURN
10    IF(NNQ(1).GT.3) GOTO 20
C     SERVER 1 BUSY, QUEUE NOT FULL
      CALL FILEM(1,ATRIB)
      RETURN
C     QUEUE FULL, JOB BALKS
20    K=K+1
      IF(K.GT.1) GOTO 30
      TLAST=TNOW
      GOTO 40
30    TBLK=TNOW-TLAST
      TLAST=TNOW
      CALL COLCT(TBLK,1)
C     CALCULATE TIME TIL BREAKDOWN, INCLUDING REPAIR
C     TIME AT SUBCONTRACTORS
40    FAIL=EXPON(40.,1)
      ATRIB(2)=FAIL+TRIAG(3.,4.,6.,5)
C     CHECK TO SEE IF NO BREAKDOWN SCHEDULED, OR IF THIS BREAKDOWN
C     WILL OCCUR BEFORE THE ONE ON THE EVENT CALENDAR
      CALL CHKCAL
      RETURN
      END
```

D11-3.15

```
                SUBROUTINE ENDS
                COMMON/SCOM1/ATRIB(100),DD(100),DDL(100),DTNOW,II,MFA,MSTOP,NCLNR
               1,NCRDR,NPRNT,NNRUN,NNSET,NTAPE,SS(100),SSL(100),TNEXT,TNOW,XX(100)
                EQUIVALENCE (XX(1),BUSF)
                EQUIVALENCE (XX(2),BUSS)
      C         CALCULATE TIME TIL NEXT BREAKDOWN
                FAIL=EXPON(40.,1)
                ATRIB(2)=FAIL
      C         CHECK TO SEE IF NO BREAKDOWN SCHEDULED, OR IF THIS BREAKDOWN
      C         WILL OCCUR BEFORE THE ONE ON THE EVENT CALENDAR
                CALL CHKCAL
      C         CALCULATE TIME IN SYSTEM AND COLLECT STATISTICS
                TSYS=TNOW-ATRIB(1)
                CALL COLCT(TSYS,2)
                IF(NNQ(2).NE.0) GOTO 10
      C         NO JOBS WAITING FOR SERVER 2, SET TO IDLE
                BUSS=0
                RETURN
      10        IF(NNQ(2).GT.1) GOTO 20
      C          ONE JOB WAITING FOR SERVER 2, REMOVE FORM FILE AND
      C         SCHEDULE END OF SERVICE
                CALL RMOVE(1,2,ATRIB)
                CALL SCHDL(3,EXPON(.5,3),ATRIB)
                RETURN
      C         TWO JOBS WAITING, REMOVE FIRST AND SCHEDULE END OF
      C         SERVICE, CHECK IF SERVER 1 BLOCKED
      20        CALL RMOVE(1,2,ATRIB)
                CALL SCHDL(3,EXPON(.5,3),ATRIB)
                IF(NNQ(3).NE.0) GOTO 30
                RETURN
      C         SERVER 1 WAS BLOCKED, REMOVE BLOCKED JOB FROM FILE 3,
      C         PLACE IN SERVER 2 QUEUE, SET BLOCKING INDICATOR TO 0
      30        CALL RMOVE(1,3,ATRIB)
                CALL FILEM(2,ATRIB)
                XX(3)=0
                IF(NNQ(1).GT.0) GOTO 40
      C         NO JOBS WAITING FOR SERVER 1, SET TO IDLE
                BUSF=0
                RETURN
      C         JOBS WAITING FOR SERVER 1, REMOVE FIRST FROM QUEUE
      C          AND SCHEDULE END OF SERVICE
      40        CALL RMOVE(1,1,ATRIB)
                CALL SCHDL(2,EXPON(.25,2),ATRIB)
                RETURN
                END
```

D11-3.16

11-4.   The simulation of a PERT network is described in Example 4 of the GASP IV book (Pritsker, A.A.B., The GASP IV Simulation Language, New York: John Wiley & Sons, Inc., 1974). In D11-4.1, a flow chart of the one event that is required when simulating PERT networks is given.

Subroutine INTLC and subroutine EVENT(I) are shown in D11-4.2 and D11-4.3, respectively. The listing of the input statements is shown in D11-4.4 .

When obtaining samples for the activity durations, stream number 9 was employed which is the stream number assigned when no value is given for random sampling on activities in the network (actually the default stream is defined by NNSTR-1 where NNSTR is the SLAM II variable defining the largest stream number which currently is set at 10). The outputs obtained are identical to those presented in Figure 7-4.

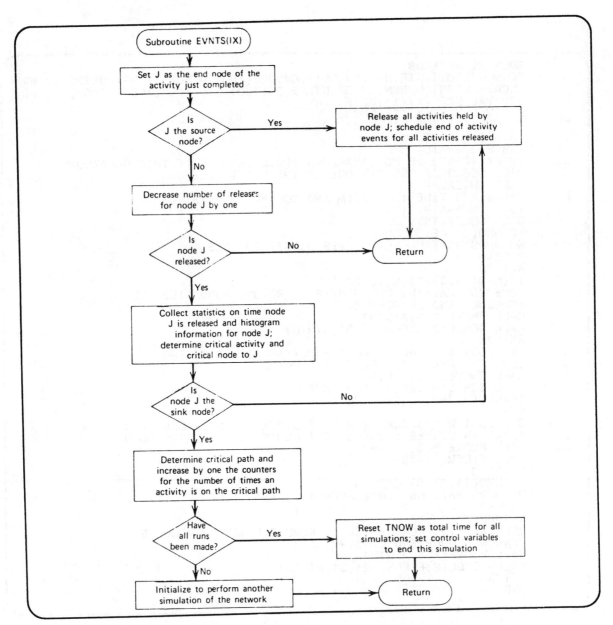

D11-4.1

```
        SUBROUTINE INTLC
        COMMON/SCOM1/ATRIB(100),DD(100),DDL(100),DTNOW,II,MFA,MSTOP,NCLNR
       1,NCRDR,NPRNT,NNRUN,NNSET,NTAPE,SS(100),SSL(100),TNEXT,TNOW,XX(100)
        COMMON/UCOM1/NOD(6),KOUNT,NSINK,TIME
        TIME=0
C       INITIALIZE NUMBER OF ACTIVITIES TO RELEASE EACH NODE
        NOD(1)=0
        NOD(2)=1
        NOD(3)=2
        NOD(4)=2
        NOD(5)=1
        NOD(6)=3
        NSINK=6
        KOUNT=0
        ATRIB(2)=1
C       SCHEDULE ARRIVAL TO SOURCE NODE TO START SIMULATION
        CALL SCHDL(1,0.0,ATRIB)
        RETURN
        END
```

D11-4.2

```
      SUBROUTINE EVENT(I)
      COMMON/SCOM1/ATRIB(100),DD(100),DDL(100),DTNOW,II,MFA,MSTOP,NCLNR
     1,NCRDR,NPRNT,NNRUN,NNSET,NTAPE,SS(100),SSL(100),TNEXT,TNOW,XX(100)
      COMMON/UCOM1/NOD(6),KOUNT,NSINK,TIME
C     SET J AS THE END NODE OF THE ACTIVITY JUST COMPLETED
      J=ATRIB(2)
      IF(J.EQ.1) GOTO 10
C     J IS NOT THE SOURCE NODE, DECREASE NUMBER OF RELEASES
C     FOR NODE J BY ONE
      NOD(J)=NOD(J)-1
C     IS NODE J RELEASED?
      IF(NOD(J).NE.0) RETURN
C     COLLECT STATISTICS ON TIME NODE J IS RELEASED
      TREL=TNOW-TIME
      CALL COLCT(TREL,J)
C     IS NODE J THE SINK NODE?
      IF(J.EQ.NSINK) GOTO 20
C     NODE J IS NOT THE SINK NODE, SCHEDULE END OF ACTIVITY EVENT
C     FOR ALL ACTIVITIES HELD IN NODE J FILE
10    K=NNQ(J)
      DO 30 I=1,K
      CALL COPY(I,J,ATRIB)
      DT=TRIAG(ATRIB(3),ATRIB(4),ATRIB(5),9)
      CALL SCHDL(1,DT,ATRIB)
30    CONTINUE
      RETURN
C     CHECK FOR COMPLETION OF 400 RUNS
C     IF NOT, REINITIALIZE VARIABLES AND SCHEDULE EVENT TO
C     RESTART SIMULATION
20    KOUNT=KOUNT+1
      IF(KOUNT.EQ.400) GOTO 40
      NOD(1)=0
      NOD(2)=1
      NOD(3)=2
      NOD(4)=2
      NOD(5)=1
      NOD(6)=3
      TIME=TNOW
      ATRIB(2)=1
      CALL SCHDL(1,0.0,ATRIB)
      RETURN
40    MSTOP=-1
      RETURN
      END
```

D11-4.3

```
1    GEN,ROLSTON,PROBLEM 11.4,7/25/80,1;
2    LIMITS,5,5,20;
3    STAT,2,NODE 2;
4    STAT,3,NODE 3;
5    STAT,4,NODE 4;
6    STAT,5,NODE 5;
7    STAT,6,PROJ. COMPLETE,20/15/.5;
8    ENTRY/1,1,2,1,3,5;
9    ENTRY/1,1,3,3,6,9;
10   ENTRY/1,1,4,10,13,19;
11   ENTRY/2,2,5,3,9,12;
12   ENTRY/2,2,3,1,3,8;
13   ENTRY/5,5,6,3,6,9;
14   ENTRY/4,4,6,1,3,8;
15   ENTRY/3,3,6,8,9,16;
16   ENTRY/3,3,4,4,7,13;
17   FIN;
```

D11-4.4

11-4,Embellishment. The code required for the embellishment in which the criticality index is computed is shown in D11-4.5 through D11-4.9. The outputs for this example are shown in D11-4.10 and D11-4.11.

```
        SUBROUTINE EVENT(I)
        COMMON/SCOM1/ATRIB(100),DD(100),DDL(100),DTNOW,II,MFA,MSTOP,NCLNR
       1,NCRDR,NPRNT,NNRUN,NNSET,NTAPE,SS(100),SSL(100),TNEXT,TNOW,XX(100)
        COMMON/UCOM1/NOD(6),KOUNT,NSINK,TIME,NCA(9),NCN(6),NUMC(9)
C       SET J AS THE END NODE OF THE ACTIVITY JUST COMPLETED
        J=ATRIB(2)
        IF(J.EQ.1) GOTO 10
C       DECREASE THE NUMBER OF RELEASES FOR NODE J BY ONE
        NOD(J)=NOD(J)-1
C       IS NODE J RELEASED?
        IF(NOD(J).NE.0) RETURN
C       YES, SET NCA(J) EQUAL TO THE NUMBER OF THE ACTIVITY WHICH
C       CAUSED THE RELEASE, NCN(J) EQUAL TO THE SOURCE NODE OF
C       THAT ACTIVITY
        NCA(J)=ATRIB(6)
        NCN(J)=ATRIB(1)
C       COLLECT STATISTICS ON TIME NODE J RELEASED
        TREL=TNOW-TIME
        CALL COLCT(TREL,J)
        IF(J.EQ.NSINK) GOTO 20
C       IF J IS NOT THE SINK NODE, SCHEDULE END OF ACTIVITY EVENTS
C       FOR ALL ACTIVITIES IN NODE J FILE
10      K=NNQ(J)
        DO 30 I=1,K
        CALL COPY(I,J,ATRIB)
        DT=TRIAG(ATRIB(3),ATRIB(4),ATRIB(5),9)
        CALL SCHDL(1,DT,ATRIB)
30      CONTINUE
        RETURN
20      KOUNT=KOUNT+1
C       CALL ROUTINE TO DETERMINE CRITICAL PATH
        CALL CRIT
        IF(KOUNT.EQ.400) GOTO 40
C       REINITIALIZE VARIABLES FOR ANOTHER RUN
        NOD(1)=0
        NOD(2)=1
        NOD(3)=2
        NOD(4)=2
        NOD(5)=1
        NOD(6)=3
        TIME=TNOW
C       SCHEDULE EVENT TO RESTART THE SIMULATION
        ATRIB(2)=1
        CALL SCHDL(1,0.0,ATRIB)
        RETURN
40      MSTOP=-1
        RETURN
        END
```

D11-4.5

```
      SUBROUTINE INTLC
      COMMON/SCOM1/ATRIB(100),DD(100),DDL(100),DTNOW,II,MFA,MSTOP,NCLNR
     1,NCRDR,NPRNT,NNRUN,NNSET,NTAPE,SS(100),SSL(100),TNEXT,TNOW,XX(100)
      COMMON/UCOM1/NOD(6),KOUNT,NSINK,TIME,NCA(9),NCN(6),NUMC(9)
      TIME=0
C     INITIALIZE NUMBER OF RELEASES FOR EACH NODE
      NOD(1)=0
      NOD(2)=1
      NOD(3)=2
      NOD(4)=2
      NOD(5)=1
      NOD(6)=3
      NSINK=6
      KOUNT=0
      DO 10 I=1,9
      NUMC(I)=0
10    CONTINUE
C     SCHEDULE EVENT TO START SIMULATION
      ATRIB(2)=1
      CALL SCHDL(1,0.0,ATRIB)
      RETURN
      END
```

D11-4.6

```
      SUBROUTINE CRIT
      COMMON/UCOM1/NOD(6),KOUNT,NSINK,TIME,NCA(9),NCN(6),NUMC(9)
C     THIS SUBROUTINE DETERMINES THE CRITICAL PATH, AND INCRE-
C     MENTS THE NUMBER OF TIMES AN ACTIVITY WAS ON THE CRITICAL
C     PATH. STARTING WITH THE SINK NODE, THE ACTIVITY WHICH
C     CAUSED THE NODE TO BE RELEASED IS DETERMINED (THE NO. OF
C     THE ACTIVITY IS STORED IN NCA(J)). THE SOURCE NODE OF THIS
C     ACTIVITY IS STORED IN NCN(J), WHICH ENABLES THE ROUTINE TO
C     TRACE BACK THROUGH THE NETWORK.
      LCN=6
      LCA=NCA(6)
10    NUMC(LCA)=NUMC(LCA)+1
      LCN=NCN(LCN)
      IF(LCN.EQ.1) RETURN
      LCA=NCA(LCN)
      GOTO 10
      END
```

D11-4.7

```
      SUBROUTINE OTPUT
      COMMON/SCOM1/ATRIB(100),DD(100),DDL(100),DTNOW,II,MFA,MSTOP,NCLNR
     1,NCRDR,NPRNT,NNRUN,NNSET,NTAPE,SS(100),SSL(100),TNEXT,TNOW,XX(100)
      COMMON/UCOM1/NOD(6),KOUNT,NSINK,TIME,NCA(9),NCN(6),NUMC(9)
      DIMENSION XNUMC(9)
      DO 10 I=1,9
      XNUMC(I)=FLOAT(NUMC(I))/400.
      WRITE(NPRNT,20) I,XNUMC(I)
10    CONTINUE
20    FORMAT(//,35X,9HACT. NO. ,I1,19H CRITICALITY INDEX ,F5.4,/)
      RETURN
      END
```

D11-4.8

```
 1   GEN,ROLSTON,PROBLEM 11.4A,7/25/80,1;
 2   LIMITS,5,6,20;
 3   STAT,2,NODE 2;
 4   STAT,3,NODE 3;
 5   STAT,4,NODE 4;
 6   STAT,5,NODE 5;
 7   STAT,6,PROJ. COMPLETE,20/15/.5;
 8   ENTRY/1,1,2,1,3,5,1;
 9   ENTRY/1,1,3,3,6,9,2;
10   ENTRY/1,1,4,10,13,19,3;
11   ENTRY/2,2,5,3,9,12,4;
12   ENTRY/2,2,3,1,3,8,5;
13   ENTRY/5,5,6,3,6,9,8;
14   ENTRY/4,4,6,1,3,8,9;
15   ENTRY/3,3,6,8,9,16,6;
16   ENTRY/3,3,4,4,7,13,7;
17   FIN;
```

D11-4.9

INTERMEDIATE RESULTS

ACT. NO. 1 CRITICALITY INDEX .6150

ACT. NO. 2 CRITICALITY INDEX .1900

ACT. NO. 3 CRITICALITY INDEX .1950

ACT. NO. 4 CRITICALITY INDEX .1200

ACT. NO. 5 CRITICALITY INDEX .4950

ACT. NO. 6 CRITICALITY INDEX .2500

ACT. NO. 7 CRITICALITY INDEX .4350

ACT. NO. 8 CRITICALITY INDEX .1200

ACT. NO. 9 CRITICALITY INDEX .6300

D11-4.10

```
                    S L A M   I I   S U M M A R Y   R E P O R T

          SIMULATION PROJECT PROBLEM 11.4A          BY ROLSTON

          DATE  7/25/1980                           RUN NUMBER   1 OF   1


          CURRENT TIME   0.8313E+04
          STATISTICAL ARRAYS CLEARED AT TIME  0.0000E+00



            **STATISTICS FOR VARIABLES BASED ON OBSERVATION**

                  MEAN        STANDARD    COEFF. OF   MINIMUM     MAXIMUM     NUMBER OF
                  VALUE       DEVIATION   VARIATION   VALUE       VALUE       OBSERVATIONS

                            NO VALUES RECORDED
   NODE 2         0.2992E+01  0.7788E+00  0.2603E+00  0.1270E+01  0.4851E+01     400
   NODE 3         0.7451E+01  0.1300E+01  0.1744E+00  0.4456E+01  0.1198E+02     400
   NODE 4         0.1596E+02  0.2037E+01  0.1276E+00  0.1182E+02  0.2245E+02     400
   NODE 5         0.1102E+02  0.2041E+01  0.1852E+00  0.5562E+01  0.1571E+02     400
   PROJ. COMPLETE 0.2078E+02  0.2137E+01  0.1028E+00  0.1577E+02  0.2697E+02     400



              **FILE STATISTICS**

   FILE                 AVERAGE     STANDARD    MAXIMUM   CURRENT   AVERAGE
   NUMBER  LABEL/TYPE   LENGTH      DEVIATION   LENGTH    LENGTH    WAITING TIME

     1                  3.0000      0.0000         3        3       8312.8711
     2                  2.0000      0.0000         2        2       8312.8711
     3                  2.0000      0.0000         2        2       8312.8711
     4                  1.0000      0.0000         1        1       8312.8711
     5                  1.0000      0.0000         1        1       8312.8711
     6      CALENDAR    3.0854      0.9706         4        0          6.4121
```

D11-4.11

11-5.   For a complete description of the simulation of the inventory situation presented in this exercise see pages 182-200 of Reference 5. The solution for this exercise is contained in D11-5.1 through D11-5.6.

11-5,Embellishment(a). In subroutine REVEW, change the second argument in the call to subroutine SCHDL to RLOGN(3.,1.,2).

```
  1    GEN,ROLSTON,PROBLEM 11.5,8/28/80,1;
  2    LIMITS,1,2,100;
  3    TIMST,XX(1),INV POSITION;
  4    TIMST,XX(2),STOCK;
  5    STAT,1,TB LOST SALES;
  6    STAT,2,SAFETY STOCK;
  7    INIT,0,312;
  8    MONTR,CLEAR,52;
  9    FIN;
```

D11-5.1

```
      SUBROUTINE DMAND
      COMMON/SCOM1/ATRIB(100),DD(100),DDL(100),DTNOW,II,MFA,MSTOP,NCLNR
     1,NCRDR,NPRNT,NNRUN,NNSET,NTAPE,SS(100),SSL(100),TNEXT,TNOW,XX(100)
      COMMON/UCOM1/TLOST,R,SCL
      EQUIVALENCE (POS,XX(1))
      EQUIVALENCE (STOCK,XX(2))
C *** SCHEDULE NEXT CUSTOMER ARRIVAL
      CALL SCHDL(1,EXPON(.2,1),ATRIB)
C *** SATISFY CUSTOMER IF POSSIBLE
      IF(STOCK.GT.0.) GOTO 20
      IF(DRAND(1).LE..2) GOTO 10
C *** LOST SALE
      TBL=TNOW-TLOST
      CALL COLCT(TBL,1)
      TLOST=TNOW
      RETURN
C *** BACKORDER
10    POS=POS-1
      CALL FILEM(1,ATRIB)
      RETURN
C *** SALE
20    POS=POS-1
      STOCK=STOCK-1
      RETURN
      END
```

D11-5.2

```
      SUBROUTINE RECPT
      COMMON/SCOM1/ATRIB(100),DD(100),DDL(100),DTNOW,II,MFA,MSTOP,NCLNR
     1,NCRDR,NPRNT,NNRUN,NNSET,NTAPE,SS(100),SSL(100),TNEXT,TNOW,XX(100)
      COMMON/UCOM1/TLOST,R,SCL
      EQUIVALENCE (POS,XX(1))
      EQUIVALENCE (STOCK,XX(2))
      CALL COLCT(STOCK,2)
      STOCK=STOCK+ATRIB(1)
C *** SATISFY ANY BACKORDERS
5     IF(NNQ(1)) 20,20,10
10    CALL RMOVE(1,1,ATRIB)
      STOCK=STOCK-1.
      IF(STOCK) 20,20,5
20    RETURN
      END
```

D11-5.3

```
      SUBROUTINE REVEW
      COMMON/SCOM1/ATRIB(100),DD(100),DDL(100),DTNOW,II,MFA,MSTOP,NCLNR
     1,NCRDR,NPRNT,NNRUN,NNSET,NTAPE,SS(100),SSL(100),TNEXT,TNOW,XX(100)
      COMMON/UCOM1/TLOST,R,SCL
      EQUIVALENCE (POS,XX(1))
      EQUIVALENCE (STOCK,XX(2))
C *** SCHEDULE NEXT REVIEW
      CALL SCHDL(3,4.,ATRIB)
C *** DETERMINE IF ORDER TO BE PLACED
      IF(POS.GT.R) RETURN
C *** PLACE ORDER WITH LEAD TIME OF 3 WEEKS
C *** AND AMOUNT UP TO THE STOCK CONTROL LEVEL
      ATRIB(1)=SCL-POS
      CALL SCHDL(2,3.,ATRIB)
      POS=SCL
      RETURN
      END
```

D11-5.4

```
      SUBROUTINE INTLC
      COMMON/SCOM1/ATRIB(100),DD(100),DDL(100),DTNOW,II,MFA,MSTOP,NCLNR
     1,NCRDR,NPRNT,NNRUN,NNSET,NTAPE,SS(100),SSL(100),TNEXT,TNOW,XX(100)
      COMMON/UCOM1/TLOST,R,SCL
      EQUIVALENCE (POS,XX(1))
      EQUIVALENCE (STOCK,XX(2))
      POS=72
      R=18
      SCL=72
      STOCK=72
      TLOST=0
C *** SCHEDULE FIRST CUSTOMER DEMAND
      CALL SCHDL(1,0.,ATRIB)
C *** SCHEDULE FIRST REVIEW EVENT
      CALL SCHDL(3,4.,ATRIB)
      RETURN
      END
```

D11-5.5

```
                    SUBROUTINE EVENT(IX)
                    GOTO (1,2,3),IX
      1             CALL DMAND
                    RETURN
      2             CALL RECPT
                    RETURN
      3             CALL REVEW
                    RETURN
                    END
```

D11-5.6

11-5,Embellishment(b).   Change subroutine DMAND shown in D11-5.2 to that shown in D11-5.7.   Also change subroutine RECPT shown in D11-5.3 to the one shown in D11-5.8.

11-5,Embellishment(c).   Change subroutine DMAND given in D11-5.2 by adding calls to subroutine REVEW before each return statement.   In subroutine REVEW, delete the call to subroutine SCHDL which schedules the next review.   The scheduling of the first review event in subroutine INTLC is permissible but not required, as REVEW is called when the first demand occurs which is at time 0.

11-5,Embellishment(d).   In subroutine INTLC, set XLOST to 0 to represent the current number of lost sales and XBO to 0 to indicate the current number of backorders.   In subroutine DMAND, replace the first executable statement of subroutine DMAND of D11-5.2 by the following statements:

```
      TBD=.2
      IF(TNOW.LE.0.) GOTO 5
      ALB=(XLOST+XBO)/TNOW
      IF(ALB.LT.1.) TBD=.18
      IF(ALB.GT.2.) TBD=.22
    5 CALL SCHDL(1,EXPON(TBD,1),ATRIB)
```

11-5,Embellishment(e). There are two approaches to including a profit structure: update profit as each event occurs; or maintain a running count of the number of sales, number of reviews, number of orders placed, and number of backorders and compute the profit at the end of a run. There is no clear cut choice as to which way it should be done. If the cost figures are included in each subroutine then they should be read in or included in a DATA statement and in COMMON statements for use throughout the program. If the profit is computed in subroutine OTPUT then the economic data need only be included in a DATA statement in subroutine OTPUT. To compute the average number of units in inventory, function TTAVG(2) could be used but this requires knowledge from Chapter 12. Alternatively, the time integrated number of units in stock can be calculated by integrating the value of STOCK each time it changes (summing up the rectangles representing stock-on-hand). For the data values presented in the example, and using the random number generator on the VAX 11/780, a total profit over the 312 weeks of $24,218 was obtained.

```
        SUBROUTINE DMAND
        COMMON/SCOM1/ATRIB(100),DD(100),DDL(100),DTNOW,II,MFA,MSTOP,NCLNR
       1,NCRDR,NPRNT,NNRUN,NNSET,NTAPE,SS(100),SSL(100),TNEXT,TNOW,XX(100)
        COMMON/UCOM1/TLOST,R,SCL
        EQUIVALENCE (POS,XX(1))
        EQUIVALENCE (STOCK,XX(2))
C ***   SCHEDULE NEXT CUSTOMER ARRIVAL
        CALL SCHDL(1,EXPON(.2,1),ATRIB)
C ***   SATISFY CUSTOMER IF POSSIBLE
        ADMND=NPSSN(2.,3)+1
        IF(STOCK.GT.ADMND) GOTO 20
        IF(DRAND(1).LE..2) GOTO 10
C ***   LOST SALE
        TBL=TNOW-TLOST
        CALL COLCT(TBL,1)
        TLOST=TNOW
        RETURN
C ***   BACKORDER
10      POS=POS-ADMND
        ATRIB(1)=ADMND
        CALL FILEM(1,ATRIB)
        RETURN
C ***   SALE
20      POS=POS-ADMND
        STOCK=STOCK-ADMND
        RETURN
        END
```

D11-5.7

```
        SUBROUTINE RECPT
        COMMON/SCOM1/ATRIB(100),DD(100),DDL(100),DTNOW,II,MFA,MSTOP,NCLNR
       1,NCRDR,NPRNT,NNRUN,NNSET,NTAPE,SS(100),SSL(100),TNEXT,TNOW,XX(100)
        COMMON/UCOM1/TLOST,R,SCL
        EQUIVALENCE (POS,XX(1))
        EQUIVALENCE (STOCK,XX(2))
        CALL COLCT(STOCK,2)
        STOCK=STOCK+ATRIB(1)
C ***   SATISFY ANY BACKORDERS
5       IF(NNQ(1)) 20,20,10
10      CALL COPY(1,1,ATRIB)
        IF(STOCK.LT.ATRIB(1)) GOTO 20
        CALL RMOVE(1,1,ATRIB)
        STOCK=STOCK-ATRIB(1)
        IF(STOCK) 20,20,5
20      RETURN
        END
```

D11-5.8

11-6. The input listing and FORTRAN subroutines other than the main program and subroutine EVENT for the conveyor discrete event model are shown in D11-6.1 through D11-6.4.

In D11-6.1, a definition of the XX variables is given. In subroutine INTLC, D11-6.2, each server is set idle and the first arrival is scheduled to occur at time 0. In subroutine ARVL, event 1, the next arrival to the conveyor system is scheduled to occur at .25 time units if ATRIB(3) is less than or equal to 1. Attribute 3 is used to distinguish between new arrivals and old arrivals. This could have been accomplished with ATRIB(2) if initial arrivals were assigned an ATRIB(2) value of 0. ATRIB(2) is used to define the station number or server number to which the unit is arriving. By using a separate attribute for this purpose, a clearer coding of the problem is obtained. In ARVL, a test is made to see if the server is busy and, if so, another ARVL event is scheduled but to the next server by indexing ATRIB(2) by 1. If the item is to be recirculated then ATRIB(2) is set to 1. Whenever the server is free when a unit arrives, an end-of-service event is scheduled to occur. The end-of-service event is listed in D11-6.4. The event is similar to other end-of-service event subroutines presented in the text. In general, once a modeler masters the basic logic used in arrival event routines and end-of-service event routines, complex queueing situations are easily modeled. In an arrival event, the next arrival is usually scheduled and the disposition of the current arrival is determined. In an end-of-service event, the disposition of the entity served and the server are modeled.

The embellishments for Exercise 9-5 are easily incorporated into the discrete event model with minor changes. For embellishment (a) the second argument in subroutine SCHDL is modified to include a time delay of 2 minutes rather than 1 minute and 10 minutes rather than 5 minutes. In general, it may be preferred to use a variable for this value and then to define the time delay in a DATA statement. Embellishment (b) involves adding a queue and storing entities when they arrive to server 5. Embellishment (c) incorporates queues before each server and tests on the number in the queue to determine the disposition of the entity arriving to the server.

```
 1    GEN,ROLSTON,PROBLEM 11.6,7/24/80,1;
 2    LIMITS,0,3,200;
 3    STAT,1,TIME IN SYSTEM;
 4    TIMST,XX(1),SERVER 1;
 5    TIMST,XX(2),SERVER 2;
 6    TIMST,XX(3),SERVER 3;
 7    TIMST,XX(4),SERVER 4;
 8    TIMST,XX(5),SERVER 5;
 9    TIMST,XX(6),NO. OF BUSY SERV;
10    INIT,0,100;
11    FIN;
```

D11-6.1

```
      SUBROUTINE INTLC
      COMMON/SCOM1/ATRIB(100),DD(100),DDL(100),DTNOW,II,MFA,MSTOP,NCLNR
     1,NCRDR,NPRNT,NNRUN,NNSET,NTAPE,SS(100),SSL(100),TNEXT,TNOW,XX(100)
      DIMENSION BUSY(5)
      EQUIVALENCE (XX(1),BUSY(1))
      DO 5 I=1,5
      BUSY(I)=0
    5 CONTINUE
      XX(6)=0
      ATRIB(3)=1
      CALL SCHDL(1,0.0,ATRIB)
      RETURN
      END
```

D11-6.2

```
      SUBROUTINE ARVL
      COMMON/SCOM1/ATRIB(100),DD(100),DDL(100),DTNOW,II,MFA,MSTOP,NCLNR
     1,NCRDR,NPRNT,NNRUN,NNSET,NTAPE,SS(100),SSL(100),TNEXT,TNOW,XX(100)
      DIMENSION BUSY(5)
      EQUIVALENCE (XX(1),BUSY(1))
      IF(ATRIB(3).GT.1.) GOTO 10
C     IF NEW ARRIVAL, SCHEDULE NEXT ARRIVAL
      CALL SCHDL(1,.25,ATRIB)
C     SAVE ARRIVAL TIME
      ATRIB(1)=TNOW
C     SET ATRIB(2) EQUAL TO THE SERVER NUMBER ARRIVING TO
      ATRIB(2)=1
10    I=ATRIB(2)
C     IS THE SERVER THE UNIT IS ARRIVING TO BUSY?
      IF(BUSY(I).GT.0.)GOTO 20
C     NO, SET SERVER TO BUSY, INCREMENT NO. OF BUSY SERVERS,
C     SCHEDULE END OF SERVICE
      BUSY(I)=1
      XX(6)=XX(6)+1
      CALL SCHDL(2,EXPON(1.0,1),ATRIB)
      RETURN
C     YES, SET ATRIB(2) TO NEXT SERVER NO, SET ATRIB(3) TO 2
C     TO INDICATE RETURNING UNIT, SCHEDULE ARRIVAL TO NEXT
C     SERVER
20    IF(I.GT.4) GOTO 30
      ATRIB(2)=I+1
      ATRIB(3)=2
      CALL SCHDL(1,1.0,ATRIB)
      RETURN
30    ATRIB(2)=1
      ATRIB(3)=2
      CALL SCHDL(1,5.0,ATRIB)
      RETURN
      END
```

D11-6.3

```
      SUBROUTINE ENDSV
      COMMON/SCOM1/ATRIB(100),DD(100),DDL(100),DTNOW,II,MFA,MSTOP,NCLNR
     1,NCRDR,NPRNT,NNRUN,NNSET,NTAPE,SS(100),SSL(100),TNEXT,TNOW,XX(100)
      DIMENSION BUSY(5)
      EQUIVALENCE (XX(1),BUSY(1))
C     CALCULATE TIME IN SYSTEM AND COLLECT STATISTICS
      TSYS=TNOW-ATRIB(1)
      CALL COLCT(TSYS,1)
C     SET STATUS OF SERVER COMPLETING SERVICE TO IDLE
C     DECREMENT NO. OF BUSY SERVERS
      I=ATRIB(2)
      BUSY(I)=0
      XX(6)=XX(6)-1.
      RETURN
      END
```

D11-6.4

11-7.   The events used to model the banking system are: (1) customer arrival to the system (ARVL); and (2) end of service (ENDSV). Customers are modeled as entities with two attributes. The first attribute is used to record the arrival time of the customer to the system and is used tocompute the time in the system for the customer. The second attribute is used when scheduling the end-of-service event and denotes the teller number from which the customer is obtaining service. A customer entity waiting for teller I is stored in file I where I equals 1 or 2. The following variables are employed in the simulation:

Variable	Definition	Initial Value
XX(I)	$\begin{cases} 0 & \text{if teller I is idle} \\ 1 & \text{if teller I is busy} \end{cases}$	(1.,1.)
NBALK	Number of customers that balk	0
NCUST	Number of customers that arrive	6
NJOCK	Number of customers that jockey	0
TLAST	Time of last customer departure from the system	0

Subroutine EVENT is shown in D11-7.1. The EVENT "arrival at the bank" is flowcharted in Figure D11-7.2 and coded in D11-7.3. Comments are given throughout the code which describe the statements. The flow chart for the end-of-service event is given in Dll-7.4 and the coding in D11-7.5.

```
      SUBROUTINE EVENT(I)
      GO TO (1,2),I
    1 CALL ARVL
      RETURN
    2 CALL ENDSV
      RETURN
      END
```

D11-7.1

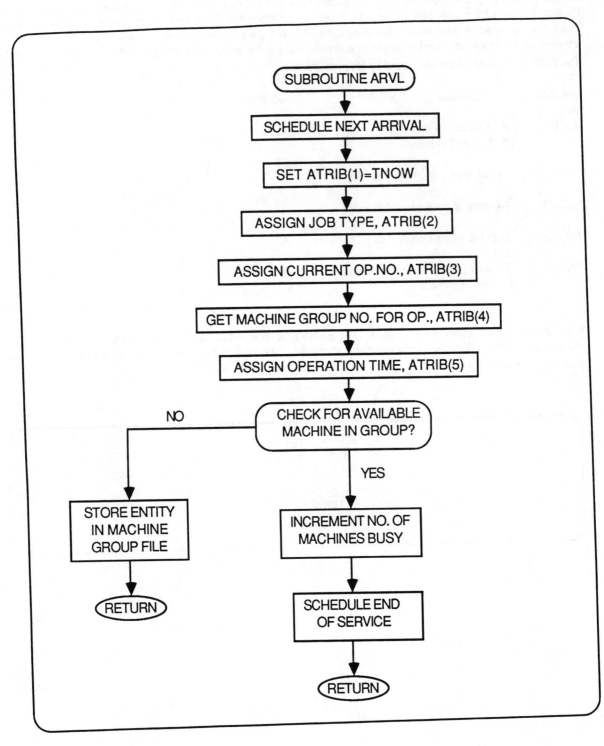

D11-7.2

```
      SUBROUTINE ARVL
      COMMON/SCOM1/ ATRIB(100),DD(100),DDL(100),DTNOW,II,MFA,MSTOP,NCLNR
     1,NCRDR,NPRNT,NNRUN,NNSET,NTAPE,SS(100),SSL(100),TNEXT,TNOW,XX(100)
      COMMON/UCOM1/ NBALK,NCUST,NJOCK,TLAST
C*****CAUSE NEXT ARRIVAL,MARK ARRIVAL TIME, AND INCREMENT NCUST
      CALL SCHDL(1,EXPON(.5,1),ATRIB)
      ATRIB(1)=TNOW
      NCUST=NCUST+1
C*****IF THE SYSTEM IS FULL
      IF(NNQ(1)+NNQ(2).LT.6) GO TO 10
C*****THEN BALK
      NBALK=NBALK+1
      RETURN
C*****OTHERWISE IF A TELLER IS FREE
   10 IF(XX(1)+XX(2).GT.1.0) GO TO 20
C*****THEN SET ITLR TO FIRST FREE TELLER
      ITLR=1
      IF(XX(1).EQ.1.) ITLR=2
C*****AND SET TELLER TO BUSY,SCHEDULE END OF SERVICE, AND RETURN
      XX(ITLR)=1.
      ATRIB(2)=ITLR
      CALL SCHDL(2,RNORM(1.,0.3,2),ATRIB)
      RETURN
C*****ELSE PLACE CUSTOMER IN SHORTEST LINE
   20 IFILE=1
      IF(NNQ(1).GT.NNQ(2)) IFILE=2
      CALL FILEM(IFILE,ATRIB)
      RETURN
      END
```

D11-7.3

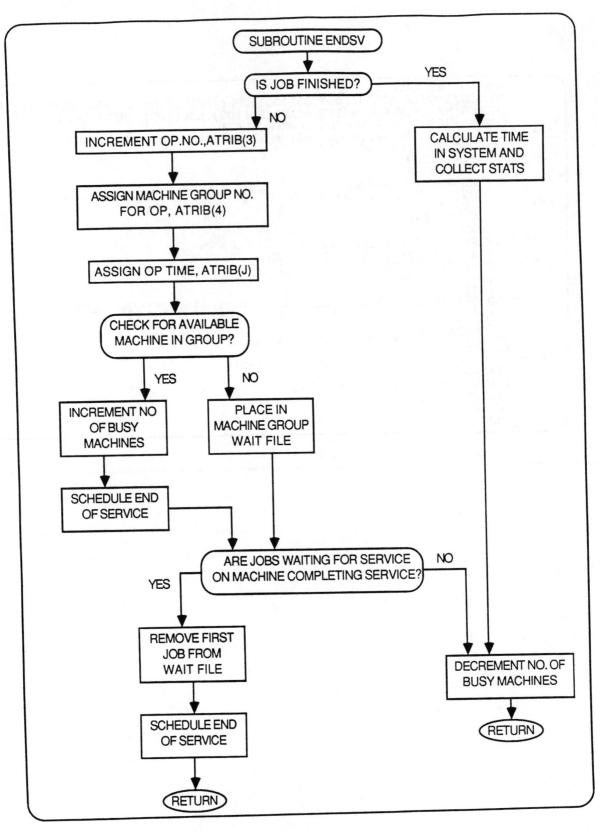

D11-7.4

```
      SUBROUTINE ENDSV
      COMMON/SCOM1/ ATRIB(100),DD(100),DDL(100),DTNOW,II,MFA,MSTOP,NCLNR
     1,NCRDR,NPRNT,NNRUN,NNSET,NTAPE,SS(100),SSL(100),TNEXT,TNOW,XX(100)
      COMMON/UCOM1/ NBALK,NCUST,NJOCK,TLAST
C*****COLLECT STATISTICS ON DEPARTING CUSTOMERS
      TSYS=TNOW-ATRIB(1)
      TBD=TNOW-TLAST
      TLAST=TNOW
      CALL COLCT(TSYS,1)
      CALL COLCT(TBD,2)
C*****SET I TO TELLER ENDING SERVICE, J TO OTHER TELLER
      I=ATRIB(2)
      J=1
      IF(I.EQ.1) J=2
C*****TEST NUMBER OF WAITING CUSTOMERS
      IF(NNQ(I).GT.0) GO TO 20
      IF(NNQ(J).GT.0) GO TO 10
C*****BOTH LANES ARE EMPTY,THEREFORE SET TELLER I TO IDLE AND RETURN
      XX(I)=0.
      RETURN
C*****LANE I IS EMPTY AND LANE J IS OCCUPIED, THEREFORE JOCKEY J TO I
   10 CALL RMOVE(NNQ(J),J,ATRIB)
      ATRIB(2)=I
      CALL SCHDL(2,RNORM(1.,0.3,2),ATRIB)
      NJOCK=NJOCK+1
      RETURN
C*****LANE I IS OCCUPIED, THEREFORE PROCESS THE FIRST CUSTOMER IN LANE I
   20 CALL RMOVE(1,I,ATRIB)
      ATRIB(2)=I
      CALL SCHDL(2,RNORM(1.,0.3,2),ATRIB)
C*****IF THE NUMBER IN LANE J EXCEEDS THE NUMBER IN LANE I BY TWO
      IF(NNQ(J).LT.NNQ(I)+2) RETURN
C*****THEN JOCKEY LAST CUSTOMER IN LANE J TO LANE I
      CALL RMOVE(NNQ(J),J,ATRIB)
      CALL FILEM(I,ATRIB)
      NJOCK=NJOCK+1
      RETURN
      END
```

D11-7.5

The initial conditions for the simulation are established in subroutine INTLC which is given in D11-7.6. Subroutine OTPUT is employed to obtain the specialized output on number of arrivals and jockeying and is presented in D11-7.7. The input statements for this problem are shown in D11-7.8. The user function employed in the TIMST statement is identical to the one presented in Example 9-1 and is shown in D11-7.9.

```
      SUBROUTINE INTLC
      COMMON/SCOM1/ ATRIB(100),DD(100),DDL(100),DTNOW,II,MFA,MSTOP,NCLNR
     1,NCRDR,NPRNT,NNRUN,NNSET,NTAPE,SS(100),SSL(100),TNEXT,TNOW,XX(100)
      COMMON/UCOM1/ NBALK,NCUST,NJOCK,TLAST
      ATRIB(1)=0.
      NBALK=0
      NCUST=6
      NJOCK=0
      TLAST=0.
      DO 10 K=1,2
      XX(K)=1.
      ATRIB(2)=K
      CALL SCHDL(2,RNORM(1.,0.3,2),ATRIB)
      CALL FILEM(K,ATRIB)
   10 CALL FILEM(K,ATRIB)
      CALL SCHDL(1,.1,ATRIB)
      RETURN
      END
```

D11-7.6

```
      SUBROUTINE OTPUT
      COMMON/SCOM1/ ATRIB(100),DD(100),DDL(100),DTNOW,II,MFA,MSTOP,NCLNR
     1,NCRDR,NPRNT,NNRUN,NNSET,NTAPE,SS(100),SSL(100),TNEXT,TNOW,XX(100)
      COMMON/UCOM1/ NBALK,NCUST,NJOCK,TLAST
      PBALK=100.*FLOAT(NBALK)/FLOAT(NCUST)
      WRITE(NPRNT,10) NCUST,PBALK,NJOCK
   10 FORMAT(//35X,30HNUMBER OF ARRIVING CUSTOMERS =,I8,/35X,30HPERCENT
     1OF CUSTOMERS BALKING =,F8.5,/35X,19HNUMBER OF JOCKEYS =,11X,I8)
      RETURN
      END
```

D11-7.7

```
    1    GEN,OREILLY,PROBLEM 11.7,7/25/83,1;
    2    LIMITS,2,2,10;
    3    TIMST,USERF(1),NO. OF CUST.;
    4    SEEDS,4367651(1),6121137(2);
    5    STAT,1,TIME IN SYSTEM;
    6    STAT,2,TIME BET. DEPART;
    7    TIMST,XX(1),TELLER 1 UTIL;
    8    TIMST,XX(2),TELLER 2 UTIL;
    9    INIT,0,1000;
   10    MONTR,TRACE,0,10,ATRIB(1),NNQ(1),XX(1),NNQ(2),XX(2);
   11    FIN;
```

D11-7.8

```
      FUNCTION USERF(I)
C*****
C  CALCULATES THE TOTAL NUMBER OF CUSTOMERS IN THE SYSTEM
C*****
      USERF=NNQ(1)+NNACT(1)+NNQ(2)+NNACT(2)
      RETURN
      END
```

D11-7.9

11-7,Embellishment. In this embellishment, a new event called GAP is written. GAP will be event 3 and will simulate the arrival of gaps to process cars waiting to go into street traffic. Subroutine EVENT and subroutine GAP are shown in D11-7.10 and D11-7.11.

In subroutine GAP, the next gap is scheduled and then it is determined if a car is waiting to go into the street. If not, a return is made. If a car is waiting, it is removed from file 3 which is a file for cars waiting to go into the street. Next, a test is made on the number in file 4. to see if a teller is blocked. Cars that block the tellers are stored in file 4. If a teller is blocked, the car blocking the teller is removed from file 4 and subroutine ENDSV is called to allow the car to leave the banking system. This illustrates that event routines can be invoked directly by the modeler. The code for subroutine ENDSV is shown in D11-7.12. The only changes from D11-7.5 involve the disposition of the car departing the tellers in the first 6 statements of the subroutine.

```
      SUBROUTINE EVENT(I)
      GO TO (1,2,3),I
1     CALL ARVL
      RETURN
2     CALL ENDSV
      RETURN
3     CALL GAP
      RETURN
      END
```

D11-7.10

```
      SUBROUTINE GAP
      COMMON/SCOM1/ ATRIB(100),DD(100),DDL(100),DTNOW,II,MFA,MSTOP,NCLNR
     1,NCRDR,NPRNT,NNRUN,NNSET,NTAPE,SS(100),SSL(100),TNEXT,TNOW,XX(100)
      CALL SCHDL(3,UNFRM(0.3,0.5,2),ATRIB)
      IF(NNQ(3).EQ.0) RETURN
      CALL RMOVE(1,3,ATRIB)
      IF(NNQ(4).EQ.0) RETURN
      CALL RMOVE(1,4,ATRIB)
      CALL ENDSV
      RETURN
      END
```

D11-7.11

```
      SUBROUTINE ENDSV
      COMMON/SCOM1/ ATRIB(100),DD(100),DDL(100),DTNOW,II,MFA,MSTOP,NCLNR
     1,NCRDR,NPRNT,NNRUN,NNSET,NTAPE,SS(100),SSL(100),TNEXT,TNOW,XX(100)
      COMMON/UCOM1/ NBALK,NCUST,NJOCK,TLAST
      IF(NNQ(3).EQ.3) THEN
      CALL FILEM(4,ATRIB)
      RETURN
      ELSE
      CALL FILEM(3,ATRIB)
      ENDIF
C*****COLLECT STATISTICS ON DEPARTING CUSTOMERS
      TSYS=TNOW-ATRIB(1)
      TBD=TNOW-TLAST
      TLAST=TNOW
      CALL COLCT(TSYS,1)
      CALL COLCT(TBD,2)
C*****SET I TO TELLER ENDING SERVICE, J TO OTHER TELLER
      I=ATRIB(2)
      J=1
      IF(I.EQ.1) J=2
C*****TEST NUMBER OF WAITING CUSTOMERS
      IF(NNQ(I).GT.0) GO TO 20
      IF(NNQ(J).GT.0) GO TO 10
C*****BOTH LANES ARE EMPTY,THEREFORE SET TELLER I TO IDLE AND RETURN
      XX(I)=0.
      RETURN
C*****LANE I IS EMPTY AND LANE J IS OCCUPIED, THEREFORE JOCKEY J TO I
   10 CALL RMOVE(NNQ(J),J,ATRIB)
      ATRIB(2)=I
      CALL SCHDL(2,RNORM(1.,0.3,2),ATRIB)
      NJOCK=NJOCK+1
      RETURN
C*****LANE I IS OCCUPIED, THEREFORE PROCESS THE FIRST CUSTOMER IN LANE I
   20 CALL RMOVE(1,I,ATRIB)
      ATRIB(2)=I
      CALL SCHDL(2,RNORM(1.,0.3,2),ATRIB)
C*****IF THE NUMBER IN LANE J EXCEEDS THE NUMBER IN LANE I BY TWO
      IF(NNQ(J).LT.NNQ(I)+2) RETURN
C*****THEN JOCKEY LAST CUSTOMER IN LANE J TO LANE I
      CALL RMOVE(NNQ(J),J,ATRIB)
      CALL FILEM(I,ATRIB)
      NJOCK=NJOCK+1
      RETURN
      END
```

D11-7.12

11-8.   To model the clerk waiting one minute before traveling to the warehouse to serve a single customer, a third event is added to the solution. This event is scheduled to occur one minute after a single customer is assigned to a clerk. Thus, when a clerk ends service and there is only one unit in file 4, the one unit is removed from file 4 and placed in the file for clerks waiting the one minute before proceeding to the warehouse. A GO event is then scheduled to occur in one minute. Additional changes are required in subroutine ARVL so that a newly arriving customer checks to see if a clerk is waiting in file 6 and, if one is waiting, the newly arriving customer is assigned to the clerk as long as no more than six customers are assigned to the clerk. Subroutine ARVL is shown in D11-8.1 and subroutine GO which is event 3 is shown in D11-8.2. The changes required in subroutine ENDSV are shown in D11-8.3. A listing of the input statements is given in D11-8.4.

```
      SUBROUTINE ARVL
      COMMON/SCOM1/ATRIB(100),DD(100),DDL(100),DTNOW,II,MFA,MSTOP,NCLNR
     1,NCRDR,NPRNT,NNRUN,NNSET,NTAPE,SS(100),SSL(100),TNEXT,TNOW,XX(100)
      DIMENSION CLERK(3)
C     SCHEDULE NEXT ARRIVAL
      CALL SCHDL(1,EXPON(2.,1),ATRIB)
C     MARK ARRIVAL TIME FOR CURRENT CUSTOMER
      ATRIB(1)=TNOW
C      IS THERE A CLERK IN FILE 6 WAITING FOR ADDITIONAL CUSTOMERS?
      IF(NNQ(6).NE.0) GOTO 20
C     NO, ARE ALL CLERKS BUSY?
      IF(NNQ(5).GT.0) GOTO 10
C     IF ALL BUSY, PLACE ARRIVAL IN WAIT FILE AND RETURN
      CALL FILEM(4,ATRIB)
      RETURN
C     IF NOT, REMOVE FIRST FREE CLERK
10    CALL RMOVE(1,5,CLERK)
      ICLRK=CLERK(1)
C     PLACE THE CUSTOMER IN FILE ICLRK
      CALL FILEM(ICLRK,ATRIB)
C     PLACE THE CLERK IN FILE 6 TO AWAIT ADDITIONAL CUSTOMERS
      CALL FILEM(6,CLERK)
C     SCHEDULE THE EVENT TO GO TO THE WAREHOUSE IN 1 MINUTE
      CALL SCHDL(3,1.0,CLERK)
      RETURN
C     COPY ATTRIBUTES OF CLERK IN FILE 6
20    CALL COPY(1,6,CLERK)
      ICLRK=CLERK(1)
C     PLACE CUSTOMER IN FILE ICLRK
      CALL FILEM(ICLRK,ATRIB)
C     CHECK NO. OF CUSTOMERS IN CLERK'S FILE
      IF(NNQ(ICLRK).LT.6) RETURN
C     IF EQUAL TO SIX, REMOVE CLERK FROM FILE 6
      CALL RMOVE(1,6,CLERK)
C     REMOVE GO EVENT FROM EVENT CALENDAR
      NRANK=NFIND(1,NCLNR,2,0,3.0,0.0)
      CALL RMOVE(NRANK,NCLNR,CLERK)
C     COLLECT STATS ON NUMBER OF CUSTOMERS SERVED
      CALL COLCT(6.,2)
C     SCHEDULE END OF SERVICE FOR FIRST CUSTOMER
      DT=UNFRM(.5,1.5,1)+RNORM(18.,3.6,1)+UNFRM(.5,1.5,1)+UNFRM(1.,3.,1)
      CALL SCHDL(2,DT,CLERK)
      RETURN
      END
```

D11-8.1

```
      SUBROUTINE GO
      COMMON/SCOM1/ATRIB(100),DD(100),DDL(100),DTNOW,II,MFA,MSTOP,NCLNR
     1,NCRDR,NPRNT,NNRUN,NNSET,NTAPE,SS(100),SSL(100),TNEXT,TNOW,XX(100)
      DIMENSION CLERK(3)
C     REMOVE CLERK FROM FILE 6
      CALL RMOVE(1,6,CLERK)
C     COLLECT STATS ON NUMBER OF CUSTOMERS SERVED
      ICLRK=CLERK(1)
      XCUST=NNQ(ICLRK)
      CALL COLCT(XCUST,2)
C     SCHEDULE END OF SERVICE FOR FIRST CUSTOMER
      XMN=XCUST*3
      STD=.2*XMN
      DT=UNFRM(.5,1.5,1)+RNORM(XMN,STD,1)+UNFRM(.5,1.5,1)+UNFRM(1.,3.,1)
      CALL SCHDL(2,DT,CLERK)
      RETURN
      END
```

D11-8.2

```
      SUBROUTINE ENDSV
      COMMON/SCOM1/ATRIB(100),DD(100),DDL(100),DTNOW,II,MFA,MSTOP,NCLNR
     1,NCRDR,NPRNT,NNRUN,NNSET,NTAPE,SS(100),SSL(100),TNEXT,TNOW,XX(100)
      DIMENSION CLERK(3)
C     REMOVE THE CUSTOMER COMPLETING SERVICE FROM THE CLERK'S WAIT FILE
      CLERK(1)=ATRIB(1)
      ICLRK=CLERK(1)
      CALL RMOVE(1,ICLRK,ATRIB)
C     COLLECT STATS ON TIME IN SYSTEM
      TSYS=TNOW-ATRIB(1)
      CALL COLCT(TSYS,1)
C     IF MORE CUSTOMERS REMAIN IN THE CLERK'S WAIT FILE,
      IF(NNQ(ICLRK).EQ.0) GOTO 10
C     SCHEDULE END OF SERVICE FOR NEXT CUSTOMER AND RETURN
      CALL SCHDL(2,UNFRM(1.,3.,1),CLERK)
      RETURN
10    CONTINUE
      IF(NNQ(4).EQ.0) GOTO 30
      IF(NNQ(4).EQ.1) GOTO 40
C     ELSE, IF MORE THAN ONE CUSTOMER IS WAITING FOR SERVICE
C     THEN SERVICE UP TO SIX CUSTOMERS
      NCUST=NNQ(4)
      IF(NCUST.GT.6) NCUST=6
      XCUST=NCUST
      CALL COLCT(XCUST,2)
      DO 20 J=1,NCUST
      CALL RMOVE(1,4,ATRIB)
20    CALL FILEM(ICLRK,ATRIB)
C     SCHEDULE END OF SERVICE FOR FIRST CUSTOMER AND RETURN
      XMN=NCUST*3
      STD=.2*XMN
      DT=UNFRM(.5,1.5,1)+RNORM(XMN,STD,1)+UNFRM(.5,1.5,1)+UNFRM(1.,3.,1)
      CALL SCHDL(2,DT,CLERK)
      RETURN
C     IF NO CUSTOMERS ARE WAITING, PLACE CLERK IN FILE OF FREE
C     CLERKS AND RETURN
30    CALL FILEM(5,CLERK)
      RETURN
C     IF ONLY ONE CUSTOMER IS WAITING, REMOVE HIM FROM THE WAIT
C     FILE AND PLACE IN THE CLERK'S FILE
40    CALL RMOVE(1,4,ATRIB)
      CALL FILEM(ICLRK,ATRIB)
C     PLACE CLERK IN FILE 6 TO WAIT FOR MORE CUSTOMERS
      CALL FILEM(6,CLERK)
C     SCHEDULE EVENT TO GO TO WAREHOUSE IN 1 MINUTE
      CALL SCHDL(3,1.0,CLERK)
      RETURN
      END
```

D11-8.3

```
1    GEN,ROLSTON,PROBLEM 11.8,8/1/80,1;
2    LIMITS,6,2,50;
3    STAT,1,TIME IN SYSTEM;
4    STAT,2,NUM OF REQUEST;
5    INIT,0,1000;
6    MONTR,TRACE,0,50;
7    FIN;
```

D11-8.4

11-9.   The changes required for this embellishment to Example 11-1 involve the probabilistic determination of courses of action. These are modeled by generating a random number using DRAND and testing it against prescribed probabilities. A third collect variable is required before obtaining observations on time between balks. The time of the last balk is initialized at zero in subroutine INTLC, that is, TLAST = 0. The changes required in subroutine ARVL are shown in D11-9.1 and the changes required in subroutine ENDSV are shown in D11-9.2.

```
      SUBROUTINE ARVL
      COMMON/SCOM1/ATRIB(100),DD(100),DDL(100),DTNOW,II,MFA,MSTOP,NCLNR
     1,NCRDR,NPRNT,NNRUN,NNSET,NTAPE,SS(100),SSL(100),TNEXT,TNOW,XX(100)
      COMMON/UCOM1/TLAST
      DIMENSION CLERK(3)
C     SCHEDULE NEXT ARRIVAL
      CALL SCHDL(1,EXPON(2.,1),ATRIB)
C     WILL THE CUSTOMER REQUEST SOMETHING NOT IN INVENTORY?
      IF(DRAND(2).GT..10) GOTO 5
C     YES, WILL HE REORDER?
      IF(DRAND(2).LE..25) GOTO 5
C     NO, COLLECT STATS ON TIME BETWEEN BALKS
      TBD=TNOW-TLAST
      TLAST=TNOW
      CALL COLCT(TBD,3)
      RETURN
C     MARK ARRIVAL TIME
5     ATRIB(1)=TNOW
C     IF ALL CLERKS ARE BUSY
      IF(NNQ(5).GT.0) GOTO 10
C     THEN PLACE ARRIVAL IN WAIT FILE AND RETURN
      CALL FILEM(4,ATRIB)
      RETURN
C     ELSE REMOVE THE FIRST FREE CLERK
10    CALL RMOVE(1,5,CLERK)
      ICLRK=CLERK(1)
C     PLACE THE CUSTOMER IN FILE ICLRK AND COLLECT STATS ON NO. SERVED
      CALL FILEM(ICLRK,ATRIB)
      CALL COLCT(1.,2)
C     AND SCHEDULE END OF SERVICE FOR FIRST CUSTOMER
      DT=UNFRM(.5,1.5,1)+RNORM(3.,.6,1)+UNFRM(.5,1.5,1)+UNFRM(1.,3.,1)
      CALL SCHDL(2,DT,CLERK)
      RETURN
      END
```

D11-9.1

```
        SUBROUTINE ENDSV
        COMMON/SCOM1/ATRIB(100),DD(100),DDL(100),DTNOW,II,MFA,MSTOP,NCLNR
       1,NCRDR,NPRNT,NNRUN,NNSET,NTAPE,SS(100),SSL(100),TNEXT,TNOW,XX(100)
        DIMENSION CLERK(3)
        CLERK(1)=ATRIB(1)
        ICLRK=CLERK(1)
C       IS ITEM WRONG?
        IF(DRAND(2).GT..15) GOTO 5
C       YES, RESCHEDULE END OF SERVICE AT TIME TO WAREHOUSE AND
C       BACK PLUS TIME TO PICK
        DT=UNFRM(.5,1.5,1)+RNORM(3.,.6,1)+UNFRM(.5,1.5,1)
        CALL SCHDL(2,DT,CLERK)
        RETURN
C       NO, REMOVE THE CUSTOMER COMPLETING SERVICE FROM THE WAIT FILE
5       CALL RMOVE(1,ICLRK,ATRIB)
C       COLLECT STATS ON TIME IN SYSTEM
        TSYS=TNOW-ATRIB(1)
        CALL COLCT(TSYS,1)
C       IF MORE CUSTOMERS REMAIN IN THE WAIT FILE
        IF(NNQ(ICLRK).EQ.0) GOTO 10
C       SCHEDULE END OF SERVICE FOR THE NEXT CUSTOMER AND RETURN
        CALL SCHDL(2,UNFRM(1.,3.,1),CLERK)
        RETURN
C       ELSE
10      CONTINUE
C       IF CUSTOMERS ARE WAITING FOR SERVICE
        IF(NNQ(4).EQ.0) GOTO 30
C       THEN SERVICE UP TO SIX CUSTOMERS
        NCUST=NNQ(4)
        IF(NCUST.GT.6) NCUST=6
        XCUST=NCUST
        CALL COLCT(XCUST,2)
        DO 20 J=1,NCUST
        CALL RMOVE(1,4,ATRIB)
20      CALL FILEM(ICLRK,ATRIB)
C       AND SCHEDULE END OF SERVICE FOR FIRST CUSTOMER AND RETURN
        XMN=NCUST*3
        STD=.2*XMN
        DT=UNFRM(.5,1.5,1)+RNORM(XMN,STD,1)+UNFRM(.5,1.5,1)+UNFRM(1.,3.,1)
        CALL SCHDL(2,DT,CLERK)
        RETURN
C       ELSE PLACE THE CLERK IN THE FILE OF FREE CLERKS AND RETURN
30      CALL FILEM(5,CLERK)
        RETURN
        END
```

D11-9.2

11-10(a). It is desired to set N, the number of runs, such that

$$P[\mu_X - \gamma\sigma_X \leq \bar{x} \leq \mu_X + \gamma\sigma_X] \geq 1-\alpha$$

Subtracting μ_X and dividing by $\sigma_X/\sqrt{N}$ yields

$$P[-\gamma\sqrt{N} \leq z \leq \gamma\sqrt{N}] \geq 1-\alpha$$

where $z = \dfrac{\bar{x}-\mu_X}{\sigma_X/\sqrt{N}}$ is approximately normally distributed by the central limit theorem.

Rearranging terms and using $P[z \geq \gamma\sqrt{N}] = P[z \leq -\gamma\sqrt{N}]$ yields

$$P[z \leq -\gamma\sqrt{N}] \leq \frac{\alpha}{2}$$

The smallest N to satisfy this inequality occurs when the equality holds. For $\alpha = 0.05$ and $\gamma = 0.1$, we have from normal tables that

$$P[z \leq -1.96] = 0.025$$

and, hence, $-0.1\sqrt{N} = -1.96$

or $N \doteq 400$

11-10(b). The requirement here is

$$P[(1-\delta)\sigma_X^2 \leq s^2 \leq (1+\delta)\sigma_X^2] \geq 1-\alpha$$

where s^2 is the estimate for σ_X^2. The random variable $(N-1)s^2/\sigma_X^2$ is Chi-square distributed with $(N-1)$ degrees of freedom, χ^2_{N-1}. We rearrange terms to obtain

$$P[(1-\delta)(N-1) \leq \frac{s^2(N-1)}{\sigma_X^2} \leq (1+\delta)(N-1)] \geq \alpha$$

For large N, χ^2_{N-1} approaches a normal distribution with mean $(N-1)$ and variance $2(N-1)$.

Subtracting $(N-1)$ and dividing by $\sqrt{2(N-1)}$ and using the same analysis as in solution 11-10(a) yields

$$P[z \leq -\frac{\delta(N-1)}{\sqrt{2(N-1)}}] \leq \frac{\alpha}{2}$$

For $\delta = 0.1$ and $\alpha = 0.05$, the smallest N satisfying the inequality is approximately 800.

11–10(c).The criticality index is an average based on a binomial distributed random variable.

Let

$$X_i = \begin{cases} 0 & \text{if not critical on run i} \\ 1 & \text{if critical on run i} \end{cases}$$

$$p = P[X_i = 1]$$

$$Y = \sum_{i=1}^{N} X_i \text{ is binomial distributed with}$$

$$\mu = Np \text{ and } \sigma^2 = Np(1-p)$$

If $\bar{X} = \dfrac{Y}{N}$ then $\dfrac{\bar{X}-p}{\dfrac{p(1-P)}{N}} = z$ is approximately normal with $\mu=0$ and $\sigma=1$.

The condition specified in the exercise is

$$P[-\delta \leq \bar{X} - p \leq \delta] \geq 1-\alpha$$

Rearranging yields

$$P[\frac{-\delta\sqrt{N}}{\sqrt{p(1-p)}} \leq z \leq \frac{\delta\sqrt{N}}{\sqrt{p(1-p}}] \geq 1-\alpha$$

$$\text{or } P[z \leq \frac{-\delta\sqrt{N}}{\sqrt{p(1-p)}}] \leq \frac{\alpha}{2}$$

For $\delta = 0.005$ and $\alpha = 0.05$, we have

$$\frac{-.005\sqrt{N}}{\sqrt{p(1-p)}} = -1.96$$

Solving yields $N = (400)^2 p(1-p)$

The worst case (largest N) is for $p = \dfrac{1}{2}$ in which

case $N = 40,000$. For $\delta = 0.05$, $N = 400$.

Based on these results, care is required in specifying desired differences from the mean.

CHAPTER 12

Advanced Discrete Event Concepts
And Subprograms

12-1(a). The following code searches through file 2 starting with the first entry and proceeding until an entry is found with the value of attribute 3 equal to 10, or until the last entry in the file is reached.

```
              NEXT=MMFE(2)
C****   GO TO 100 IF NO ENTRY FOUND WITH ATTRIBUTE 3 EQUAL TO 10
     10   IF(NEXT.EQ.0) GO TO 100
          CALL COPY(-NEXT,2,ATRIB)
          IF(ATRIB(3).EQ.10.0) GO TO 20
          NEXT=NSUCR(NEXT)
          GO TO 10
     20   CONTINUE
```

This could also be done as follows:

```
          NEXT=NFIND(-MMFE(2),2,3,0,10.0,0.0)
          IF(NEXT.EQ.0)   GO TO 100
```

To search the file from the last entry to the first until an entry is found with attribute 3 equal to 10, change the first statement to set NEXT to MMLE(2). If the entry does not satisfy the condition, the predecessor entry pointer is set to NEXT by the statement NEXT = NPRED(NEXT).

12-1(b). Use NFIND as shown below

```
          LOC=NFIND(-MMFE(1),1,4,2,0.0,0.0)
```

Either of the following statements will remove the entry:

```
          CALL RMOVE(-LOC,1,ATRIB)
            or
          CALL ULINK(-LOC,1)
```

If ULINK is used, then LINK must be called immediately or the entry will be lost to the system.

12-1(c). LOC=NFIND(-MMFE(3),3,3,-2,10.0,0.0)

12-1(d). LOC=NFIND(-MMFE(3),3,3,1,10.0,0.0)

12-1(e). LOC=NFIND(-MMFE(4),4,2,2,3.0,0.0)

For the solutions to 12-1(c),(d),(e), coding similar to that given for 12-1(a) could be used.

12-2.   The statements requested are:

>   IF(NNQ(NCLNR).LT.2) GO TO 50
>   CALL COPY(2,NCLNR,ATRIB)

These statements will cause the attributes of the second entry in the event file to be copied into the array ATRIB. If less than two entries exist in the event file, a transfer to statement 50 is made.

On the LIMITS statement, the user specifies a value of MATR, the maximum number of attributes per entry. Thus, the values of ATRIB(I), I=1 MATR are reserved for the user's attributes. If a value is not assigned to an attribute, the last value in ATRIB before filing will be used. In the event file, the value of ATRIB(MATR+1) is the event code (for network events it is the negative of a pointer to NSET/QSET), and the value of ATRIB(MATR+2) is the time at which the event is scheduled to occur.

12-3.   a) COLCT; b) COLCT; c) COLCT; d) TIMST; e) TIMST; f) COLCT; g)TIMST.

12-4.   The cells of the histogram are: [-∞ to 5], (5 to 6], (6 to 7], (7 to 8], (8 to 9], (9 to 10], (10 to 11], (11 to 12], (12 to 13], (13 to 14], (14 to 15], and (15 to ∞].

The specified values would be placed in the cells as follows:

> 7.2 -> fourth cell; 9.1 -> sixth cell; 5.0 -> first cell; 4.1 -> first cell; 27.7 -> twelfth cell; and 3.3 -> first cell.

12-5.   The entire program can be written in subroutine INTLC as shown below.

```
        SUBROUTINE INTLC
        COMMON/SCOM1/ATRIB(100),DD(100),DDL(100),DTNOW,II,MFA,MSTOP,NCLNR
       1NCRDR,NPRNT,NNRUN,NNSET,NTAPE,SS(100),SSL(100),TNEXT,TNOW,XX(100)
        DO 10 I=1,400
        XVALUE=RNORM(50.,10.,1)
        IF(XVALUE.LT.0)XVALUE=0.
        IF(XVALUE.GT.100.)XVALUE=100.
        CALL COLCT(XVALUE,1)
    10  CONTINUE
        MSTOP=-1
        RETURN
        END
```

The following STAT statement should be included in the SLAM II input statements.

    STAT,1,NORMAL DIST.,10/30/4;

12-6.   Assuming that entries in the event file are not included, Function NTOTE counts the total entries in all files.

```
      FUNCTION NTOTE(IDUM)
      COMMON/SCON1/ATRIB(100),DD(100),DDL(100),DTNOW,II,MFA,MSTOP,NCLNR,
     1NCRDR,NPRNT,NNRUN,NNSET,NTAPE,SS(100),SSL(100),TNEXT,TNOW,XX(100)
      NFIL=NCLNR-1
      NTOTE=0
      IF(NFIL.LT.1)RETURN
      DO 20 I=1,NFIL
      NTOTE=NTOTE+NNQ(I)
   20 CONTINUE
      RETURN
      END
```

12-7.   The subroutine FIND3 is given in D12-7.1 where it is assumed that an entry's location is desired and that search is to begin at the first entry in the file.

An example of the use of FIND3 is shown below.

```
      X(1)=3.0
      X(2)=3.0
      X(3)=5.0
      CALL FIND3(NTRY,4,X)
```

If a -1 is returned as the value of NTRY, no entity was found in file 4 that has the desired attribute values.

```
      SUBROUTINE FIND3(NTRY,J,X)
      COMMON/SCOM1/ATRIB(100),DD(100),DDL(100),DTNOW,II,MFA,MSTOP,NCLNR,
     1NCRDR,NPRNT,NNRUN,NNSET,NTAPE,SS(100),SSL(100),TNEXT,TNOW,XX(100)
      DIMENSION X(3)
C****  SOLUTION TO EXRCISE 12-7
      NEXT=MMFE(J)
   30 IF(NEXT.EQ.0) GO TO 40
      CALL COPY(-NEXT,J,ATRIB)
      IF(ATRIB(1).GT.X(1)) GO TO 10
      IF(ATRIB(2).GT.X(2)) GO TO 10
      IF(ATRIB(3).GT.X(3)) GO TO 10
      NTRY=NEXT
      GO TO 20
   10 NEXT=NSUCR(NEXT)
      GO TO 30
   40 NTRY=-1
   20 RETURN
      END
```

D12-7.1

12-8.   The function SUMPRO is given in D12-8.1. It calculates the sum of the product of attributes KATT and NATT for the entries in file J.

```
      FUNCTION SUMPRO(J,KATT,NATT)
      COMMON/SCOM1/ATRIB(100),DD(100),DDL(100),DTNOW,II,MFA,MSTOP,NCLNR,
     1NCRDR,NPRNT,NNRUN,NNSET,NTAPE,SS(100),SSL(100),TNEXT,TNOW,XX(100)
      DIMENSION A(100)
      SUMPRO=0.0
      NIF=NNQ(J)
      IF(NIF.EQ.0) RETURN
C**** COPY ATTRIBUTES BY RANK IN FILE J
      DO 10 I=1,NIF
      CALL COPY(I,J,A)
      SUMPRO=SUMPRO+A(KATT)*A(NATT)
   10 CONTINUE
      RETURN
      END
```

D12-8.1

12-9.   In the code presented in D12-9.1, the following definitions are used.

> X   Array of independent variables
> Y   Array of dependent variables
> XVALUE   Independent value of variable
> NVAL   Dimension of X and Y

Linear extrapolation is used based on the slopes at the ends of the table. If an error is detected, subroutine ERROR is called with a code of 18.

```
      FUNCTION GGTBLE(X,Y,XVALUE,NVAL)
C
C****  INPUT PARAMETERS
C
C      X = VALUES OF INDEPENDENT VARIABLE
C      Y = VALUES OF DEPENDENT VARIABLE
C      XVALUE = VALUE OF X FOR WHICH VALUE OF Y IS DESIRED
C      NVAL = NUMBER OF VALUES OF X AND Y
C
C****  OUTPUT
C
C      GGTBLE - TABLE LOOK UP FOR VALUE CORRESPONDING
C               TO XVALUE WITH EXTRAPOLATION BEYOND
C               X(1) AND X(NVAL) AND INTERPOLATION OTHERWISE
C
      DIMENSION X(NVAL),Y(NVAL)
      IF (NVAL.LE.0) CALL ERROR(18)
C
C****  EXTRAPOLATION FORMULAS
C
      IF (XVALUE.LT.X(1)) THEN
          GGTBLE=Y(1)-(X(1)-XVALUE)*(Y(2)-Y(1))/(X(2)-X(1))
          RETURN
      ENDIF
      IF (XVALUE.GT.X(NVAL)) THEN
          GGTBLE=Y(NVAL)+(XVALUE-X(NVAL))*(Y(NVAL)-Y(NVAL-1))/
     &          (X(NVAL)-X(NVAL-1))
          RETURN
      ENDIF
C
C****  INTERPOLATION FORMULA
C
      DO 10 I=1,NVAL-1
          IF (XVALUE.GT.X(I).AND.XVALUE.LE.X(I+1)) THEN
              GGTBLE=Y(I)+(XVALUE-X(I))*(Y(I+1)-Y(I))/(X(I+1)-X(I))
              RETURN
          ENDIF
   10 CONTINUE
      RETURN
      END
```

D12-9.1

12-10. A function to obtain a sample from a regression equation is shown in D12-10.1.

```
      FUNCTION REGRES(X,A,XMEAN,STD,ISTRM)
      DIMENSION A(6)
      REGRES=A(1)*SQRT(X)+A(2)*X+A(3)*X*X
     1      +A(4)*X*X*X+A(5)*EXP(X)+A(6)*ALOG(X)
     2      +RNORM(XMEAN,STD,ISTRM)
      RETURN
      END
```

D12-10.1

12-11.   The code for this exercise is shown in D12-11.1.

12-12.   This exercise is taken from Schriber (Schriber, T., <u>Simulation Using GPSS</u>, John Wiley, 1974).   This particular problem can also be solved using only SLAM II network concepts which could be an added embellishment to the problem.

Basically, the problem is a queueing problem with arrival and end-of- service events. The main difference is that there are job types and machine groups to be considered. Flow charts for subroutine ARVL and ENDSV are given in D12-12.1 and D12-12.2 respectively.

There are many data structures that can be used for storing the route information and the one used in this example is established in subroutine INTLC shown in D12-12.3. In INTLC, the two dimensional array ROUT is used with the first column indicating machine group number and the second column the mean operation time. The route for job type 1 is contained in the first three rows of ROUT; for job type 2, in rows 4-7; and for job type 3, in rows 8-12. The attribute definitions for jobs are given in comment statements in subroutine ARVL which is shown in D12-12.4. Files 1 through 6 are used for storing jobs waiting for machine groups 1 through 6, respectively.   Subroutine ENDSV is shown in D12-12.5 and contains the code corresponding to the flow chart given in D12-12.2. The input statements are shown in D12-12.6. A monitor statement with the SUMRY option is used to obtain reports after every 2400 minutes corresponding to one week. Cumulative summary reports for each 2400 minutes are shown in Figures D12-12.7 through D12-12.11. Although these summary reports were not requested in the problem statement, they do portray system operation over the five week period. If separate weekly reports (40 hour) are desired, then statistics can be cleared directly by the user after each 40 hour week after subroutine SUMRY is called. Summary statistics can be obtained for the five separate runs by calling subroutine COLCT at the end of each week.

```
C****  SET NOW TO THE LOCATION OF THE EVENT
       NOW=MFA
C****  SET NEXT TO THE LOCATION WHERE THE FILE3 ENTRY WILL BE PLACED
       NEXT=NSUCR(MFA)
       ATRIB(1)=NEXT
       CALL SCHDL(7,UNFRM(10.,20.,1),ATRIB)
C****  PLACE ENTITY IN FILE 3
       ATRIB(1)=127.
       ATRIB(2)=TNOW
C****  SET ATRIB(3) TO LOCATION OF THE CORRESPONDING EVENT
       ATRIB(3)=NOW
       CALL FILEM(3,ATRIB)
```

D12-11.1

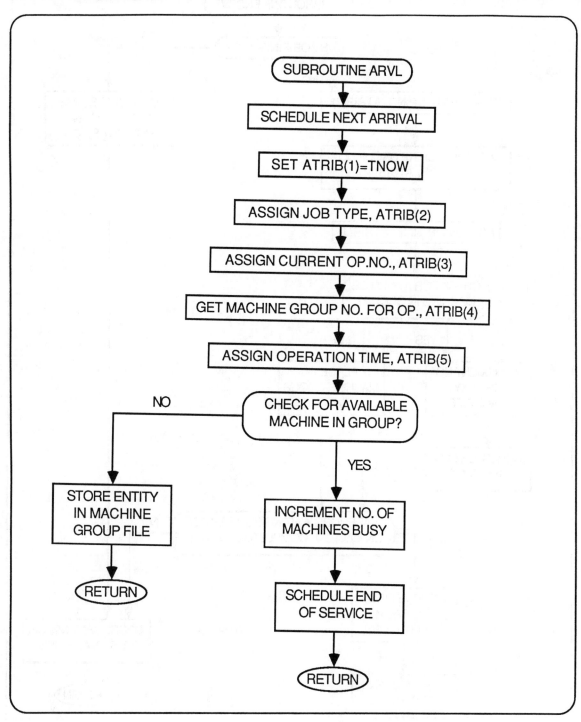

D12-12.1

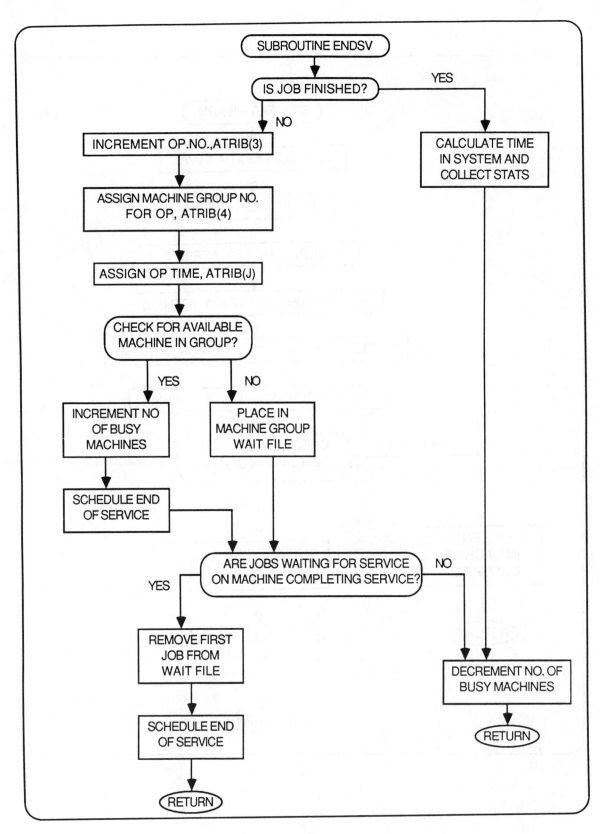

D12-12.2

```
      SUBROUTINE INTLC
      COMMON/SCOM1/ATRIB(100),DD(100),DDL(100),DTNOW,II,MFA,MSTOP,NCLNR
     1,NCRDR,NPRNT,NNRUN,NNSET,NTAPE,SS(100),SSL(100),TNEXT,TNOW,XX(100)
      COMMON/UCOM1/ROUT(12,2),NUMM(6),XVAL(3),FX(3)
      DO 10 I=1,6
      XX(I)=0
10    CONTINUE
      NUMM(1)=14
      NUMM(2)=5
      NUMM(3)=4
      NUMM(4)=8
      NUMM(5)=16
      NUMM(6)=4
      XVAL(1)=1
      XVAL(2)=2
      XVAL(3)=3
      FX(1)=.24
      FX(2)=.68
      FX(3)=1.0
      ROUT(1,1)=1
      ROUT(1,2)=125
      ROUT(2,1)=3
      ROUT(2,2)=35
      ROUT(3,1)=2
      ROUT(3,2)=20
      ROUT(4,1)=6
      ROUT(4,2)=60
      ROUT(5,1)=5
      ROUT(5,2)=105
      ROUT(6,1)=4
      ROUT(6,2)=90
      ROUT(7,1)=2
      ROUT(7,2)=65
      ROUT(8,1)=1
      ROUT(8,2)=235
      ROUT(9,1)=5
      ROUT(9,2)=250
      ROUT(10,1)=4
      ROUT(10,2)=50
      ROUT(11,1)=3
      ROUT(11,2)=30
      ROUT(12,1)=6
      ROUT(12,2)=25
      CALL SCHDL(1,0.0,ATRIB)
      RETURN
      END
```

D12-12.3

```
      SUBROUTINE ARVL
      COMMON/SCOM1/ATRIB(100),DD(100),DDL(100),DTNOW,II,MFA,MSTOP,NCLNR
     1,NCRDR,NPRNT,NNRUN,NNSET,NTAPE,SS(100),SSL(100),TNEXT,TNOW,XX(100)
      COMMON/UCOM1/ROUT(12,2),NUMM(6),XVAL(3),FX(3)
C     SCHEDULE NEXT ARRIVAL
      CALL SCHDL(1,EXPON(9.6,1),ATRIB)
C     ATRIB(1) IS ARRIVAL TIME
C     ATRIB(2) IS JOB TYPE
C     ATRIB(3) IS OPERATION NUMBER
C     ATRIB(4) IS MACHINE NUMBER
C     ATRIB(5) IS OPERATION TIME
      ATRIB(1)=TNOW
      ATRIB(2)=DPROB(FX,XVAL,3,2)
      ATRIB(3)=1
      NTYP=ATRIB(2)
      IF(NTYP.EQ.1) NSTR=1
      IF(NTYP.EQ.2) NSTR=5
      IF(NTYP.EQ.3) NSTR=8
      ATRIB(4)=ROUT(NSTR,1)
      XMN=ROUT(NSTR,2)
      ATRIB(5)=EXPON(XMN,3)
      I=ATRIB(4)
      NBUSY=XX(I)
C     CHECK FOR AVAILABLE MACHINE
      IF(NBUSY.LT.NUMM(I)) GOTO 10
C     NO, PLACE IN QUEUE OF MACHINE GROUP
      CALL FILEM(I,ATRIB)
      RETURN
C     YES, INCREMENT NO. OF BUSY MACHINES
10    XX(I)=XX(I)+1.0
C     SCHEDULE END OF SERVICE
      CALL SCHDL(2,ATRIB(5),ATRIB)
      RETURN
      END
```

D12-12.4

```
        SUBROUTINE ENDSV
        COMMON/SCOM1/ATRIB(100),DD(100),DDL(100),DTNOW,II,MFA,MSTOP,NCLNR
       1,NCRDR,NPRNT,NNRUN,NNSET,NTAPE,SS(100),SSL(100),TNEXT,TNOW,XX(100)
        COMMON/UCOM1/ROUT(12,2),NUMM(6),XVAL(3),FX(3)
        NTYP=ATRIB(2)
        IF(NTYP.EQ.1) NMAX=4
        IF(NTYP.EQ.2) NMAX=3
        IF(NTYP.EQ.3) NMAX=5
        NOP=ATRIB(3)
        NMACH=ATRIB(4)
C       CHECK TO SEE IF CURRENT OPERATION IS THE LAST
        IF(NOP.LT.NMAX) GOTO 10
C       YES, COLLECT STATS ON TIME IN SYSTEM
        TSYS=TNOW-ATRIB(1)
        CALL COLCT(TSYS,NTYP)
        GOTO 20
10      IF(NTYP.EQ.1) NSTR=1
        IF(NTYP.EQ.2) NSTR=5
        IF(NTYP.EQ.3) NSTR=8
        NROW=NSTR+NOP
C       NO, INCREMENT OPERATION NO., GET NEXT MACHINE NO. AND TIME
        ATRIB(3)=ATRIB(3)+1.0
        ATRIB(4)=ROUT(NROW,1)
        XMN=ROUT(NROW,2)
        ATRIB(5)=EXPON(XMN,3)
        I=ATRIB(4)
        NBUSY=XX(I)
C       CHECK TO SEE IF NEXT MACHINE TYPE AVAILABLE
        IF(NBUSY.LT.NUMM(I)) GOTO 30
C       NO, PLACE JOB IN QUEUE BEFORE MACHINE GROUP
        CALL FILEM(I,ATRIB)
        GOTO 20
C       YES, INCREMENT NO. OF BUSY MACHINES
C       SCHEDULE END OF SERVICE
30      XX(I)=XX(I)+1.0
        CALL SCHDL(2,ATRIB(5),ATRIB)
C       ARE JOBS WAITING IN QUEUE OF MACHINE COMPLETING SERVICE?
20      IF(NNQ(NMACH).EQ.0) GOTO 40
C       YES, REMOVE FIRST ONE AND SCHEDULE END OF SERVICE
        CALL RMOVE(1,NMACH,ATRIB)
        CALL SCHDL(2,ATRIB(5),ATRIB)
        RETURN
C       NO, DECREMENT NO. OF BUSY MACHINES
40      XX(NMACH)=XX(NMACH)-1.
        RETURN
        END
```

D12-12.5

```
    1   GEN,ROLSTON,PROBLEM 12.12,8/1/80,1;
    2   LIMITS,6,5,250;
    3   STAT,1,TIME 1;
    4   STAT,2,TIME 2;
    5   STAT,3,TIME 3;
    6   TIMST,XX(1),CAST;
    7   TIMST,XX(2),LATHES;
    8   TIMST,XX(3),PLANERS;
    9   TIMST,XX(4),DRILL PRESSES;
   10   TIMST,XX(5),SHAPERS;
   11   TIMST,XX(6),POLISHERS;
   12   INIT,0,12000;
   13   MONTR,SUMRY,2400,2400;
   14   FIN;
```

D12-12.6

```
                    S L A M   I I   S U M M A R Y   R E P O R T

         SIMULATION PROJECT PROBLEM 12.12          BY ROLSTON

         DATE  8/ 1/1980                           RUN NUMBER    1 OF    1


         CURRENT TIME   0.2400E+04
         STATISTICAL ARRAYS CLEARED AT TIME  0.0000E+00
```

STATISTICS FOR VARIABLES BASED ON OBSERVATION

	MEAN VALUE	STANDARD DEVIATION	COEFF. OF VARIATION	MINIMUM VALUE	MAXIMUM VALUE	NUMBER OF OBSERVATIONS
TIME 1	0.3119E+03	0.1349E+03	0.4324E+00	0.7811E+02	0.6343E+03	66
TIME 2	0.3314E+03	0.1668E+03	0.5034E+00	0.6865E+02	0.8112E+03	97
TIME 3	0.5902E+03	0.3387E+03	0.5739E+00	0.1424E+03	0.1728E+04	66

STATISTICS FOR TIME-PERSISTENT VARIABLES

	MEAN VALUE	STANDARD DEVIATION	MINIMUM VALUE	MAXIMUM VALUE	TIME INTERVAL	CURRENT VALUE
CAST	0.1045E+02	0.4026E+01	0.0000E+00	0.1400E+02	0.2400E+04	0.1000E+02
LATHES	0.3745E+01	0.1670E+01	0.0000E+00	0.5000E+01	0.2400E+04	0.5000E+01
PLANERS	0.1648E+01	0.1312E+01	0.0000E+00	0.4000E+01	0.2400E+04	0.0000E+00
DRILL PRESSES	0.5191E+02	0.2230E+01	0.0000E+00	0.8000E+01	0.2400E+04	0.8000E+01
SHAPERS	0.1369E+02	0.3376E+01	0.0000E+00	0.1600E+02	0.2400E+04	0.1600E+02
POLISHERS	0.2320E+01	0.1316E+01	0.0000E+00	0.4000E+01	0.2400E+04	0.3000E+01

FILE STATISTICS

FILE NUMBER	LABEL/TYPE	AVERAGE LENGTH	STANDARD DEVIATION	MAXIMUM LENGTH	CURRENT LENGTH	AVERAGE WAITING TIME
1		2.6381	4.0384	13	0	89.1765
2		2.5693	3.2408	13	2	59.2916
3		0.1460	0.5630	5	0	14.5990
4		0.3483	0.8900	6	0	19.9018
5		1.0247	2.3079	11	0	33.6883
6		0.3339	0.8698	6	0	18.6355
7	CALENDAR	39.0453	9.5166	49	44	75.0271

D12-12.7

```
                    S L A M   I I   S U M M A R Y   R E P O R T

           SIMULATION PROJECT PROBLEM 12.12          BY ROLSTON

           DATE  8/ 1/1980                           RUN NUMBER   1 OF   1


           CURRENT TIME   0.4800E+04
           STATISTICAL ARRAYS CLEARED AT TIME  0.0000E+00


               **STATISTICS FOR VARIABLES BASED ON OBSERVATION**

                    MEAN      STANDARD    COEFF. OF    MINIMUM    MAXIMUM    NUMBER OF
                    VALUE     DEVIATION   VARIATION    VALUE      VALUE      OBSERVATIONS

   TIME 1      0.2803E+03   0.1480E+03   0.5281E+00   0.6889E+02   0.8015E+03      118
   TIME 2      0.3277E+03   0.1649E+03   0.5033E+00   0.5598E+02   0.8790E+03      217
   TIME 3      0.5997E+03   0.3172E+03   0.5290E+00   0.1088E+03   0.1730E+04      148


               **STATISTICS FOR TIME-PERSISTENT VARIABLES**

                    MEAN      STANDARD    MINIMUM      MAXIMUM      TIME        CURRENT
                    VALUE     DEVIATION   VALUE        VALUE        INTERVAL    VALUE

   CAST           0.1091E+02   0.3505E+01   0.0000E+00   0.1400E+02   0.4800E+04   0.1400E+02
   LATHES         0.3826E+01   0.1511E+01   0.0000E+00   0.5000E+01   0.4800E+04   0.4000E+01
   PLANERS        0.1626E+01   0.1249E+01   0.0000E+00   0.4000E+01   0.4800E+04   0.4000E+01
   DRILL PRESSES  0.5915E+01   0.2046E+01   0.0000E+00   0.8000E+01   0.4800E+04   0.7000E+01
   SHAPERS        0.1343E+02   0.3220E+01   0.0000E+00   0.1600E+02   0.4800E+04   0.1200E+02
   POLISHERS      0.2179E+01   0.1269E+01   0.0000E+00   0.4000E+01   0.4800E+04   0.4000E+01


               **FILE STATISTICS**

   FILE              AVERAGE     STANDARD    MAXIMUM    CURRENT    AVERAGE
   NUMBER  LABEL/TYPE  LENGTH    DEVIATION   LENGTH     LENGTH     WAITING TIME

   1                   1.6092     3.1969       13         8        65.4594
   2                   1.6353     2.6122       13         0        43.3679
   3                   0.0821     0.4142        5         0        12.7179
   4                   0.5352     1.0883        6         0        19.9129
   5                   1.0196     2.0232       11         0        30.9756
   6                   0.1837     0.6470        6         0        16.3260
   7      CALENDAR    39.8904     7.1113       50        47        76.8663
```

D12-12.8

```
                    S L A M   I I   S U M M A R Y   R E P O R T

         SIMULATION PROJECT PROBLEM 12.12          BY ROLSTON

         DATE  8/ 1/1980                           RUN NUMBER   1 OF   1


         CURRENT TIME   0.7200E+04
         STATISTICAL ARRAYS CLEARED AT TIME  0.0000E+00


              **STATISTICS FOR VARIABLES BASED ON OBSERVATION**

              MEAN        STANDARD      COEFF. OF     MINIMUM      MAXIMUM      NUMBER OF
              VALUE       DEVIATION     VARIATION     VALUE        VALUE        OBSERVATIONS

   TIME 1     0.2896E+03  0.1489E+03    0.5142E+00    0.6889E+02   0.8015E+03   191
   TIME 2     0.3242E+03  0.1587E+03    0.4894E+00    0.5598E+02   0.8790E+03   344
   TIME 3     0.6478E+03  0.3341E+03    0.5157E+00    0.1088E+03   0.1791E+04   225


              **STATISTICS FOR TIME-PERSISTENT VARIABLES**

              MEAN        STANDARD      MINIMUM       MAXIMUM      TIME         CURRENT
              VALUE       DEVIATION     VALUE         VALUE        INTERVAL     VALUE

   CAST           0.1157E+02  0.3156E+01  0.0000E+00  0.1400E+02  0.7200E+04  0.1400E+02
   LATHES         0.3792E+01  0.1501E+01  0.0000E+00  0.5000E+01  0.7200E+04  0.5000E+01
   PLANERS        0.1711E+01  0.1243E+01  0.0000E+00  0.4000E+01  0.7200E+04  0.0000E+00
   DRILL PRESSES  0.6017E+01  0.1961E+01  0.0000E+00  0.8000E+01  0.7200E+04  0.1000E+01
   SHAPERS        0.1361E+02  0.3122E+01  0.0000E+00  0.1600E+02  0.7200E+04  0.1000E+02
   POLISHERS      0.2332E+01  0.1251E+01  0.0000E+00  0.4000E+01  0.7200E+04  0.4000E+01


              **FILE STATISTICS**

   FILE                  AVERAGE     STANDARD    MAXIMUM    CURRENT    AVERAGE
   NUMBER  LABEL/TYPE    LENGTH      DEVIATION   LENGTH     LENGTH     WAITING TIME

     1                   1.7429      2.9901        13         6        62.4322
     2                   1.5893      2.6589        13         2        42.0708
     3                   0.0803      0.3928         5         0        10.9096
     4                   0.6118      1.3583         8         0        23.0623
     5                   1.6127      2.7299        14         0        41.7685
     6                   0.1946      0.6644         6         2        15.2296
     7     CALENDAR     41.0382      6.4379        53        36        77.6748
```

D12-12.9

```
                         S L A M   I I   S U M M A R Y   R E P O R T

          SIMULATION PROJECT PROBLEM 12.12          BY ROLSTON

          DATE  8/ 1/1980                           RUN NUMBER   1 OF   1


          CURRENT TIME   0.9600E+04
          STATISTICAL ARRAYS CLEARED AT TIME  0.0000E+00


            **STATISTICS FOR VARIABLES BASED ON OBSERVATION**

                  MEAN        STANDARD     COEFF. OF    MINIMUM      MAXIMUM      NUMBER OF
                  VALUE       DEVIATION    VARIATION    VALUE        VALUE        OBSERVATIONS

   TIME 1         0.2947E+03  0.1464E+03   0.4966E+00   0.4652E+02   0.8015E+03   261
   TIME 2         0.3301E+03  0.1578E+03   0.4781E+00   0.5598E+02   0.8906E+03   460
   TIME 3         0.6553E+03  0.3178E+03   0.4850E+00   0.1088E+03   0.1791E+04   319


            **STATISTICS FOR TIME-PERSISTENT VARIABLES**

                  MEAN        STANDARD     MINIMUM      MAXIMUM      TIME         CURRENT
                  VALUE       DEVIATION    VALUE        VALUE        INTERVAL     VALUE

   CAST           0.1185E+02  0.2949E+01   0.0000E+00   0.1400E+02   0.9600E+04   0.1400E+02
   LATHES         0.3916E+01  0.1449E+01   0.0000E+00   0.5000E+01   0.9600E+04   0.2000E+01
   PLANERS        0.1802E+01  0.1295E+01   0.0000E+00   0.4000E+01   0.9600E+04   0.1000E+01
   DRILL PRESSES  0.6155E+01  0.1963E+01   0.0000E+00   0.8000E+01   0.9600E+04   0.4000E+01
   SHAPERS        0.1401E+02  0.2914E+01   0.0000E+00   0.1600E+02   0.9600E+04   0.1600E+02
   POLISHERS      0.2428E+01  0.1236E+01   0.0000E+00   0.4000E+01   0.9600E+04   0.1000E+01


            **FILE STATISTICS**

   FILE                    AVERAGE      STANDARD     MAXIMUM   CURRENT   AVERAGE
   NUMBER  LABEL/TYPE      LENGTH       DEVIATION    LENGTH    LENGTH    WAITING TIME

     1                     1.9168       2.9971       13        0         62.1649
     2                     1.5234       2.4512       13        0         37.8872
     3                     0.1510       0.6382        7        0         14.4939
     4                     1.1452       2.3661       13        0         37.3930
     5                     2.0151       2.8930       14        8         43.8659
     6                     0.2931       0.9030        7        0         19.1410
     7       CALENDAR     42.1680       6.1663       53       40         78.1190
```

D12-12.10

```
                    S L A M   I I   S U M M A R Y   R E P O R T

         SIMULATION PROJECT PROBLEM 12.12            BY ROLSTON

         DATE  8/ 1/1980                     RUN NUMBER    1 OF    1


         CURRENT TIME   0.1200E+05
         STATISTICAL ARRAYS CLEARED AT TIME  0.0000E+00


                 **STATISTICS FOR VARIABLES BASED ON OBSERVATION**

                   MEAN       STANDARD     COEFF. OF    MINIMUM      MAXIMUM      NUMBER OF
                   VALUE      DEVIATION    VARIATION    VALUE        VALUE        OBSERVATIONS

        TIME 1    0.2855E+03  0.1449E+03   0.5077E+00   0.2584E+02   0.8015E+03      330
        TIME 2    0.3214E+03  0.1576E+03   0.4903E+00   0.2660E+02   0.8906E+03      571
        TIME 3    0.6503E+03  0.3330E+03   0.5121E+00   0.1088E+03   0.1977E+04      396


                     **STATISTICS FOR TIME-PERSISTENT VARIABLES**

                   MEAN       STANDARD     MINIMUM      MAXIMUM      TIME         CURRENT
                   VALUE      DEVIATION    VALUE        VALUE        INTERVAL     VALUE

        CAST          0.1182E+02  0.2784E+01   0.0000E+00   0.1400E+02   0.1200E+05   0.1200E+02
        LATHES        0.3789E+01  0.1478E+01   0.0000E+00   0.5000E+01   0.1200E+05   0.2000E+01
        PLANERS       0.1803E+01  0.1277E+01   0.0000E+00   0.4000E+01   0.1200E+05   0.4000E+01
        DRILL PRESSES 0.6119E+01  0.1969E+01   0.0000E+00   0.8000E+01   0.1200E+05   0.2000E+01
        SHAPERS       0.1379E+02  0.2977E+01   0.0000E+00   0.1600E+02   0.1200E+05   0.1500E+02
        POLISHERS     0.2435E+01  0.1252E+01   0.0000E+00   0.4000E+01   0.1200E+05   0.4000E+01


                        **FILE STATISTICS**

        FILE                 AVERAGE     STANDARD    MAXIMUM   CURRENT    AVERAGE
        NUMBER  LABEL/TYPE   LENGTH      DEVIATION   LENGTH    LENGTH     WAITING TIME

          1                   1.6103      2.7928        13        0        59.0938
          2                   1.2958      2.2810        13        0        36.0782
          3                   0.1348      0.5917         7        3        13.8246
          4                   1.1177      2.2709        13        0        36.1518
          5                   1.8324      2.7401        14        0        43.0303
          6                   0.2903      0.8702         7        0        18.6264
          7      CALENDAR    41.7583      5.9562        53       41        77.7140
```

D12-12.11

12-12,Embellishment(a). Change the Priority statement for each file to lowest value first using attribute 5, LVF(5).

12-12,Embellishment(b). Use the PRIORITY statement for each file which specifies a highest value first ranking rule based on attribute 2, HVF(2).

12-12,Embellishment(c). Change the mean arrival rate as specified in subroutine ARVL from 9.6 to 9 in the call to subroutine SCHDL.

12-12,Embellishment(d). The cost structure should include the cost of in-process inventory, the cost of new machines and labor, and the profitability associated with increased throughput through the machine shop.

12-13. To change job sequences and processing times, the array ROUT would need to be changed. As long as the number of operations for each job type remains the same, ROUT is a convenient form for storing the route information. If a variation in the number of operations is to be investigated, then a good data structure would be to store route information by job type in a two dimensional array with the job type being the row number and the sequence of machine groups stored as the columns of the array. (A singly dimensioned array with pointers could also be used for storing this information. If the number of job types is large, this would save on storage space.) A second two dimensional array organized in the same manner could then be used to store the operation times.

12-14. There have been many articles which describe generalized job shop simulations. It is recommended that one of these articles be assigned and the exercise be directed toward producing a SLAM II program which corresponds to the program used in the article.

12-15. A system flow diagram showing a sketch with system parameters is given in D12-15.1. There are five discrete events included in the model which are listed below:

1) Customer arrival to the system, ARRIVE;

2) Placement of request for information, REQUEST;

3) Arrival of scanner at a terminal point, SCAN;

4) Arrival of an answer at a station, ANSWER;

5) End of customer service, LEAVE.

The definitions of the non-SLAM II variables and subroutine INTLC are given in D12-15.2. Subroutine EVENT is shown in D12-15.3. The event subroutines, ARRIVE, REQUEST, SCAN, ANSWER and LEAVE are shown in D12- 15.4 through D12-15.8.

A special subroutine output reporting on system statistics can be generated by coding OTPUT which is shown in D12-15.9. The SLAM II input statements are shown in D12-15.10. The output from subroutine OTPUT is shown in D12- 15.11 and the SLAM II Summary Report in D12-15.12.

12-15,Embellishment(a). Buffer storage appears to be one of the critical resources in this exercise.

12-15,Embellishment(b). The concept of reneging can be applied directly in this exercise. A RENEGE event is added to the model as event 6. Subroutine ARRIVE to include the scheduling of RENEGE event upon customer arrival is shown in D12-15.13. Subroutine RENEGE is shown in D12-15.14.

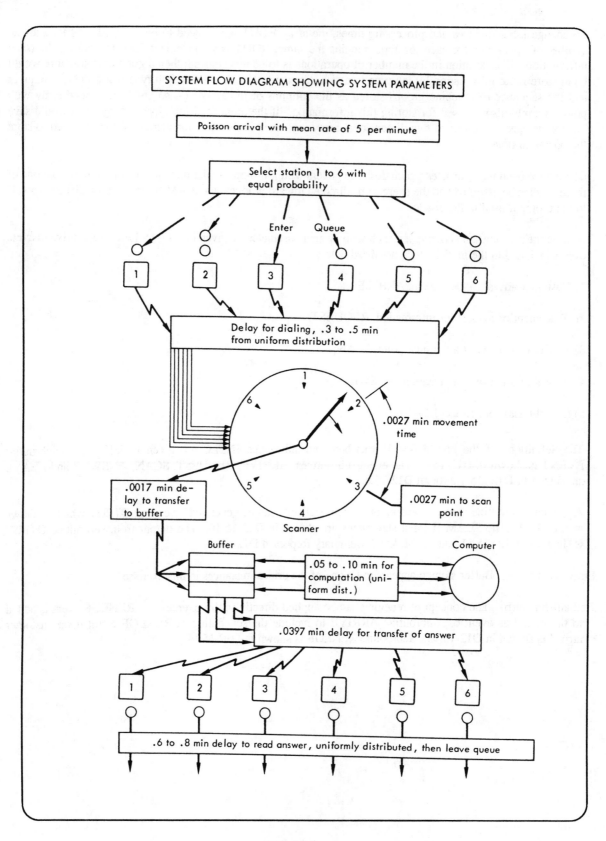

D12.15.1

```
***********************************************************************
*                                                                     *
*        NARC = NUMBER OF ARRIVING CUSTOMERS                          *
*        NBUFF = BUFFER UNIT FILE                                     *
*        NSTA(I) = NUMBER OF PEOPLE WAITING AT STATION I              *
*        JBUFF = 1 - BUFFER IS FULL AND THE SCANNER IS STOPPED        *
*                0 - OTHERWISE                                        *
*        LTSTATUS(I) = STATUS OF LINE I                               *
*                    1 - LINE IS OPEN AND READY                       *
*                    2 - CALL HAS BEEN PLACED                         *
*                    3 - CALL HAS BEEN TRANSFERRED TO THE BUFFER AND   *
*                        THE COMPUTER IS WORKING ON THE REQUEST        *
*                    4 - ANSWER HAS BEEN SENT AND IS BEING READ        *
*                                                                     *
***********************************************************************
       SUBROUTINE INTLC
       COMMON/SCOM1/ATRIB(100),DD(100),DDL(100),DTNOW,II,MFA,MSTOP,NCLNR
      1,NCRDR,NPRNT,NNRUN,NNSET,NTAPE,SS(100),SSL(100),TNEXT,TNOW,XX(100)
       COMMON/UCOM1/NSCAN,LTSTATUS(8),JBUFF,NARC,TRESPON,NBUFF
       REAL NSTA(6)
       EQUIVALENCE(XX(1),NSTA(1))
       NARC=0
       NSCAN=1
       JBUFF=0
       NBUFF=7
       DO 10 J=1,6
         NSTA(J)=0.0
   10    LTSTATUS(J)=1
       ATRIB(1)=TNOW
       CALL SCHDL(1,0.0,ATRIB)
       CALL SCHDL(3,0.0054,ATRIB)
       RETURN
       END
```

D12-15.2

```
       SUBROUTINE EVENT(I)
       GO TO (1,2,3,4,5),I
    1  CALL ARRIVE
       RETURN
    2  CALL REQUEST
       RETURN
    3  CALL SCAN
       RETURN
    4  CALL ANSWER
       RETURN
    5  CALL LEAVE
       RETURN
       END
```

D12-15.3

```
      SUBROUTINE ARRIVE
      COMMON/SCOM1/ATRIB(100),DD(100),DDL(100),DTNOW,II,MFA,MSTOP,NCLNR
     1,NCRDR,NPRNT,NNRUN,NNSET,NTAPE,SS(100),SSL(100),TNEXT,TNOW,XX(100)
      COMMON/UCOM1/NSCAN,LTSTATUS(8),JBUFF,NARC,TRESPON,NBUFF
      REAL NSTA(6)
      EQUIVALENCE (XX(1),NSTA(1))
      NARC=NARC+1
      J=UNFRM(1.0,7.0,1)
      IF (NSTA(J).NE.0) GO TO 10
      ATRIB(1)=TNOW
      ATRIB(2)=J
      CALL SCHDL(2,UNFRM(0.3,0.5,1),ATRIB)
   10 NSTA(J)=NSTA(J)+1
      CALL SCHDL(1,EXPON(0.2,1),ATRIB)
      RETURN
      END
```

D12-15.4

```
      SUBROUTINE REQUEST
      COMMON/SCOM1/ATRIB(100),DD(100),DDL(100),DTNOW,II,MFA,MSTOP,NCLNR
     1,NCRDR,NPRNT,NNRUN,NNSET,NTAPE,SS(100),SSL(100),TNEXT,TNOW,XX(100)
      COMMON/UCOM1/NSCAN,LTSTATUS(8),JBUFF,NARC,TRESPON,NBUFF
      J=ATRIB(2)
      CALL FILEM(J,ATRIB)
      LTSTATUS(J)=2
      RETURN
      END
```

D12-15.5

```
      SUBROUTINE SCAN
      COMMON/SCOM1/ATRIB(100),DD(100),DDL(100),DTNOW,II,MFA,MSTOP,NCLNR
     1,NCRDR,NPRNT,NNRUN,NNSET,NTAPE,SS(100),SSL(100),TNEXT,TNOW,XX(100)
      COMMON/UCOM1/NSCAN,LTSTATUS(8),JBUFF,NARC,TRESPON,NBUFF
      K=LTSTATUS(NSCAN)
      GO TO (1,2,1,1),K
    1 STIME=0.0054
   10 CALL SCHDL(3,STIME,ATRIB)
      NSCAN=NSCAN+1
      IF (NSCAN.GT.6) NSCAN=1
      RETURN
    2 IF (NNQ(NBUFF).LT.3) GO TO 20
      JBUFF=1
      RETURN
   20 CALL RMOVE(1,NSCAN,ATRIB)
      CALL FILEM(NBUFF,ATRIB)
      LTSTATUS(NSCAN)=3
      CALL SCHDL(4,0.0514+UNFRM(0.05,0.1,1),ATRIB)
      STIME=0.0171
      GO TO 10
      END
```

D12-15.6

```
SUBROUTINE ANSWER
COMMON/SCOM1/ATRIB(100),DD(100),DDL(100),DTNOW,II,MFA,MSTOP,NCLNR
1,NCRDR,NPRNT,NNRUN,NNSET,NTAPE,SS(100),SSL(100),TNEXT,TNOW,XX(100)
COMMON/UCOM1/NSCAN,LTSTATUS(8),JBUFF,NARC,TRESPON,NBUFF
NRANK=NFIND(1,NBUFF,2,0,ATRIB(2),0.001)
CALL RMOVE(NRANK,NBUFF,ATRIB)
TRESPON=TNOW-ATRIB(1)
CALL COLCT(TRESPON,1)
J=ATRIB(2)
LTSTATUS(J)=4
CALL SCHDL(5,UNFRM(0.6,0.8,1),ATRIB)
IF (JBUFF.EQ.0) RETURN
JBUFF=0
CALL SCAN
RETURN
END
```

D12-15.7

```
SUBROUTINE LEAVE
COMMON/SCOM1/ATRIB(100),DD(100),DDL(100),DTNOW,II,MFA,MSTOP,NCLNR
1,NCRDR,NPRNT,NNRUN,NNSET,NTAPE,SS(100),SSL(100),TNEXT,TNOW,XX(100)
COMMON/UCOM1/NSCAN,LTSTATUS(8),JBUFF,NARC,TRESPON,NBUFF
REAL NSTA(6)
EQUIVALENCE (XX(1),NSTA(1))
J=ATRIB(2)
NSTA(J)=NSTA(J)-1
LTSTATUS(J)=1
IF (NSTA(J).GT.0) THEN
ATRIB(1)=TNOW
CALL SCHDL(2,UNFRM(0.3,0.5,1),ATRIB)
ENDIF
RETURN
END
```

D12-15.8

```
      SUBROUTINE OTPUT
      COMMON/SCOM1/ATRIB(100),DD(100),DDL(100),DTNOW,II,MFA,MSTOP,NCLNR
     1,NCRDR,NPRNT,NNRUN,NNSET,NTAPE,SS(100),SSL(100),TNEXT,TNOW,XX(100)
      COMMON/UCOM1/NSCAN,LTSTATUS(8),JBUFF,NARC,TRESPON,NBUFF
      WRITE(6,3)
    3 FORMAT(//)
      WRITE(6,4)
    4 FORMAT(17X,13(2H *),' INFORMATION SYSTEM SIMULATION ',13(2H* ))
      WRITE(6,5)
    5 FORMAT(1X)
      WRITE(6,9)
    9 FORMAT(17X,'*',81X,'*')
      WRITE(6,8)
    8 FORMAT(40X,'OUTPUT REPORT',20X,'BY VASEK')
      WRITE(6,9)
      WRITE(6,30),NARC
      WRITE(6,9)
      WRITE(6,20)
      WRITE(6,9)
      WRITE(6,25)
      WRITE(6,80)
      WRITE(6,5)
      DO 100 I=1,6
         WRITE(6,50),I,TTAVG(I)
         WRITE(6,5)
  100 CONTINUE
   50 FORMAT(17X,'*',29X,I1,18X,F4.1,29X,'*')
      WRITE(6,9)
      WRITE(6,10),CCAVG(1)
      WRITE(6,9)
      WRITE(6,5)
      WRITE(6,60)
   10 FORMAT(30X,'THE AVERAGE RESPONSE TIME WAS   ',F5.1,' MINUTES.')
   20 FORMAT(30X,'THE AVERAGE NUMBER OF CUSTOMERS WAITING IN EACH
     ;STATION:')
   25 FORMAT(42X,'STATION #',10X,'NUMBER OF CUSTOMERS')
   30 FORMAT(30X,'THE NUMBER OF CUSTOMERS THAT ARRIVED TO THE
     ;SYSTEM WAS ',I5)
   60 FORMAT(17X,41(2H *))
   80 FORMAT(17X,'*',24X,9('-'),10X,19('-'),19X,'*')
      RETURN
      END
```

D12-15.9

```
    1   GEN,VASEK,PROBLEM 12.15,7/3/86,1;
    2   LIMITS,7,2,400;
    3   TIMST,XX(1),AVG LNGTH LINE 1;
    4   TIMST,XX(2),AVG LNGTH LINE 2;
    5   TIMST,XX(3),AVG LNGTH LINE 3;
    6   TIMST,XX(4),AVG LNGTH LINE 4;
    7   TIMST,XX(5),AVG LNGTH LINE 5;
    8   TIMST,XX(6),AVG LNGTH LINE 6;
    9   STAT,1,RESPONSE TIME;
   10   INIT,0,480;
   11   FIN;
```

D12-15.10

```
                        **INTERMEDIATE RESULTS**



* * * * * * * * * * * INFORMATION SYSTEM SIMULATION * * * * * * * * * * * *
*                                                                          *
*              OUTPUT REPORT                    BY VASEK                    *
*                                                                          *
*     THE NUMBER OF CUSTOMERS THAT ARRIVED TO THE SYSTEM WAS   2495         *
*                                                                          *
*     THE AVERAGE NUMBER OF CUSTOMERS WAITING IN EACH STATION:             *
*                                                                          *
*                STATION #        NUMBER OF CUSTOMERS                       *
*                _____        _____                       *
*                                                                          *
*                    1                  12.7                               *
*                                                                          *
*                    2                  35.9                               *
*                                                                          *
*                    3                  35.7                               *
*                                                                          *
*                    4                  32.1                               *
*                                                                          *
*                    5                   5.7                               *
*                                                                          *
*                    6                  21.5                               *
*                                                                          *
*        THE AVERAGE RESPONSE TIME WAS    0.5 MINUTES.                      *
*                                                                          *
* * * * * * * * * * * * * * * * * * * * * * * * * * * * * * * * * * * * * * *
```

D12–15.11

```
                          S L A M   I I   S U M M A R Y   R E P O R T


                 SIMULATION PROJECT PROBLEM 12.15          BY VASEK

                 DATE  7/ 3/1986                           RUN NUMBER   1 OF    1


                 CURRENT TIME   0.4800E+03
                 STATISTICAL ARRAYS CLEARED AT TIME  0.0000E+00



                 **STATISTICS FOR VARIABLES BASED ON OBSERVATION**

                   MEAN        STANDARD     COEFF. OF     MINIMUM       MAXIMUM      NUMBER OF
                   VALUE       DEVIATION    VARIATION     VALUE         VALUE        OBSERVATIONS

   RESPONSE TIME   0.5450E+00  0.6025E-01   0.1106E+00    0.4092E+00    0.6863E+00   2278



                 **STATISTICS FOR TIME-PERSISTENT VARIABLES**

                   MEAN        STANDARD     MINIMUM       MAXIMUM       TIME         CURRENT
                   VALUE       DEVIATION    VALUE         VALUE         INTERVAL     VALUE

   AVG LNGTH LINE 1  0.1274E+02  0.5530E+01  0.0000E+00    0.2300E+02    0.4800E+03   0.1600E+02
   AVG LNGTH LINE 2  0.3589E+02  0.1238E+02  0.0000E+00    0.5800E+02    0.4800E+03   0.5700E+02
   AVG LNGTH LINE 3  0.3568E+02  0.1615E+02  0.0000E+00    0.5800E+02    0.4800E+03   0.5200E+02
   AVG LNGTH LINE 4  0.3208E+02  0.1376E+02  0.0000E+00    0.5200E+02    0.4800E+03   0.4600E+02
   AVG LNGTH LINE 5  0.5700E+01  0.4466E+01  0.0000E+00    0.1800E+02    0.4800E+03   0.1000E+02
   AVG LNGTH LINE 6  0.2149E+02  0.8519E+01  0.0000E+00    0.4100E+02    0.4800E+03   0.3800E+02



                 **FILE STATISTICS**

   FILE                   AVERAGE     STANDARD     MAXIMUM    CURRENT    AVERAGE
   NUMBER  LABEL/TYPE     LENGTH      DEVIATION    LENGTH     LENGTH     WAITING TIME

     1                    0.0138      0.1166       1          0          0.0175
     2                    0.0131      0.1138       1          0          0.0164
     3                    0.0144      0.1191       1          0          0.0182
     4                    0.0140      0.1175       1          0          0.0176
     5                    0.0134      0.1151       1          0          0.0177
     6                    0.0140      0.1175       1          0          0.0174
     7                    0.6004      0.7287       3          1          0.1265
     8      CALENDAR      7.8309      0.4377       8          8          0.0403
```

D12-15.12

```
      SUBROUTINE ARRIVE
      COMMON/SCOM1/ATRIB(100),DD(100),DDL(100),DTNOW,II,MFA,MSTOP,NCLNR
     1,NCRDR,NPRNT,NNRUN,NNSET,NTAPE,SS(100),SSL(100),TNEXT,TNOW,XX(100)
      COMMON/UCOM1/NSCAN,LTSTATUS(8),JBUFF,NARC,TRESPON,NBUFF
      REAL NSTA(6)
      EQUIVALENCE (XX(1),NSTA(1))
      NARC=NARC+1
      J=UNFRM(1.0,7.0,1)
      ATRIB(2)=J
      MPREN=MFA
      MPCQ=NSUCR(MFA)
      ATRIB(3)=MPCQ
      CALL SCHDL(6,5.0,ATRIB)
      ATRIB(4)=MPREN
      CALL FILEM(J+7,ATRIB)
      IF (NSTA(J).NE.0) GO TO 10
      CALL RMOVE(1,J+7,ATRIB)
      MPREN=ATRIB(4)
      CALL RMOVE(-MPREN,NCLNR,ATRIB)
      ATRIB(1)=TNOW
      CALL SCHDL(2,UNFRM(0.3,0.5,1),ATRIB)
   10 NSTA(J)=NSTA(J)+1
      CALL SCHDL(1,EXPON(0.2,1),ATRIB)
      RETURN
      END
```

D12-15.13

```
      SUBROUTINE RENEGE
      COMMON/SCOM1/ATRIB(100),DD(100),DDL(100),DTNOW,II,MFA,MSTOP,NCLNR
     1,NCRDR,NPRNT,NNRUN,NNSET,NTAPE,SS(100),SSL(100),TNEXT,TNOW,XX(100)
      REAL NSTA(6)
      EQUIVALENCE (XX(1),NSTA(1))
      J=ATRIB(2)
      MPCQ=ATRIB(3)
      CALL RMOVE(-MPCQ,J+7,ATRIB)
      NSTA(J)=NSTA(J)-1
      RETURN
      END
```

D12-15.14

CHAPTER 13

Continuous Modeling

13-1.   The statements are as given in the text and experimentation should be directed to ascertaining the effects of changing parameter values.

13-1,Embellishment(a).   The variable of XX(1) should be added to the SS(6) in subroutine STATE. The value of XX(1) is changed every 0.025 years by the statements:

```
NETWORK;
        CREATE,0.025, 0.025;
        ASSIGN,XX(1)=RNORM(0.0,3.0,1);
        TERM;
        ENDNETWORK;
```

These statements would be added to the SLAM II input statements. It is not correct to add the RNORM value directly to SS(6) as subroutine STATE is called repeatedly within a step and new samples would be drawn at each call.

13-1,Embellishment(b).   The equation for SS(6) in subroutine STATE should be multiplied by 1.20. The remainder of the problem formulation is the same as given in the text.

13-1,Embellishment(c).   A simple addition in the equation for DD(1) of .2*SS(4) is the appropriate modeling change.

13-1,Embellishment(d).   The solution of this problem involves two time events and a state event. A solution using discrete events and a SEVNT statement is shown in subroutine EVENT in D13-1.1. A list of network statements to accomplish the modeling for the embellishment is given in D13-1.2.

```
 1   GEN,ROLSTON,PROBLEM 13.1D,8/11/80,1;
 2   LIMITS,0,2,10;
 3   CONTINUOUS,5,1,.00025,.025,.025;
 4   INTLC,SS(1)=.83,SS(2)=.003,SS(3)=.0001,SS(4)=0.0,SS(5)=0.0;
 5   RECORD,TNOW,TIME,0,P,.025;
 6   VAR,SS(1),P,PLANTS;
 7   VAR,SS(2),H,HERBIVORES;
 8   VAR,SS(3),C,CARNIVORES;
 9   VAR,SS(4),O,ORGANIC;
10   VAR,SS(5),E,ENVIRONMENT;
11   VAR,SS(6),S,SOLAR ENERGY;
12   SEVENT,3,SS(3),XP,.6,.01;
13   INIT,0,2.0;
14   FIN;

     SUBROUTINE STATE
     COMMON/SCOM1/ATRIB(100),DD(100),DDL(100),DTNOW,II,MFA,MSTOP,NCLNR
    1,NCRDR,NPRNT,NNRUN,NNSET,NTAPE,SS(100),SSL(100),TNEXT,TNOW,XX(100)
     DATA PI/3.14159/
     SS(6)=95.9*(1.+0.635*SIN(2.*PI*TNOW))
     DD(1)=SS(6)-4.03*SS(1)
     DD(2)=.48*SS(1)-17.87*SS(2)
     DD(3)=4.85*SS(2)-4.65*SS(3)
     DD(4)=2.55*SS(1)+6.12*SS(2)+1.95*SS(3)
     DD(5)=SS(1)+6.9*SS(2)+2.7*SS(3)
     RETURN
     END

     SUBROUTINE INTLC
     COMMON/SCOM1/ATRIB(100),DD(100),DDL(100),DTNOW,II,MFA,MSTOP,NCLNR
    1,NCRDR,NPRNT,NNRUN,NNSET,NTAPE,SS(100),SSL(100),TNEXT,TNOW,XX(100)
     CALL SCHDL(1,.1,ATRIB)
     CALL SCHDL(2,.5,ATRIB)
     RETURN
     END

     SUBROUTINE EVENT(I)
     COMMON/SCOM1/ATRIB(100),DD(100),DDL(100),DTNOW,II,MFA,MSTOP,NCLNR
    1,NCRDR,NPRNT,NNRUN,NNSET,NTAPE,SS(100),SSL(100),TNEXT,TNOW,XX(100)
     GOTO (1,2,3),I
 1   CALL SCHDL(1,.1,ATRIB)
     SS(3)=SS(3)+.3
     RETURN
 2   CALL SCHDL(2,.5,ATRIB)
     SS(1)=.3*SS(1)
     RETURN
 3   SS(2)=SS(2)+.2
     RETURN
     END
```

208

```
NETWORK;
        CREATE,.1,.1;
        ASSIGN,SS(3) = SS(3)+0.3;        STOCK CARNIVORES
        TERM;
        CRATE,.5,.5;
        ASSIGN,SS(1)=.3*SS(1);           SPRAY LAKE
        TERM;
        DETECT,SS(3),XP,0.6,0.01;
        ASSIGN,SS(2)=SS(2)+0.2;          REPLENISH HER.
        TERM;
        ENDNETWORK;
```

D13-1.2

13-2.   The integral is the expected value of the maximum of eight uniformly distributed random variables in the interval from 3 to 6. It is related to Exercises 2-11 and 7-14 which is concerned with a rotary index table which indexes only after eight operations are performed with each operation time being uniformly distributed between 3 and 6. To evaluate a definite integral using SLAM II, DD(1) is set equal to the integrand in subroutine STATE. The value of SS(1) will then be the integral of DD(1). The independent variable is set to TNOW and the integration is performed by letting time start at 3 and end at 6. The coding of subroutine STATE is shown in D13-2.1. The input statements are shown in D13-2.2. The summary report shows the value of SS(1) to be 5.667 in D13-2.3. The definite integral can be integrated by parts to yield a value of y of 5 2/3.

```
SUBROUTINE STATE
COMMON/SCOM1/ATRIB(100),DD(100),DDL(100),DTNOW,II,MFA,MSTOP,NCLNR
1,NCRDR,NPRNT,NNRUN,NNSET,NTAPE,SS(100),SSL(100),TNEXT,TNOW,XX(100)
DD(1)=0.0012193*TNOW*(TNOW-3.0)**7
RETURN
END
```

D13-2.1

```
1   GEN,PRITSKER,PROBLEM 13.2,7/1/86,1;
2   CONTINUOUS,1,,,,0.1;
3   INIT,3.0,6.0;
4   FIN;
```

D13-2.2

```
                    S L A M   I I   S U M M A R Y   R E P O R T

        SIMULATION PROJECT PROBLEM 13.2          BY PRITSKER

        DATE  7/ 1/1986                          RUN NUMBER    1 OF    1


        CURRENT TIME    0.6000E+01
        STATISTICAL ARRAYS CLEARED AT TIME   0.3000E+01


        **STATE AND DERIVATIVE VARIABLES**


          (1)        SS(1)          DD(1)
           1       0.5667E+01     0.1600E+02
```

D13-2.3

13-3.   Adding a second pilot to Example 13-1 involves defining four new state variables to represent the second pilot pushing the eject button. When this occurs the equations for DD(5) through DD(8) are included in the model. In this situation it is necessary to schedule a discrete event. A test on TNOW in subroutine STATE should not be made. The reason for this is that the value of TNOW is changed by SLAM II within a step and having an augmented model invoked based on TNOW may result in solving different models within a single step. By including a discrete event at time 1, SLAM II prohibits a step from exceeding the value of 1 until the discrete event has been enacted. In the discrete event, XX(3) is set to 1, and it is this value that should be tested in subroutine STATE. The main program and subroutine STATE are shown in D13-3.1; subroutine EVENT is shown in D13-3.2; subroutine INTLC is shown in D13-3.3, and the input statements are shown in D13-3.4. Note that a LIMITS statement must now be included since we have an entry on the event calendar. A plot of the position of the two pilots is shown in D13-3.5 when the aircraft speed is 500 feet per second.

```
      DIMENSION NSET(1000)
      COMMON/SCOM1/ATRIB(100),DD(100),DDL(100),DTNOW,II,MFA,MSTOP,NCLNR
     1,NCRDR,NPRNT,NNRUN,NNSET,NTAPE,SS(100),SSL(100),TNEXT,TNOW,XX(100)
      COMMON/UCOM1/CD,G,RHO,THED,VA,VE,XM,XS,Y1
      COMMON QSET(1000)
      EQUIVALENCE (NSET(1),QSET(1))
      NNSET=1000
      NCRDR=5
      NPRNT=6
      NTAPE=7
      READ(NCRDR,101) XM,G,CD,XS,Y1,VE,THED,RHO
      CALL SLAM
      STOP
101   FORMAT(7F5.0,E10.4)
      END

      SUBROUTINE STATE
      COMMON/SCOM1/ATRIB(100),DD(100),DDL(100),DTNOW,II,MFA,MSTOP,NCLNR
     1,NCRDR,NPRNT,NNRUN,NNSET,NTAPE,SS(100),SSL(100),TNEXT,TNOW,XX(100)
      COMMON/UCOM1/CD,G,RHO,THED,VA,VE,XM,XS,Y1
      DD(1)=SS(3)*COS(SS(4))-VA
      DD(2)=SS(3)*SIN(SS(4))
      IF(XX(1).LT.1.0) GOTO 101
      XD=.5*RHO*CD*XS*SS(3)*SS(3)
      DD(3)=-XD/XM-G*SIN(SS(4))
      DD(4)=-G*COS(SS(4))/SS(3)
101   IF(XX(3).LT.1.0) GOTO 102
      DD(5)=SS(7)*COS(SS(8))-VA
      DD(6)=SS(7)*SIN(SS(8))
      IF(XX(2).LT.1.0) GOTO 102
      XD=.5*RHO*CD*XS*SS(7)*SS(7)
      DD(7)=-XD/XM-G*SIN(SS(8))
      DD(8)=-G*COS(SS(8))/SS(7)
102   XX(4)=ABS(SS(1)-SS(5))
      XX(5)=ABS(SS(2)-SS(6))
      XX(6)=SQRT(XX(4)**2+XX(5)**2)
      RETURN
      END
```

D13-3.1

```
      SUBROUTINE EVENT(I)
      COMMON/SCOM1/ATRIB(100),DD(100),DDL(100),DTNOW,II,MFA,MSTOP,NCLNR
     1,NCRDR,NPRNT,NNRUN,NNSET,NTAPE,SS(100),SSL(100),TNEXT,TNOW,XX(100)
      GOTO (1,2,3,4),I
1     MSTOP=-1
      RETURN
2     XX(1)=1
      RETURN
3     XX(2)=1
      RETURN
4     XX(3)=1.0
      RETURN
      END
```

D13-3.2

```
      SUBROUTINE INTLC
      COMMON/SCOM1/ATRIB(100),DD(100),DDL(100),DTNOW,II,MFA,MSTOP,NCLNR
     1,NCRDR,NPRNT,NNRUN,NNSET,NTAPE,SS(100),SSL(100),TNEXT,TNOW,XX(100)
      COMMON/UCOM1/CD,G,RHO,THED,VA,VE,XM,XS,Y1
      READ(NCRDR,101) VA
      THE=THED/57.3
      VX=VA-VE*SIN(THE)
      VY=VE*COS(THE)
      SS(1)=0
      SS(2)=0
      SS(3)=SQRT(VX*VX+VY*VY)
      SS(4)=ATAN(VY/VX)
      XX(1)=0
      XX(2)=0.0
      XX(3)=0.0
      SS(5)=-7.0
      SS(6)=0
      SS(7)=SS(3)
      SS(8)=SS(4)
      CALL SCHDL(4,1.0,ATRIB)
      RETURN
101   FORMAT(F10.0)
      END
```

D13-3.3

```
      7. 32.2   1.  10.    4.  40.  15.   .0023769
   GEN,ROLSTON,PROBLEM 13.3,8/21/80,1;
   LIMITS,0,0,4;
   INIT,0,4;
   CONT,8,0,.0001,.01,.01,W,0,.000005
   RECORD,TNOW,TIME,1,B,.02;
   VAR,SS(1),1,X POS 1,-100,30;
   VAR,SS(2),2,Y POS 1,0,20;
   VAR,SS(5),3,X POS 2,-100,30;
   VAR,SS(6),4,Y POS 2,0,20;
   RECORD,TNOW,TIME,2,B,.02;
   VAR,XX(4),X,SEPARX,0,500;
   VAR,XX(5),Y,SEPARY,0,100;
   VAR,XX(6),T,SEPART,0,500;
   SEVNT,1,SS(5),XN,-60.0,0.0;
   SEVNT,2,SS(2),XP,4.0,0.0;
   SEVNT,3,SS(6),XP,4.0,0.0;
   SIMULATE;
    500.
   FIN;
```

D13-3.4

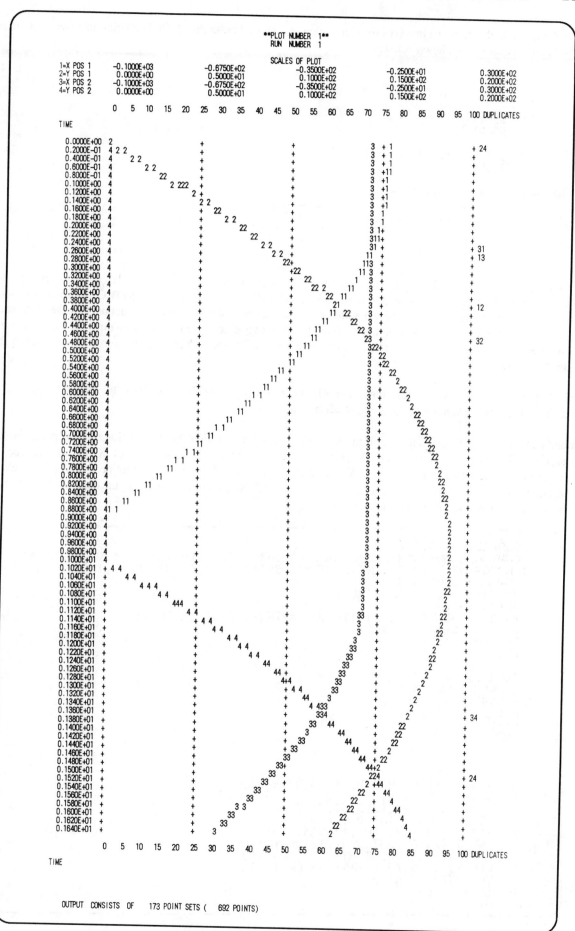

13-4.   The solution to this exercise involves the conversion of a second order differential equation into two first order differential equations as shown below:

$$\ddot{y} + 0.3\dot{y} + y = 1$$

```
Let SS(1)=y
    SS(2)=ẏ
```

```
then DD(2)=1.-0.3*SS(2) - SS(1)
     DD(1)=SS(2)
```

The analytical solution is

$$y(t)=1-e^{-.15t}(\cos At + B \sin At) \; ; \; A = \sqrt{(1-.15^2)} \; \text{and} \; B = \frac{.15}{A}$$

Subroutine STATE and the input statements for this exercise are shown in D13-4.1. It is assumed in the solution that the mass is originally at rest and SS(1) and SS(2) are set to zero in an INTLC statement. Included in the solution is an evaluation of the solution to the second order differential equation and the difference between the values computed using the RKF numerical integration package and evaluating the solution at the same points in time. A table of these values is shown in D13-4.2. Note that the error is well within the specified amount of .001 + .0001*SS(1).

13-4,Embellishment(a).   The coding for subroutine STATE is shown in D13-4.3 without the analytic solution. No additional changes are required for this embellishment.

13-4,Embellishment(b).   The complete coding to model this situation in which the vertical force is a sample from a uniform distribution is shown in D13-4.4. An event is used to obtain a value of XX(1) at five second intervals. In the equation for DD(2), XX(1) is used as the vertical force. In subroutine INTLC, the event is scheduled at time 0 to get a first value for XX(1). A LIMITS statement is included in the data input since there is an event which will be on the event calendar.

```
      SUBROUTINE STATE
      COMMON/SCOM1/ATRIB(100),DD(100),DDL(100),DTNOW,II,MFA,MSTOP,NCLNR
     1,NCRDR,NPRNT,NNRUN,NNSET,NTAPE,SS(100),SSL(100),TNEXT,TNOW,XX(100)
      COMMON/UCOM1/A,B
      DD(2)=1.-0.3*SS(2)-SS(1)
      DD(1)=SS(2)
      SS(3)=1.-EXP(-.15*TNOW)*(COS(A*TNOW)+B*SIN(A*TNOW))
      SS(4)=ABS(SS(3)-SS(1))
      RETURN
      END

      SUBROUTINE INTLC
      COMMON/UCOM1/A,B
      A=SQRT(1.-(.15**2))
      B=.15/A
      RETURN
      END


    1    GEN,ROLSTON,PROBLEM 13.4,8/28/80,1;
    2    CONTINUOUS,2,2,.01,.5,1.0,W,.001,.0001;
    3    INTLC,SS(1)=0,SS(2)=0,SS(3)=0;
    4    RECORD,TNOW,TIME,0,B,2.0;
    5    VAR,SS(1),Y,Y,MIN(.1),MAX(.1);
    6    VAR,SS(2),D,YDOT,MIN(.1),MAX(.1);
    7    VAR,DD(2),2,YDDOT,MIN(.1),MAX(.1);
    8    VAR,SS(3),3,ANAL,MIN(.1),MAX(.1);
    9    VAR,SS(4),0,DIFF;
   10    INIT,0,60;
   11    FIN;
```

TIME	Y	YDOT	YDDOT	ANAL	DIFF
0.0000E+00	0.0000E+00	0.0000E+00	0.1000E+01	0.0000E+00	0.0000E+00
0.0000E+00	0.0000E+00	0.0000E+00	0.1000E+01	0.0000E+00	0.0000E+00
0.1000E+01	0.4177E+00	0.7272E+00	0.3641E+00	0.4177E+00	0.5573E-05
0.2000E+01	0.1190E+01	0.6882E+00	-0.3962E+00	0.1190E+01	0.3147E-04
0.3000E+01	0.1611E+01	0.1126E+00	-0.6448E+00	0.1611E+01	0.5710E-04
0.4000E+01	0.1438E+01	-0.4033E+00	-0.3167E+00	0.1438E+01	0.2551E-04
0.5000E+01	0.9616E+00	-0.4651E+00	0.1780E+00	0.9616E+00	0.4041E-04
0.6000E+01	0.6394E+00	-0.1414E+00	0.4031E+00	0.6394E+00	0.7415E-04
0.7000E+01	0.6872E+00	0.2108E+00	0.2496E+00	0.6872E+00	0.4005E-04
0.8000E+01	0.9711E+00	0.3042E+00	-0.6238E-01	0.9711E+00	0.2998E-04
0.9000E+01	0.1204E+01	0.1318E+00	-0.2440E+00	0.1204E+01	0.7010E-04
0.1000E+02	0.1215E+01	-0.1007E+00	-0.1847E+00	0.1215E+01	0.4578E-04
0.1100E+02	0.1052E+01	-0.1929E+00	0.5965E-02	0.1052E+01	0.1562E-04
0.1200E+02	0.8899E+00	-0.1080E+00	0.1425E+00	0.8900E+00	0.5698E-04
0.1300E+02	0.8574E+00	0.4071E-01	0.1304E+00	0.8574E+00	0.4470E-04
0.1400E+02	0.9466E+00	0.1185E+00	0.1788E-01	0.9466E+00	0.3636E-05
0.1500E+02	0.1055E+01	0.8202E-01	-0.7969E-01	0.1055E+01	0.4184E-04
0.1600E+02	0.1092E+01	-0.1019E-01	-0.8867E-01	0.1092E+01	0.3910E-04
0.1700E+02	0.1046E+01	-0.7041E-01	-0.2487E-01	0.1046E+01	0.4292E-05
0.1800E+02	0.9756E+00	-0.5909E-01	0.4215E-01	0.9756E+00	0.2849E-04
0.1900E+02	0.9428E+00	-0.3758E-02	0.5832E-01	0.9428E+00	0.3189E-04
0.2000E+02	0.9640E+00	0.4022E-01	0.2397E-01	0.9640E+00	0.8285E-05
0.2100E+02	0.1008E+01	0.4085E-01	-0.2052E-01	0.1008E+01	0.1776E-04
0.2200E+02	0.1035E+01	0.8863E-02	-0.3718E-01	0.1034E+01	0.2444E-04
0.2300E+02	0.1027E+01	-0.2188E-01	-0.1998E-01	0.1027E+01	0.9656E-05
0.2400E+02	0.9995E+00	-0.2727E-01	0.8632E-02	0.9996E+00	0.1019E-04
0.2500E+02	0.9799E+00	-0.9602E-02	0.2297E-01	0.9799E+00	0.1782E-04
0.2600E+02	0.9813E+00	0.1112E-01	0.1535E-01	0.9813E+00	0.9358E-05
0.2700E+02	0.9972E+00	0.1763E-01	-0.2495E-02	0.9972E+00	0.5007E-05
0.2800E+02	0.1011E+01	0.8454E-02	-0.1373E-01	0.1011E+01	0.1240E-04
0.2900E+02	0.1013E+01	-0.5063E-02	-0.1115E-01	0.1013E+01	0.8225E-05
0.3000E+02	0.1004E+01	-0.1105E-01	-0.3769E-03	0.1004E+01	0.1907E-05
0.3100E+02	0.9941E+00	-0.6711E-02	0.7902E-02	0.9941E+00	0.8404E-05
0.3200E+02	0.9917E+00	0.1838E-02	0.7758E-02	0.9917E+00	0.6676E-05
0.3300E+02	0.9965E+00	0.6712E-02	0.1488E-02	0.9965E+00	0.5960E-07
0.3400E+02	0.1003E+01	0.4990E-02	-0.4339E-02	0.1003E+01	0.5364E-05
0.3500E+02	0.1005E+01	-0.2497E-03	-0.5209E-02	0.1005E+01	0.5126E-05
0.3600E+02	0.1003E+01	-0.3933E-02	-0.1715E-02	0.1003E+01	0.8345E-06
0.3700E+02	0.9988E+00	-0.3538E-02	0.2236E-02	0.9988E+00	0.3219E-05
0.3800E+02	0.9967E+00	-0.4340E-03	0.3387E-02	0.9967E+00	0.3815E-05
0.3900E+02	0.9978E+00	0.2210E-02	0.1549E-02	0.9978E+00	0.1192E-05
0.4000E+02	0.1000E+01	0.2413E-02	-0.1043E-02	0.1000E+01	0.1907E-05
0.4100E+02	0.1002E+01	0.6465E-03	-0.2135E-02	0.1002E+01	0.2742E-05
0.4200E+02	0.1002E+01	-0.1176E-02	-0.1248E-02	0.1002E+01	0.1192E-05
0.4300E+02	0.1000E+01	-0.1592E-02	0.4009E-03	0.1000E+01	0.9537E-06
0.4400E+02	0.9989E+00	-0.6354E-03	0.1304E-02	0.9989E+00	0.1848E-05
0.4500E+02	0.9989E+00	0.5781E-03	0.9368E-03	0.9989E+00	0.1073E-05
0.4600E+02	0.9998E+00	0.1018E-02	-0.7927E-04	0.9998E+00	0.4172E-06
0.4700E+02	0.1001E+01	0.5350E-03	-0.7691E-03	0.1001E+01	0.1192E-05
0.4800E+02	0.1001E+01	-0.2477E-03	-0.6691E-03	0.1001E+01	0.8345E-06
0.4900E+02	0.1000E+01	-0.6308E-03	-0.6354E-04	0.1000E+01	0.1192E-06
0.5000E+02	0.9997E+00	-0.4134E-03	0.4355E-03	0.9997E+00	0.7749E-06
0.5100E+02	0.9995E+00	0.7599E-04	0.4593E-03	0.9995E+00	0.5960E-06
0.5200E+02	0.9998E+00	0.3782E-03	0.1120E-03	0.9998E+00	0.5960E-07
0.5300E+02	0.1000E+01	0.3017E-03	-0.2342E-03	0.1000E+01	0.4768E-06
0.5400E+02	0.1000E+01	0.5353E-05	-0.3048E-03	0.1000E+01	0.4768E-06
0.5500E+02	0.1000E+01	-0.2185E-03	-0.1149E-03	0.1000E+01	0.2384E-06
0.5600E+02	0.9999E+00	-0.2108E-03	0.1171E-03	0.9999E+00	0.1192E-06
0.5700E+02	0.9998E+00	-0.3766E-04	0.1960E-03	0.9998E+00	0.3576E-06
0.5800E+02	0.9999E+00	0.1206E-03	0.9871E-04	0.9999E+00	0.1788E-06
0.5900E+02	0.1000E+01	0.1420E-03	-0.5168E-04	0.1000E+01	0.0000E+00
0.6000E+02	0.1000E+01	0.4512E-04	-0.1220E-03	0.1000E+01	0.1192E-06
0.6000E+02	0.1000E+01	0.4512E-04	-0.1220E-03	0.1000E+01	0.1192E-06
MINIMUM	0.0000E+00	-0.4651E+00	-0.6448E+00	0.0000E+00	0.0000E+00
MAXIMUM	0.1611E+01	0.7272E+00	0.1000E+01	0.1611E+01	0.7415E-04

D13-4.2

```
    SUBROUTINE STATE
    COMMON/SCOM1/ATRIB(100),DD(100),DDL(100),DTNOW,II,MFA,MSTOP,NCLNR
   1,NCRDR,NPRNT,NNRUN,NNSET,NTAPE,SS(100),SSL(100),TNEXT,TNOW,XX(100)
    DATA PI/3.14159/
    DD(2)=SIN(2.*PI*TNOW)-0.3*SS(2)-SS(1)
    DD(1)=SS(2)
    RETURN
    END
```

D13-4.3

```
    SUBROUTINE STATE
    COMMON/SCOM1/ATRIB(100),DD(100),DDL(100),DTNOW,II,MFA,MSTOP,NCLNR
   1,NCRDR,NPRNT,NNRUN,NNSET,NTAPE,SS(100),SSL(100),TNEXT,TNOW,XX(100)
    DD(2)=XX(1)-0.3*SS(2)-SS(1)
    DD(1)=SS(2)
    RETURN
    END

    SUBROUTINE EVENT(I)
    COMMON/SCOM1/ATRIB(100),DD(100),DDL(100),DTNOW,II,MFA,MSTOP,NCLNR
   1,NCRDR,NPRNT,NNRUN,NNSET,NTAPE,SS(100),SSL(100),TNEXT,TNOW,XX(100)
    XX(1)=UNFRM(.5,1.5,1)
    CALL SCHDL(1,5.,ATRIB)
    RETURN
    END

    SUBROUTINE INTLC
    COMMON/SCOM1/ATRIB(100),DD(100),DDL(100),DTNOW,II,MFA,MSTOP,NCLNR
   1,NCRDR,NPRNT,NNRUN,NNSET,NTAPE,SS(100),SSL(100),TNEXT,TNOW,XX(100)
    CALL SCHDL(1,0.,ATRIB)
    RETURN
    END


    1   GEN,ROLSTON,PROBLEM 13.4B,8/28/80,1;
    2   LIMITS,,,1;
    3   CONTINUOUS,2,0,.01,.5,1.0,W,.001,.0001;
    4   INTLC,SS(1)=0,SS(2)=0;
    5   RECORD,TNOW,TIME,0,B,2.0;
    6   VAR,SS(1),Y,Y,MIN(.1),MAX(.1);
    7   VAR,SS(2),D,YDOT,MIN(.1),MAX(.1);
    8   VAR,DD(2),2,YDDOT,MIN(.1),MAX(.1);
    9   INIT,0,60;
   10   FIN;
```

D13-4.4

13-5.   Let SS(I) be the probability of I-1 customers in the system. The translation of the differential, difference equations into SLAM II differential equations is direct as shown in the code for subroutine STATE given in D13-5.1 where SS(4) is the expected number in the system at the time subroutine STATE is called. The input statements for this problem are shown in D13-5.2. In the INTLC statement, SS(1) is set to 1 which prescribes that the probability is 1 of no one in the system, that is, the system is starting empty and idle. In order to observe the transient behavior, the simulation was run for 5 time units and the values of the SS variables were plotted at .005 intervals as shown on the RECORD statement. When assigning this exercise, these values should be prescribed. The ending values for the state and derivative variables as obtained on the SLAM II Summary Report are shown in D13-5.3. From the values of the DD(I) variables, it is seen that at the end of the simulation there is only a small change in the rate at which the SS(I) values are changing. This indicates that steady state is being approached. Note that DD(4) is 0 as the expected number in the system is given by an equation directly and not by a rate of change. A plot over the first 2.35 time units is shown in D13-5.4.

```
SUBROUTINE STATE
COMMON/SCOM1/ATRIB(100),DD(100),DDL(100),DTNOW,II,MFA,MSTOP,NCLNR
1,NCRDR,NPRNT,NNRUN,NNSET,NTAPE,SS(100),SSL(100),TNEXT,TNOW,XX(100)
DD(1)=-2.5*SS(1)+2.0*SS(2)
DD(2)=2.5*SS(1)-4.5*SS(2)+2.0*SS(3)
DD(3)=2.5*SS(2)-2.0*SS(3)
SS(4)=1.0*SS(2)+2.0*SS(3)
RETURN
END
```

D13-5.1

```
 1    GEN,ROLSTON,PROBLEM 13.5,8/14/80,1;
 2    CONTINUOUS,3,1,.005,.05,0,,.0001,.0001;
 3    INTLC,SS(1)=1,SS(2)=0,SS(3)=0;
 4    RECORD,TNOW,TIME,0,B,.05;
 5    VAR,SS(1),0,PROB 0,0,1;
 6    VAR,SS(2),1,PROB 1,0,1;
 7    VAR,SS(3),2,PROB 2,0,1;
 8    VAR,SS(4),N,NUMBER IN SYSTEM,0,2;
 9    INIT,0,5;
10    FIN;
```

D13-5.2

```
                    S L A M   I I   S U M M A R Y   R E P O R T


          SIMULATION PROJECT PROBLEM 13.5              BY ROLSTON

          DATE  8/14/1980                             RUN NUMBER    1 OF    1


          CURRENT TIME    0.5000E+01
          STATISTICAL ARRAYS CLEARED AT TIME  0.0000E+00



          **STATE AND DERIVATIVE VARIABLES**


              (I)         SS(I)            DD(I)
               1       0.2623E+00      -0.1514E-04
               2       0.3279E+00      -0.1788E-05
               3       0.4098E+00       0.1699E-04
               4       0.1148E+01       0.0000E+00
```

D13-5.3

```
                                        **PLOT NUMBER   1**
                                          RUN  NUMBER   1

                                          SCALES OF PLOT
    0=PROB 0      0.0000E+00           0.2500E+00           0.5000E+00           0.7500E+00           0.1000E+01
    1=PROB 1      0.0000E+00           0.2500E+00           0.5000E+00           0.7500E+00           0.1000E+01
    2=PROB 2      0.0000E+00           0.2500E+00           0.5000E+00           0.7500E+00           0.1000E+01
    N=NUMBER IN SY 0.0000E+00          0.5000E+00           0.1000E+01           0.1500E+01           0.2000E+01

              0    5   10  15   20  25   30  35  40   45   50  55   60  65   70  75   80  85   90  95  100 DUPLICATES

    TIME

    0.0000E+00 1                    +                    +                    +                    0 12 1N
    0.5000E-01 +2     N    1        +                    +                    +              0
    0.1000E+00 + 2         N        N       1+           +                    +        0
    0.1500E+00 +       2         N      1 +              +              0     +
    0.2000E+00 +       2          N   + 1               +        0     0     +
    0.2500E+00 +          2         N+   1              +     0     0        +
    0.3000E+00 +           2        + N  1          +  0                     +
    0.3500E+00 +            2      +      N1         +  0                     +
    0.4000E+00 +             2    +       1N         0                       +
    0.4500E+00 +             2  +         1  N       0 +                     +
    0.5000E+00 +              2 +        1     N   O +                       +
    0.5500E+00 +              2+        1       N O  +                       +
    0.6000E+00 +              2        1      ON      +                      +
    0.6500E+00 +             +2    1      0      N     +                     +
    0.7000E+00 +             + 2    1     0       N    +                     +
    0.7500E+00 +             +   2   1 0          N +                        +
    0.8000E+00 +             +   2  10           N +                         +
    0.8500E+00 +             +    2 0            N+                          +                 + 01
    0.9000E+00 +             +     01           N                           +                 + 02
    0.9500E+00 +             +     02          +N                           +                 + 01
    0.1000E+01 +             +    01 2         +N                           +                 +
    0.1050E+01 +             +    0 1 2        + N                          +                 +
    0.1100E+01 +             +    0 1  2       +  N                         +                 +
    0.1150E+01 +             +    0 1   2      +   N                        +                 +
    0.1200E+01 +             +    0 1   2.     +    N                       +                 +
    0.1250E+01 +             +    0 1    2     +    N                       +                 +
    0.1300E+01 +             +    0 1    2     +     N                      +                 +
    0.1350E+01 +             +    0 1     2    +     N                      +                 +
    0.1400E+01 +             +    0 1     2    +     N                      +                 +
    0.1450E+01 +             +    0 1      2   +     N                      +                 +
    0.1500E+01 +             +    0 1      2   +     N                      +                 +
    0.1550E+01 +             +    0 1      2   +      N                     +                 +
    0.1600E+01 +             +    0 1      2   +      N                     +                 +
    0.1650E+01 +             +    0 1       2  +      N                     +                 +
    0.1700E+01 +             + 0   1       2  +      N                      +                 +
    0.1750E+01 +             + 0   1       2   +      N                     +                 +
    0.1800E+01 +             + 0   1       2   +      N                     +                 +
    0.1850E+01 +             + 0   1       2   +      N                     +                 +
    0.1900E+01 +             + 0   1       2   +       N                    +                 +
    0.1950E+01 +             + 0   1       2   +       N                    +                 +
    0.2000E+01 +             + 0   1       2   +       N                    +                 +
    0.2050E+01 +             + 0   1       2   +       N                    +                 +
    0.2100E+01 +             + 0   1       2   +       N                    +                 +
    0.2150E+01 +             + 0   1        2  +       N                    +                 +
    0.2200E+01 +             + 0   1        2  +       N                    +                 +
    0.2250E+01 +             + 0   1        2  +       N                    +                 +
    0.2300E+01 +             +0    1        2  +      N                     +                 +
    0.2350E+01 +              +0   1        2  +      N                     +                 +
```

D13-5.4

13-6.   Let $P_n(t)$ be the probability of n in the system at time t for n=0,1,...N. The parameters for this exercise are $\lambda=4$, $\mu=5$, and N=20. The differential, difference equations for this exercise are shown below where a prime is used to indicate a derivative.

$$P_0'(t) = -\lambda P_0(t) + \mu P_1(t)$$

$$P_n'(t) = \lambda P_{n-1}(t) - (\lambda+\mu)P_n(t) + \mu P_{n+1}(t), \; 0 < n < N$$

$$P_N'(t) = \lambda P_{N-1}(t) - \mu P_N(t)$$

The expected number in the system, E[X(t)], can be computed by

$$E[X(t)] = \sum_{n=0}^{N} n*P_n(t)$$

The variance of the number in the system at time t can be computed from

$$E[(X(t)-E[X(t)])^2]$$
$$= E[X^2(t)] - (E[X(t)])^2$$
$$= \sum_{n=0}^{N} n^2*P_n(t) - [\sum_{n=0}^{N} n*P_n(t)]^2$$

The coding of these equations in subroutine STATE is shown in D13-6.1. The input statements are shown in D13-6.2. In subroutine STATE, the expected value and variance are computed each time subroutine STATE is called. Since subroutine STATE is called frequently, this could increase the computations required for the simulation. As an alternative, an event could be called periodically in order to compute the expected value and variance of the number in the system and subroutine GPLOT could be used to record the values for plotting. When this is done, a plot number must be prescribed on the RECORD input statement.

In subroutine INTLC, SS(1) is set equal to 1 to indicate that there is 0 in the system at time 0. SS(J) is defined as the probability of J-1 in the system. The exercise also calls for the computing of the average of the expected number in the system, SS(22), computed at 5 minute intervals. Thus, starting at time 0 and every 5 minutes thereafter a value of the expected number in the system is obtained and the average of these is what is requested. This average is computed in subroutine EVENT which is scheduled to be called every 5 minutes. Subroutine INTLC is shown in D13-6.3 and subroutine EVENT is shown in D13-6.4. Note that the average, AVE, is equivalenced to XX(1) in order to plot the value of AVE.

The final values for the state and derivative variables are shown on the SLAM II Summary Report given in D13-6.5. The plot obtained for this exercise is shown in D13-6.6. From the plot, it is seen that the probabilities stabilize faster than the expected number or variance of the number in the system. It is also seen that the average as computed by averaging the expected values lags the expected value considerably. By taking smaller intervals between samples of expected value the lag can be reduced, but it will always exist. This information is useful when considering when to clear statistical arrays during simulation studies.

```
      SUBROUTINE STATE
      COMMON/SCOM1/ATRIB(100),DD(100),DDL(100),DTNOW,II,MFA,MSTOP,NCLNR
     1,NCRDR,NPRNT,NNRUN,NNSET,NTAPE,SS(100),SSL(100),TNEXT,TNOW,XX(100)
      SS(22)=0.0
      SS(23)=0.0
      DD(1)=-4.*SS(1)+5.*SS(2)
      DO 10 J=2,20
      DD(J)=4.*SS(J-1)-9.*SS(J)+5.*SS(J+1)
      SS(22)=SS(22)+(FLOAT(J-1))*SS(J)
      SS(23)=SS(23)+(FLOAT(J-1))**2*SS(J)
   10 CONTINUE
      DD(21)=4.*SS(20)-5.*SS(21)
      SS(22)=SS(22)+20.*SS(21)
      SS(23)=SS(23)+400.*SS(21)
      SS(24)=SS(23)-SS(22)*SS(22)
      RETURN
      END
```

D13-6.1

```
    1    GEN,ROLSTON,PROBLEM 13.6,8/15/80,1;
    2    LIMITS,0,0,5;
    3    CONTINUOUS,21,3,.01,.1,.2,,.001,.001;
    4    RECORD,TNOW,TIME,0,P,.2;
    5    VAR,SS(1),0,PROB 0,0,1;
    6    VAR,SS(2),1,PROB 1,0,1;
    7    VAR,SS(3),2,PROB 2,0,1;
    8    VAR,SS(22),N,E[X],0,4;
    9    VAR,SS(24),V,VARIANCE,0,20;
   10    VAR,XX(1),A,AVE NUM,0,4;
   11    INIT,0,25;
   12    FIN;
```

D13-6.2

```
      SUBROUTINE INTLC
      COMMON/SCOM1/ATRIB(100),DD(100),DDL(100),DTNOW,II,MFA,MSTOP,NCLNR
     1,NCRDR,NPRNT,NNRUN,NNSET,NTAPE,SS(100),SSL(100),TNEXT,TNOW,XX(100)
      COMMON/UCOM1/TOT,NUM
      SS(1)=1
      DO 10 I=2,21
   10 SS(I)=0
      TOT=0
      NUM=1
      CALL SCHDL(1,5.0,ATRIB)
      RETURN
      END
```

D13-6.3

```
SUBROUTINE EVENT(I)
COMMON/SCOM1/ATRIB(100),DD(100),DDL(100),DTNOW,II,MFA,MSTOP,NCLNR
1,NCRDR,NPRNT,NNRUN,NNSET,NTAPE,SS(100),SSL(100),TNEXT,TNOW,XX(100)
COMMON/UCOM1/TOT,NUM
EQUIVALENCE (AVE,XX(1))
CALL SCHDL(1,5.0,ATRIB)
TOT=TOT+SS(22)
NUM=NUM+1
AVE=TOT/FLOAT(NUM)
RETURN
END
```

D13-6.4

S L A M   I I   S U M M A R Y   R E P O R T

SIMULATION PROJECT PROBLEM 13.6   BY ROLSTON

DATE   8/15/1980   RUN NUMBER   1 OF   1

CURRENT TIME   0.2500E+02
STATISTICAL ARRAYS CLEARED AT TIME   0.0000E+00

STATE AND DERIVATIVE VARIABLES

(I)	SS(I)	DD(I)
1	0.2030E+00	-0.1733E-03
2	0.1623E+00	-0.1332E-03
3	0.1298E+00	-0.9704E-04
4	0.1038E+00	-0.6497E-04
5	0.8292E-01	-0.3737E-04
6	0.6624E-01	-0.1413E-04
7	0.5289E-01	0.4947E-05
8	0.4221E-01	0.2004E-04
9	0.3367E-01	0.3144E-04
10	0.2684E-01	0.3961E-04
11	0.2139E-01	0.4488E-04
12	0.1703E-01	0.4770E-04
13	0.1356E-01	0.4845E-04
14	0.1079E-01	0.4756E-04
15	0.8587E-02	0.4535E-04
16	0.6831E-02	0.4218E-04
17	0.5435E-02	0.3832E-04
18	0.4326E-02	0.3405E-04
19	0.3446E-02	0.2958E-04
20	0.2748E-02	0.2508E-04
21	0.2194E-02	0.2071E-04
22	0.3765E+01	0.0000E+00
23	0.2972E+02	0.0000E+00
24	0.1555E+02	0.0000E+00

D13-6.5

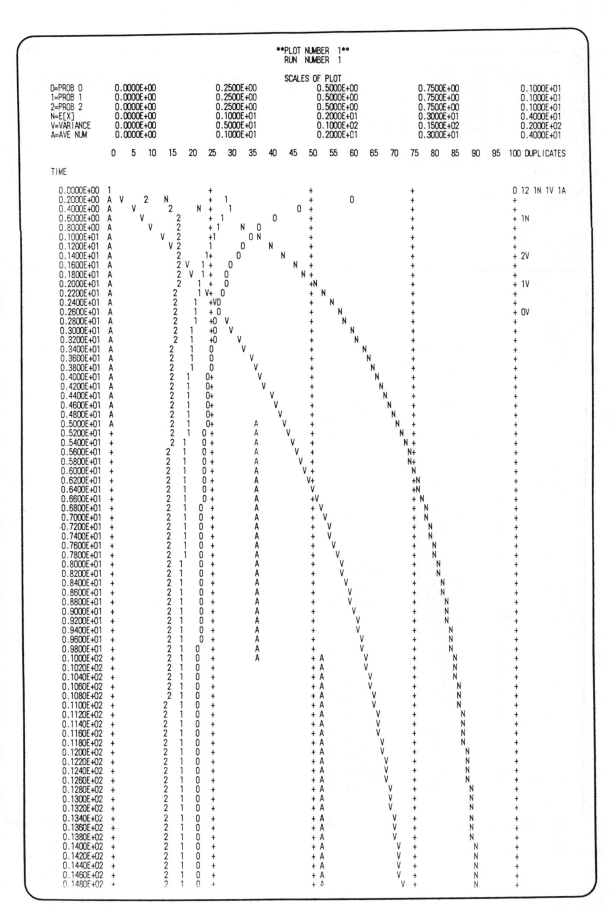

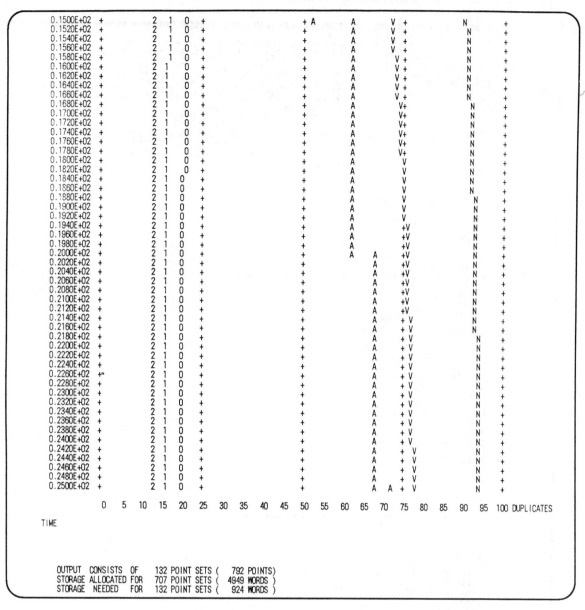

D13-6.6(2)

13-7. The problem statement and discussion for this exercise is covered in detail by Forester, J.W., _Industrial Dynamics_, John Wiley, 1961. A GASP IV model is given in Pritsker, A.A.B., _The GASP IV Simulation Language_, New York, John Wiley, 1974, pages 396-417. The SLAM II model is shown in D13-7.1 through D13-7.3.

In subroutine STATE, rates of change that are equal to a variable value (a "level" in systems dynamics terminology) over a time period are required. This requires the use of the full step size, DTFUL, in SLAM II. DTFUL is contained in labeled COMMON block GCOM2 which is included in subroutine STATE. This example illustrates the use of EQUIVALENCE statements to allow more descriptive variables to be used in defining the equations. Also, subroutine GDLAY is used to obtain delays of state variables. The outputs obtained are the same as those presented in the GASP IV book and in Forester's _Industrial Dynamics_ book.

```
      SUBROUTINE STATE
      REAL IAD,IAF,IAR,IDD,IDF,IDR,LAD,LAF,LAR,LDD,LDF,LDR,MDF,MNR,MOD,M
     1OF,MTD,MTR,MWF,NID,NIF,NIR,MPF
      COMMON/SCOM1/ATRIB(100),DD(100),DDL(100),DTNOW,II,MFA,MSTOP,NCLNR
     1,NCRDR,NPRNT,NNRUN,NNSET,NTAPE,SS(100),SSL(100),TNEXT,TNOW,XX(100)
      COMMON/GCOM2/DTFUL,DDI(100),IICU(25,2),IIDIR(25),IIEVT(25),ISEES,
     1LLSCD(25),NNEQD,NNEQS,NNEQI,NNSCD,SSI(100),TTOL(25),VVAL(25)
      COMMON/UCOM1/AID,AIF,AIR,ALF,DCD,DCF,DCR,DFD,     DFR,DHD,DHF,DHR
     1,DID,DIF,DIR,DMD,DMR,DPF,DRD,DRF,DRR,DTD,DTR,DUD,DUF,DUR,IDD,IDF,I
     2DR,LAD,LAF,LAR,LDD,LDF,LDR,MDF,MNR,MOD,MWF,NID,NIF,NIR,PDD,PDR,RDR
     3,SSD,SSF,SSR,STD,STF,STR,UND,UNF,UNR,   RRI,DMF
      EQUIVALENCE (UOR,SS(1)),(IAR,SS(2)),(RSR,SS(3)),(CPR,SS(4)),(PMR,S
     1S(5)),(MTR,SS(6)),(UOD,SS(7)),(IAD,SS(8)),(RSD,SS(9)),(CPD,SS(10))
     2,(PMD,SS(11)),(MTD,SS(12)),(UOF,SS(13)),(IAF,SS(14)),(RSF,SS(15)),
     3(CPF,SS(16)),(OPF,SS(17)),(PSR,SS(20)),(RRD,SS(23)),(SRR,SS(26)),(
     4PSD,SS(29)),(RRF,SS(32)),(SRD,SS(35)),(MOF,SS(38)),(SRF,SS(41))
      EQUIVALENCE (XX(1),RRR),(XX(2),DFF)
C
C*****INPUT TEST FUNCTIONS
C
      IF(TNOW.GT.0.0) RRR=1.1*RRI
C
C*****AUXILIARY EQUATIONS
C
      UNF=RSF*(DHF+DUF)
      LAF=CPF+OPF
      LDF=RSF*(DCF+DPF)
      IDF=AIF*RSF
      MWF=RRF+(1./DIF)*(IDF-IAF+LDF-LAF+UOF-UNF)
      DFF=DUF*IDF/IAF+DHF
      NIF=IAF/DTFUL
      STF=UOF/DFF
      UND=RSD*(DHD+DUD)
      LAD=CPD+PMD+UOF+MTD
      LDD=RSD*(DCD+DMD+DFF+DTD)
      IDD=AID*RSD
      MPF=DUF*IDF
      DFF=(MPF/IAF)+DHF
      MOD=DUD*IDD
      DFD=(MOD/IAD)+DHD
      NID=IAD/DTFUL
      STD=UOD/DFD
      UNR=RSR*(DHR+DUR)
      LAR=CPR+PMR+UOD+MTR
      LDR=RSR*(DCR+DMR+DFD+DTR)
      IDR=AIR*RSR
      DFR=DUR*IDR/IAR+DHR
      NIR=IAR/DTFUL
      STR=UOR/DFR
C
C*****RATE COMPONENTS
C
      SSR=STR
      IF(NIR.LT.STR) SSR=NIR
      PDR=RRR+(1./DIR)*(IDR-IAR+LDR-LAR+UOR-UNR)
      CALL GDLAY(18,20,PDR,DCR)
      CALL GDLAY(21,23,PSR,DMR)
      PDD=RRD+(1./DID)*(IDD-IAD+LDD-LAD+UOD-UND)
      SSD=STD
      IF(NID.LT.STD) SSD=NID
      CALL GDLAY(24,26,SSD,DTR)
      CALL GDLAY(27,29,PDD,DCD)
      CALL GDLAY(30,32,PSD,DMD)
      SSF=STF
      IF(NIF.LT.STF) SSF=NIF
      CALL GDLAY(33,35,SSF,DTD)
      MDF=MWF
      IF(ALF.LT.MWF) MDF=ALF
      CALL GDLAY(36,38,MDF,DCF)
      CALL GDLAY(39,41,MOF,DPF)
C
C*****RATE EQUATIONS
C
      DD(1)=RRR-SSR
      DD(2)=SRR-SSR
      DD(3)=(1./DRR)*(RRR-RSR)
      DD(4)=PDR-PSR
      DD(5)=PSR-RRD
      DD(6)=SSD-SRR
      DD(7)=RRD-SSD
      DD(8)=SRD-SSD
      DD(9)=(1./DRD)*(RRD-RSD)
      DD(10)=PDD-PSD
      DD(11)=PSD-RRF
      DD(12)=SSF-SRD
      DD(13)=RRF-SSF
      DD(14)=SRF-SSF
      DD(15)=(1./DRF)*(RRF-RSF)
      DD(16)=MDF-MOF
      DD(17)=MOF-SRF
      RETURN
      END
```

```
      SUBROUTINE INTLC
      REAL IAD,IAF,IAR,IDD,IDF,IDR,LAD,LAF,LAR,LDD,LDF,LDR,MDF,MNR,MOD,M
     1OF,MTD,MTR,MWF,NID,NIF,NIR,MPF
      COMMON/SCOM1/ATRIB(100),DD(100),DDL(100),DTNOW,II,MFA,MSTOP,NCLNR
     1,NCRDR,NPRNT,NNRUN,NNSET,NTAPE,SS(100),SSL(100),TNEXT,TNOW,XX(100)
      COMMON/UCOM1/AID,AIF,AIR,ALF,DCD,DCF,DCR,DFD,        DFR,DHD,DHF,DHR
     1,DID,DIF,DIR,DMD,DMR,DPF,DRD,DRF,DRR,DTD,DTR,DUD,DUF,DUR,IDD,IDF,I
     2DR,LAD,LAF,LAR,LDD,LDF,LDR,MDF,MNR,MOD,MWF,NID,NIF,NIR,PDD,PDR,RDR
     3,SSD,SSF,SSR,STD,STF,STR,UND,UNF,UNR,     RRI,DMF
      EQUIVALENCE (UOR,SS(1)),(IAR,SS(2)),(RSR,SS(3)),(CPR,SS(4)),(PMR,S
     1S(5)),(MTR,SS(6)),(UOD,SS(7)),(IAD,SS(8)),(RSD,SS(9)),(CPD,SS(10))
     2,(PMD,SS(11)),(MTD,SS(12)),(UOF,SS(13)),(IAF,SS(14)),(RSF,SS(15)),
     3(CPF,SS(16)),(OPF,SS(17)),(PSR,SS(20)),(RRD,SS(23)),(SRR,SS(26)),(
     4PSD,SS(29)),(RRF,SS(32)),(SRD,SS(35)),(MOF,SS(38)),(SRF,SS(41))
      EQUIVALENCE (XX(1),RRR),(XX(2),DFF)
C
C*****TEST INPUT CONDITIONS
C
      RRI=1000.
C
C*****PARAMETERS
C
      AID=6.
      AIF=4.
      AIR=8.
      ALF=1000.*RRI
      DCD=2.
      DCF=1.
      DCR=3.
      DHD=1.
      DHF=1.
      DHR=1.
      DID=4.
      DIF=4.
      DIR=4.
      DMD=.5
      DMR=.5
      DPF=6.
      DRD=8.
      DRF=8.
      DRR=8.
      DTD=2.
      DTR=1.
      DUD=.6
      DUF=1.
      DUR=.4
C
C*****INITIAL CONDITIONS
C
      RRR=RRI
      UOR=RRI*(DHR+DUR)
      IAR=AIR*RRI
      RSR=RRR
      CPR=DCR*RRI
      PDR=RRR
      PSR=PDR
      RRD=PSR
      PMR=DMR*RRI
      MTR=DTR*RRI
      SSD=RRD
      UOD=RRI*(DHD+DUD)
      IAD=AID*RRI
      RSD=RRI
      CPD=DCD*RRI
      PDD=RRD
      PSD=PDD
      RRF=PSD
      PMD=DMD*RRI
      MTD=DTD*RRI
      SSF=RRF
      MDF=RRF
      UOF=RRI*(DHF+DUF)
      IAF=AIF*RRI
      RSF=RRI
      CPF=DCF*RRI
      OPF=DPF*RRI
      SSR=RRR
      SRR=SSD
      SRD=SSF
      MOF=MDF
      SRF=MOF
      RETURN
      END
```

```
 1    GEN,ROLSTON,PROBLEM 13.7,8/21/80,1;
 2    CONTINUOUS,42,0,.01,1.0,1.0,W,0.0,0.0;
 3    RECORD,TNOW,TIME,0,B,1;
 4    VAR,XX(1),R,RRR,0,2000;
 5    VAR,SS(23),D,RRD,0,2000;
 6    VAR,SS(32),F,RRF,0,2000;
 7    VAR,SS(38),M,MOF,0,2000;
 8    VAR,SS(41),S,SRF,0,2000;
 9    VAR,SS(14),A,IAF,0,10000;
10    VAR,SS(13),U,UOF,0,10000;
11    VAR,SS(8),C,IAP,0,10000;
12    VAR,SS(2),B,IAR,0,10000;
13    VAR,XX(2),W,DFF,0,10;
14    INIT,-4.0,75.0;
15    FIN;
```

D13-7.3

13-8.   The incorporation of events in the world model is described by Pritsker, A.A.B. and R.E. Young in Simulation with GASP PL/I, John Wiley, 1975, pages 291-303. The solution in SLAM II follows directly from the solution presented in GASP_PL/I. Listings of the subroutines for the SLAM II solution are shown in D13-8.1 through D13-8.5. In the problem statement, thresholds and tolerances were not given for the state events (event codes 6,7, and 8). In the input statements, these are prescribed as (2.5E09,1.25E08); (.5E09,.5E08); and (.7E12,1.05E12) for the threshold and tolerances on population, pollution, and natural resources, respectively. The SLAM II solution obtained in this exercise was identical to that obtained from the GASP_PL/I program.

```
      SUBROUTINE INTLC
      REAL MSL,NREM,NRFR,NRMM,NR,NRI,NRUR,NREMT,NRMMT,LA
      COMMON/SCOM1/ATRIB(100),DD(100),DDL(100),DTNOW,II,MFA,MSTOP,NCLNR
     1,NCRDR,NPRNT,NNRUN,NNSET,NTAPE,SS(100),SSL(100),TNEXT,TNOW,XX(100)
      COMMON/UCOM1/BR,BRCM,BRFM,BRMM,BRN,BRPM,CFIFR,CIAFI,CIAFN,CIAFT,
     1CID,CIDN,CIG,CIGN,CII,CIM,CIQR,CIR,CIRA,CR,DR,DRCM,DRFM,DRMM,DRN,D
     2RPM,ECIR,ECIRN,FC,FCM,FN,FPCI,FPM,LA,NREM,NRFR,NRI,NRMM,NRUR,
     3PDN,PI,POLA,POLAT,POLCM,POLG,POLI,POLN,POLS,QLF,QLM,QLS,XNRUN
      COMMON/UCOM2/BRCMT(6),BRFMT(5),BRMMT(6),BRPMT(7),CFIFRT(5),CIMT
     1(6),CIQRT(5),DRCMT(6),DRFMT(9),DRMMT(11),DRPMT(7),FCMT(6),FPCIT(7)
     2,FPMT(7),NREMT(5),NRMMT(11),POLATT(7),POLCMT(6),QLCT(11),QLFT(5),Q
     3LMT(6),QLPT(7)
      COMMON/UCOM3/LCNSR,LZPG,LFDST,LEG
      EQUIVALENCE (P,SS(1)),(NR,SS(2)),(CI,SS(3)),(POL,SS(4)),(CIAF,SS(5
     1)),(POLR,XX(1)),(QL,XX(2)),(FR,XX(3)),(MSL,XX(4)),(QLC,XX(5)),
     2(QLP,XX(6))
C
C*****INITIAL CONDITIONS
C
      PI=1.65E9
      READ(NCRDR,101) BRN
      ECIRN=1
      NRI=900.E9
      XNRUN=1
      DRN=.028
      LA=135.E6
      PDN=26.5
      FC=1
      FN=1
      CIAFN=.3
      CII=.4E9
      CIGN=.05
      CIDN=.025
      POLS=3.6E9
      POLI=.2E9
      POLN=1
      CIAFI=.2
      CIAFT=15
      QLS=1
      P=PI
      NR=NRI
      CI=CII
      POL=POLI
      CIAF=CIAFI
      LCNSR=0
      LFDST=0
      LZPG=0
      LEG=0
      CALL SCHDL(1,20.0,ATRIB)
      CALL SCHDL(2,75.,ATRIB)
      CALL SCHDL(3,80.,ATRIB)
      RETURN
101   FORMAT(F5.2)
      END
```

```
      DIMENSION NSET(2000)
      COMMON QSET(2000)
      REAL MSL,NREM,NRFR,NRMM,NR,NRI,NRUR,NREMT,NRMMT,LA
      COMMON/SCOM1/ATRIB(100),DD(100),DDL(100),DTNOW,II,MFA,MSTOP,NCLNR
     1,NCRDR,NPRNT,NNRUN,NNSET,NTAPE,SS(100),SSL(100),TNEXT,TNOW,XX(100)
      COMMON/UCOM2/BRCMT(6),BRFMT(5),BRMMT(6),BRPMT(7),CFIFRT(5),CIMT
     1(6),CIQRT(5),DRCMT(6),DRFMT(9),DRMMT(11),DRPMT(7),FCMT(6),FPCIT(7)
     2,FPMT(7),NREMT(5),NRMMT(11),POLATT(7),POLCMT(6),QLCT(11),QLFT(5),Q
     3LMT(6),QLPT(7)
      EQUIVALENCE(NSET(1),QSET(1))
      NCRDR=5
      NPRNT=6
      NTAPE=7
      NNSET=2000
      READ(NCRDR,101) BRCMT
      READ(NCRDR,101) BRFMT
      READ(NCRDR,101) BRMMT
      READ(NCRDR,101) BRPMT
      READ(NCRDR,101) CFIFRT
      READ(NCRDR,101) CIMT
      READ(NCRDR,101) CIQRT
      READ(NCRDR,101) DRCMT
      READ(NCRDR,101) DRFMT
      READ(NCRDR,101) DRMMT
      READ(NCRDR,101) DRPMT
      READ(NCRDR,101) FCMT
      READ(NCRDR,101) FPCIT
      READ(NCRDR,101) FPMT
      READ(NCRDR,101) NREMT
      READ(NCRDR,101) NRMMT
      READ(NCRDR,101) POLATT
      READ(NCRDR,101) POLCMT
      READ(NCRDR,101) QLCT
      READ(NCRDR,101) QLFT
      READ(NCRDR,101) QLMT
      READ(NCRDR,101) QLPT
      CALL SLAM
      STOP
101   FORMAT(7F10.4)
      END
```

D13-8.2

```
      SUBROUTINE STATE
      REAL MSL,NREM,NRFR,NRMM,NR,NRI,NRUR,NREMT,NRMMT,LA
      COMMON/SCOM1/ATRIB(100),DD(100),DDL(100),DTNOW,II,MFA,MSTOP,NCLNR
     1,NCRDR,NPRNT,NNRUN,NNSET,NTAPE,SS(100),SSL(100),TNEXT,TNOW,XX(100)
      COMMON/UCOM1/BR,BRCM,BRFM,BRMM,BRN,BRPM,CFIFR,CIAFI,CIAFN,CIAFT,
     1CID,CIDN,CIG,CIGN,CII,CIM,CIQR,CIR,CIRA,CR,DR,DRCM,DRFM,DRMM,DRN,D
     1RPM,ECIR,ECIRN,FC,FCM,FN,FPCI,FPM,LA,NREM,NRFR,NRI,NRMM,NRUR,
     1PDN,PI,POLA,POLAT,POLCM,POLG,POLI,POLN,POLS,QLF,QLM,QLS,XNRUN
      COMMON/UCOM2/BRCMT(6),BRFMT(5),BRMMT(6),BRPMT(7),CFIFRT(5),CIMT
     1(6),CIQRT(5),DRCMT(6),DRFMT(9),DRMMT(11),DRPMT(7),FCMT(6),FPCIT(7)
     2,FPMT(7),NREMT(5),NRMMT(11),POLATT(7),POLCMT(6),QLCT(11),QLFT(5),Q
     3LMT(6),QLPT(7)
      COMMON/UCOM3/LCNSR,LZPG,LFDST,LEG
      EQUIVALENCE (P,SS(1)),(NR,SS(2)),(CI,SS(3)),(POL,SS(4)),(CIAF,SS(5
     1)),(POLR,XX(1)),(QL,XX(2)),(FR,XX(3)),(MSL,XX(4)),(QLC,XX(5)),
     2(QLP,XX(6))
C
C*****AUXILIARIES
C
      NRFR=NR/NRI
      CR=P/(LA*PDN)
      CIR=CI/P
      NREM=GTABL(NREMT,NRFR,0.,1.,.25)
      ECIR=CIR*(1.-CIAF)*NREM/(1.-CIAFN)
      MSL=ECIR/ECIRN
      BRMM=GTABL(BRMMT,MSL,0.,5.,1.)
      DRMM=GTABL(DRMMT,MSL,0.,5.,.5)
      DRCM=GTABL(DRCMT,CR,0.,5.,1.)
      BRCM=GTABL(BRCMT,CR,0.,5.,1.)
      CIRA=CIR*CIAF/CIAFN
      FPCI=GTABL(FPCIT,CIRA,0.,6.,1.)
      FCM=GTABL(FCMT,CR,0.,5.,1.)
      POLR=POL/POLS
      FPM=GTABL(FPMT,POLR,0.,60.,10.)
      FR=FPCI*FCM*FPM*FC/FN
      IF(LFDST.EQ.1) FR=.7*FR

      CIM=GTABL(CIMT,MSL,0.,5.,1.)
      POLCM=GTABL(POLCMT,CIR,0.,5.,1.)
      IF(LEG.EQ.1) POLCM=.75*POLCM
      POLAT=GTABL(POLATT,POLR,0.,60.,10.)
      CFIFR=GTABL(CFIFRT,FR,0.,2.,.5)
      QLM=GTABL(QLMT,MSL,0.,5.,1.)
      QLC=GTABL(QLCT,CR,0.,5.,.5)
      QLF=GTABL(QLFT,FR,0.,4.,1.)
      QLP=GTABL(QLPT,POLR,0.,60.,10.)
      NRMM=GTABL(NRMMT,MSL,0.,10.,1.)
      CIQR=GTABL(CIQRT,QLM/QLF,0.,2.,.5)
      DRPM=GTABL(DRPMT,POLR,0.,60.,10.)
      DRFM=GTABL(DRFMT,FR,0.,2.,.25)
      BRFM=GTABL(BRFMT,FR,0.,4.,1.)
      BRPM=GTABL(BRPMT,POLR,0.,60.,10.)
C
C*****RATE COMPONENTS
C
      BR=P*BRN*BRFM*BRMM*BRCM*BRPM
      DR=P*DRN*DRMM*DRPM*DRFM*DRCM
      CIG=P*CIM*CIGN
      CID=CI*CIDN
      POLG=P*POLN*POLCM
      POLA=POL/POLAT
      IF(LZPG.EQ.1) BR=.85*BR
      IF(LCNSR.EQ.1) NRMM=.9*NRMM
      NRUR=P*XNRUN*NRMM
C
C*****LEVELS
C
      QL=QLS*QLM*QLC*QLF*QLP
      P=P+DTNOW*(BR-DR)
      NR=NR-DTNOW*NRUR
      CI=CI+DTNOW*(CIG-CID)
      POL=POL+DTNOW*(POLG-POLA)
      CIAF=CIAF+DTNOW*(CFIFR*CIQR-CIAF)/CIAFT
      RETURN
      END
```

```
      SUBROUTINE EVENT(I)
      REAL MSL,NREM,NRFR,NRMM,NR,NRI,NRUR,NREMT,NRMMT,LA
      COMMON/SCOM1/ATRIB(100),DD(100),DDL(100),DTNOW,II,MFA,MSTOP,NCLNR
     1,NCRDR,NPRNT,NNRUN,NNSET,NTAPE,SS(100),SSL(100),TNEXT,TNOW,XX(100)
      COMMON/UCOM3/LCNSR,LZPG,LFDST,LEG
      EQUIVALENCE  (P,SS(1)),(NR,SS(2)),(CI,SS(3)),(POL,SS(4)),(CIAF,SS(5
     1)),(POLR,XX(1)),(QL,XX(2)),(FR,XX(3)),(MSL,XX(4)),(QLC,XX(5)),
     2(QLP,XX(6))
      GOTO (1,2,3,4,5,6,7,8),I
1     LFDST=1
      P=.9*P
      CALL SCHDL(1,20.,ATRIB)
      CALL SCHDL(5,4.,ATRIB)
      RETURN
2     NR=1.5*NR
      RETURN
3     P=.85*P
      CI=.80*CI
      RETURN
4      LEG=1
      RETURN
5     LFDST=0
      RETURN
6     LZPG=1
      RETURN
7     CALL SCHDL(4,5.,ATRIB)
      RETURN
8      LCNSR=1
      RETURN
      END
```

D13-8.4

1.05	1.	.9	.7	.6	.55	
0.	1.	1.6	1.9	2.		
1.2	1.	.85	.75	.7	.7	
1.02	.9	.7	.4	.25	.15	.1
1.	.6	.3	.15	.1		
.1	1.	1.8	2.4	2.8	3.	
.7	.8	1.	1.5	2.		
.9	1.	1.2	1.5	1.9	3.	
0.	3.	2.	1.4	1.	.7	.6
.5	.5					
3.	1.8	1.	.8	.7	.6	.53
.5	.5	.5	.5			
.92	1.3	2.	3.2	4.8	6.8	9.2
2.4	1.	.6	.4	.3	.2	
.5	1.	1.4	1.7	1.9	2.05	2.2
1.02	.9	.65	.35	.2	.1	.05
0.	.15	.5	.85	1.		
0.	1.	1.8	2.4	2.9	3.3	3.6
3.8	3.9	3.95	4.			
.6	2.5	5.	8.	11.5	15.5	20.
.05	1.	3.	5.4	7.4	8.	
2.	1.3	1.	.75	.55	.45	.38
.3	.25	.22	.2			
0.	1.	1.8	2.4	2.7		
.2	1.	1.7	2.3	2.7	2.9	
1.04	.85	.6	.3	.15	.05	.02

```
GEN,ROLSTON,PROBLEM 13.8,8/29/80,1;
LIMITS,,,10;
CONT,0,5,.2,.2,2.,W,.00001,.00001;
RECORD,TNOW,TIME,0,P,2.;
VAR,SS(1),P,POP,0,16E9;
VAR,XX(1),2,POLR,0,80;
VAR,SS(3),C,CI,0,4.0E10;
VAR,XX(2),Q,QL,0,8;
VAR,SS(2),N,NR,0,2.0E12;
VAR,XX(3),F,FR,-2,4;
VAR,XX(4),M,MSL,-20,4;
VAR,XX(5),4,QLC,-2,2;
VAR,XX(6),5,QLP,-2,2;
VAR,SS(5),A,CIAF,-.2,.6;
SEVNT,6,SS(1),XP,2.5E09,1.25E08;
SEVNT,7,SS(4),XP,.5E09,.5E08;
SEVNT,8,SS(2),X,.7E12,1.05E12
INIT,1900,2100
FIN;
   .04
```

D13-8.5

13-9.   The second order differential equation as programmed in subroutine STATE is shown below:

```
SUBROUTINE STATE
COMMON/SCOM1/...
SS(3) =(W*E*SIN(W*TNOW))/XL
DD(1)=SS(2)
DD(2)=SS(3)-R*SS(2)/XL-SS(1)/(XL*C)
RETURN
END
```

In the input statements, a state event statement is required as shown below:

```
SEVNT,1,SS(1),XP,A,.1*A;
```

13-10.   The coding of the two equations modeling the bouncing ball is shown in subroutine STATE in D13-10.1. An event is used to reverse the direction of the ball which is determined by an SEVNT statement. The coding of subroutine EVENT is shown in D13-10.2 which resets the value of SS(2) by changing its sign and multiplying by 0.8. This reverses the direction as well as reducing the ball's velocity. A printout at the event time provides the time at which the ball hits ground (SS(1)=0). When the number of bounces equals 8, MSTOP is set equal to minus one to stop the simulation. Input statements for this simulation are given in D13-10.3. A plot of the height of the ball, SS(1), is obtained from the RECORD and VAR statement specification. The times at which each bounce occurs is printed out after the intermediate results heading and shown in D13-10.4. It is seen that the values correspond to the times specified in the problem statement with a slight variation due to the use of a minimum step to detect the crossing of SS(1). A portion of the plot for this exercise is shown in D13-10.5.

```
SUBROUTINE STATE
COMMON/SCOM1/ATRIB(100),DD(100),DDL(100),DTNOW,II,MFA,MSTOP,NCLNR
1,NCRDR,NPRNT,NNRUN,NNSET,NTAPE,SS(100),SSL(100),TNEXT,TNOW,XX(100)
DATA G/32.3/
DD(1)=SS(2)
DD(2)=-G
RETURN
END
```

D13-10.1

```
SUBROUTINE EVENT(I)
COMMON/SCOM1/ATRIB(100),DD(100),DDL(100),DTNOW,II,MFA,MSTOP,NCLNR
1,NCRDR,NPRNT,NNRUN,NNSET,NTAPE,SS(100),SSL(100),TNEXT,TNOW,XX(100)
II=II+1
WRITE(6,100)II,TNOW
WRITE(6,101)
IF (II.EQ.8) MSTOP=-1
SS(2)=-0.8*SS(2)
RETURN
100 FORMAT(38X,'BOUNCE NUMBER ',I1,' OCCURRED AT TIME ',F6.4)
101 FORMAT(/)
END
```

D13-10.2

```
1   GEN,FLOSS,PROBLEM 13.10,7/1/86,1;
2   CONTINUOUS,2,0,,,0.01,N;
3   INTLC,SS(1)=4.0,SS(2)=0.0;
4   SEVNT,1,SS(1),XN,0.0,0.0;
5   RECORD,TNOW,TIME,0,B,0.05,,,N;
6   VAR,SS(1),H,HEIGHT,0.0,4.0;
7   INIT,0,10;
8   FIN;
```

D13-10.3

INTERMEDIATE RESULTS

BOUNCE NUMBER 1 OCCURRED AT TIME 0.4978

BOUNCE NUMBER 2 OCCURRED AT TIME 1.2941

BOUNCE NUMBER 3 OCCURRED AT TIME 1.9312

BOUNCE NUMBER 4 OCCURRED AT TIME 2.4408

BOUNCE NUMBER 5 OCCURRED AT TIME 2.8485

BOUNCE NUMBER 6 OCCURRED AT TIME 3.1745

BOUNCE NUMBER 7 OCCURRED AT TIME 3.4353

BOUNCE NUMBER 8 OCCURRED AT TIME 3.6439

D13-10.4

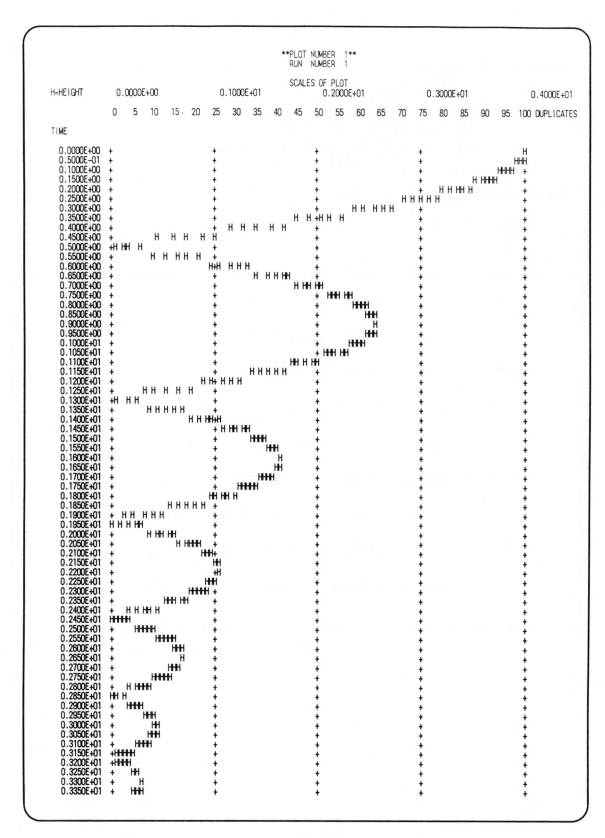

13-11. The coding of subroutines STATE and EVENT for the pendulum problem are shown in D13-11.1 and D13-11.2 respectively. The input statements for this exercise are shown in D13-11.3. The intermediate results obtained from the print statements in subroutine EVENT are shown in D13-11.4. A portion of the plot of the pendulum position overtime is shown in D13-11.5.

```
      SUBROUTINE STATE
      COMMON/SCOM1/ATRIB(100),DD(100),DDL(100),DTNOW,II,MFA,MSTOP,NCLNR
     1,NCRDR,NPRNT,NNRUN,NNSET,NTAPE,SS(100),SSL(100),TNEXT,TNOW,XX(100)
      REAL K
      DATA K/0.5/
      DD(1)=SS(2)
      DD(2)=2.0*K*K*SS(1)*SS(1)*SS(1)-(1.0+K*K)*SS(1)
      RETURN
      END
```

D13-11.1

```
      SUBROUTINE EVENT
      COMMON/SCOM1/ATRIB(100),DD(100),DDL(100),DTNOW,II,MFA,MSTOP,NCLNR
     1,NCRDR,NPRNT,NNRUN,NNSET,NTAPE,SS(100),SSL(100),TNEXT,TNOW,XX(100)
      II=II+1
      WRITE(6,100)II,TNOW
      WRITE(6,101)
      IF (II.EQ.6) MSTOP=-1
      RETURN
  100 FORMAT(38X,'MAXIMUM NUMBER ',I1,' OCCURRED AT TIME ',F7.4)
  101 FORMAT(/)
      END
```

D13-11.2

```
      1   GEN,FLOSS,PROBLEM 13.11,7/2/86,1;
      2   CONTINUOUS,2,0,,,0.1,N;
      3   INTLC,SS(1)=0.0,SS(2)=1.0;
      4   SEVNT,1,DD(1),X,0.0,0.0;
      5   RECORD,TNOW,TIME,0,P,0.1,,,N;
      6   VAR,SS(1),P,POSITION,-1.0,1.0;
      7   INIT,0,15;
      8   FIN;
```

D13-11.3

```
                    **INTERMEDIATE RESULTS**

      MAXIMUM NUMBER 1 OCCURRED AT TIME   1.6858

      MAXIMUM NUMBER 2 OCCURRED AT TIME   5.0573

      MAXIMUM NUMBER 3 OCCURRED AT TIME   8.4297

      MAXIMUM NUMBER 4 OCCURRED AT TIME  11.8010
```

D13-11.4

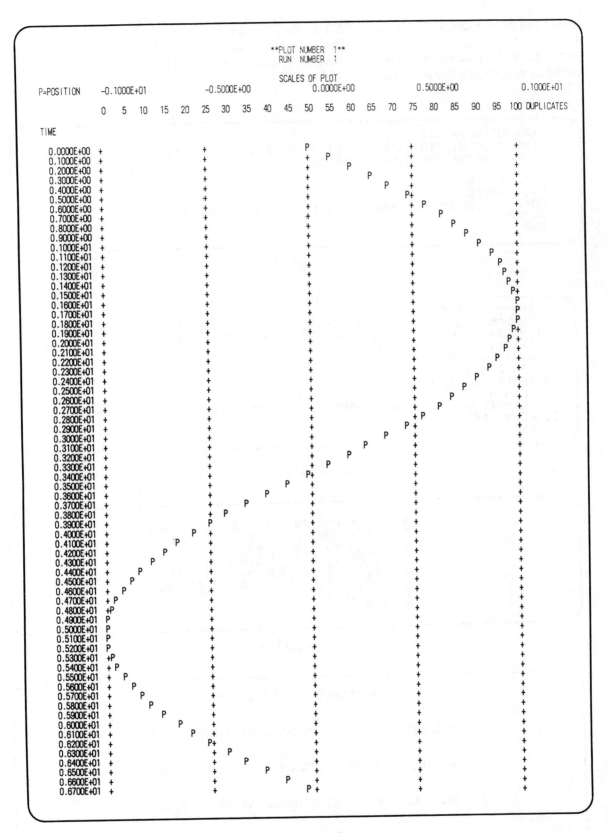

D13-11.5

CHAPTER 14

Combined Modeling

14-1. This exercise has many variations and has sufficient breadth to challenge a serious student of simulation. The solution requires the use of discrete events and the continuous features of SLAM II. The program contains many comments and should be readable. Subroutine EVENT is shown in D14-1.1. The definition of the event codes are:

1   Arrival of a drag
2   Discharge of a half drag from the pit
3   Transfer from the queue of hot ingots to the queue of cold ingots
4-7   Start of soaking for pits 1-4, respectively
8-11 Loading of half drags from the queue of cold ingots into pits 1-4, respectively

Subroutine INTLC is listed in D14-1.2 and initializes the temperatures of the pits, SS(J); initializes the value to which the pit is being driven, DRIVE(J); initializes the pit status indicator, SS(J), to 1 for a busy pit and 0 for an idle pit: sets the SLAM II variable II to 0 and the indicator for a pit in a soaking mode, IND(K) to 0 (initially none of the pits is in the soaking mode). The last statement of subroutine INTLC schedules the first arrival to occur at time 0.

The listing of subroutine ARRVL is shown in D14-1.3. The next arrival is scheduled and the temperature of the current arriving drag is established. Next, a check on XX(J) is made to determine for each half drag whether a pit is available. If not, a transfer event (event code 3) is scheduled to occur and the half drag is placed in the hot queue (file 5). When the transfer event is scheduled the location of the half drag in file 5 is maintained as ATRIB(6) by giving it the value of the successor of the pointer to the first available set of locations for storing an entry in a file. In this manner, a transfer from the hot queue to the cold queue can be made without searching the event calendar (See discussion on reneging in Chapter 12).

At statement 20, pit J is made busy, the temperature of the pit with a half drag inserted is established, the pit is put into a charging state, and the half drag is filed into the file for pit J.

```
                    SUBROUTINE EVENT(I)
                    GOTO (1,2,3,4,4,4,4,8,8,8,8),I
            1       CALL ARRVL
                    RETURN
            2       CALL DSCHR
                    RETURN
            3       CALL TRANS
                    RETURN
            4       J=I-3
                    CALL SOAK(J)
                    RETURN
            8       J=I-7
                    CALL LOAD(J)
                    RETURN
                    END
```

D14-1.1

```
      SUBROUTINE INTLC
      COMMON/SCOM1/ATRIB(100),DD(100),DDL(100),DTNOW,II,MFA,MSTOP,NCLNR
     1,NCRDR,NPRNT,NNRUN,NNSET,NTAPE,SS(100),SSL(100),TNEXT,TNOW,XX(100)
      COMMON/UCOM1/IND(4),DRIVE(4)
C     INITIALIZE TEMPERATURES
      SS(1)=1150
      SS(2)=800
      SS(3)=1500
      SS(4)=1000
C     INITIALIZE TEMPERATURE DRIVERS
      DRIVE(1)=2600
      DRIVE(2)=2600
      DRIVE(3)=600
      DRIVE(4)=600
C     INITIALIZE PIT STATUS INDICATORS
      XX(1)=1
      XX(2)=1
      XX(3)=0
      XX(4)=0
      II=0
      DO 10 K=1,4
10    IND(K)=0
C     SCHEDULE FIRST ARRIVAL
      CALL SCHDL(1,0.0,ATRIB)
      RETURN
      END
```

D14-1.2

```
      SUBROUTINE ARRVL
      COMMON/SCOM1/ATRIB(100),DD(100),DDL(100),DTNOW,II,MFA,MSTOP,NCLNR
     1,NCRDR,NPRNT,NNRUN,NNSET,NTAPE,SS(100),SSL(100),TNEXT,TNOW,XX(100)
      COMMON/UCOM1/IND(4),DRIVE(4)
C     SCHEDULE NEXT ARRIVAL
      CALL SCHDL(1,EXPON(1.75,1),ATRIB)
C     ASSIGN ARRIVAL TIME AND TEMPERATURE
      ATRIB(1)=TNOW
5     ATRIB(2)=RLOGN(400.,50.,2)
      IF(ATRIB(2).LT.300.) GOTO 5
      IF(ATRIB(2).GT.600.) GOTO 5
C     CHECK FOR UNOCCUPIED PIT FOR EACH HALF DRAG
      DO 10 K=1,2
      DO 15 J=1,4
      IF(XX(J).LT.1.) GOTO 20
15    CONTINUE
C     ALL PITS OCCUPIED, FIND TIME WHEN
C     INGOTS WILL TURN COLD
      ATRIB(3)=((150.-ATRIB(2))/157.)**2
C     ASSIGN IDENTIFIER AND SCHEDULE TRANSFER
C     OF INGOTS FROM HOT TO COLD QUEUE
      II=II+1
      ATRIB(4)=II
      CALL SCHDL(3,ATRIB(3),ATRIB)
C     PLACE INGOTS IN HOT QUEUE
      CALL FILEM(5,ATRIB)
      GOTO 10
C     UNOCCUPIED PIT FOUND
C     CHANGE PIT STATUS INDICATOR TO BUSY
20    XX(J)=1
C     SET ATTRIBUTE 5 OF THE ENTITY EQUAL TO
C     THE PIT NUMBER IT OCCUPIES
      ATRIB(5)=J
C     CHANGE THE PIT TEMPERATURE EQUATION
      SS(J)=.7*SSL(J)+.3*ATRIB(2)
      DRIVE(J)=2600.
C     PLACE ENTITY IN PIT FILE
      CALL FILEM(J,ATRIB)
10    CONTINUE
      RETURN
      END
```

D14-1.3

When the pit achieves the desired temperature which is specified through an SEVNT statement, the state event SOAK is invoked. In D14- 1.4 a listing of subroutine SOAK is shown which involves copying the attributes of the half drag that is in pit K and scheduling a discharge for that half drag to occur in two hours. The indicator IND(K) is set to 1 to indicate that the pit is in a soaking state.

The listing for the discharge event (event 2) is shown in D14-1.5. The half drag of ingots is removed from file J. In this example no further information on the half drag is requested but could easily be obtained using the attributes of the entity representing the half drag. A check is made to see if any ingots are in the hot queue and, if so, they are removed and their current temperature is established. Next, the ingots are put into a pit in the same manner as was done in the arrival subroutine. The scheduled transfer event is then removed from the event calender. If there are no hot ingots waiting, the pit is set to idle status in order that its temperature be reduced so that it can handle cold ingots.

The transfer event, TRANS, is listed in D14-1.6. NPTR is set as the pointer to the location of the entity in the hot queue. The entity is removed from the hot file by a call to subroutine RMOVE with a negative first argument to indicate a removal by pointer operation. The entity is then placed in the cold bank by filing it in file 6.

14-1,Embellishment(a). Statistics are cleared at time 100 using a MONTR statement and summary reports are obtained also using a MONTR statement. These statements are shown below:

        MONTR,CLEAR,100;
        MONTR,SUMRY,200.,100;

14-1,Embellishment(b). Events are scheduled to occur at times 200 and 300. At each of these times, a variable is set that forces all pits to cool down to process cold ingots until there are no further cold ingots waiting. When all cold ingots are processed, the variable is reset to allow for normal processing.

The loading of a cold ingot into a pit is done in subroutine LOAD which is called when state events 8, 9, 10 or 11 occur. These events occur when a pit's temperature decreases below 750^0. The listing of subroutine LOAD is given in D14-1.7. First, a check is made to see if the pit is idle. If it is busy, a return is made. This statement is required because a pit can go below 750^0 due to the insertion of a new half drag. Next, a check is made to see if there are any ingots in the cold bank and, if there are not, a return occurs. Otherwise, the first ingot in the cold bank is removed by a call to subroutine RMOVE and the pit's status is changed. The pit is then loaded as was done in subroutine ARRVL.

In subroutine STATE, the differential equations are coded for the four pits. If any of the pits are in a soak status, the rate of change of the temperature is established as zero. The code for subroutine STATE is shown in D14-1.8. The listing of the input statements for this exercise is shown in D14-1.9, and there are no special features associated with the input statements. Portions of the outputs for this exercise are shown in D14-1.10.

```
      SUBROUTINE SOAK(K)
      COMMON/SCOM1/ATRIB(100),DD(100),DDL(100),DTNOW,II,MFA,MSTOP,NCLNR
     1,NCRDR,NPRNT,NNRUN,NNSET,NTAPE,SS(100),SSL(100),TNEXT,TNOW,XX(100)
      COMMON/UCOM1/IND(4),DRIVE(4)
C     SET INDICATOR TO 1 TO KEEP TEMPERATURE CONSTANT
      IND(K)=1
C     SCHEDULE DISCHARGE
      CALL COPY(1,K,ATRIB)
      CALL SCHDL(2,2.,ATRIB)
      RETURN
      END
```

```
      SUBROUTINE DSCHR
      COMMON/SCOM1/ATRIB(100),DD(100),DDL(100),DTNOW,II,MFA,MSTOP,NCLNR
     1,NCRDR,NPRNT,NNRUN,NNSET,NTAPE,SS(100),SSL(100),TNEXT,TNOW,XX(100)
      COMMON/UCOM1/IND(4),DRIVE(4)
C     DETERMINE WHICH PIT HAS FINISHED SOAKING
      J=ATRIB(5)
C     REMOVE THE INGOTS FORM THE PIT FILE
      CALL RMOVE(1,J,ATRIB)
C     CHECK TO SEE IF HOT INGOTS ARE WAITING
      IF(NNQ(5).EQ.0) GOTO 10
C     INGOTS WAITING, REMOVE FIRST HALF DRAG FROM
C     QUEUE AND CALCULATE ITS TEMPERATURE
      CALL RMOVE(1,5,ATRIB)
      WATIME=TNOW-ATRIB(1)
      TEMPI=ATRIB(2)-(WATIME**.5)*157.
C     CHANGE THE PIT STATUS INDICATOR TO BUSY
      XX(J)=1
C     SET ATTRIBUTE 5 OF THE ENTITY EQUAL TO
C     THE PIT NUMBER IT OCCUPIES
      ATRIB(5)=J
C     CHANGE THE PIT TEMPERATURE EQUATION
      SS(J)=.7*SSL(J)+.3*TEMPI
      DRIVE(J)=2600
      IND(J)=0
C     PLACE ENTITY IN PIT FILE
      CALL FILEM(J,ATRIB)
      RETURN
C     NO HOT INGOTS WAITING
C       SET PIT STATUS INDICATOR TO IDLE
10    XX(J)=0
C     CHANGE THE PIT TEMPERATURE EQUATION
      DRIVE(J)=600
      IND(J)=0
      RETURN
      END
```

D14-1.5

```
      SUBROUTINE TRANS
      COMMON/SCOM1/ATRIB(100),DD(100),DDL(100),DTNOW,II,MFA,MSTOP,NCLNR
     1,NCRDR,NPRNT,NNRUN,NNSET,NTAPE,SS(100),SSL(100),TNEXT,TNOW,XX(100)
C     DETERMINE ENTITY IDENTIFIER
      VAL=ATRIB(4)
C     FIND RANK OF ENTITY IN HOT QUEUE
      NRANK=NFIND(1,5,4,0,VAL,0.0)
C     IF ENTITY HAS BEEN REMOVED TO PIT, CANCEL TRANSFER
      IF(NRANK.EQ.0) RETURN
C     TRANSFER ENTITY FROM HOT FILE TO COLD FILE
      CALL RMOVE(NRANK,5,ATRIB)
C     SET TEMPERATURE TO THE MINIMUM (COLD) VALUE
      ATRIB(2)=150
C     PLACE IN COLD QUEUE
      CALL FILEM(6,ATRIB)
      RETURN
      END
```

D14-1.6

```
      SUBROUTINE LOAD(K)
      COMMON/SCOM1/ATRIB(100),DD(100),DDL(100),DTNOW,II,MFA,MSTOP,NCLNR
     1,NCRDR,NPRNT,NNRUN,NNSET,NTAPE,SS(100),SSL(100),TNEXT,TNOW,XX(100)
      COMMON/UCOM1/IND(4),DRIVE(4)
C     CHECK TO SEE IF PIT OCCUPIED
      IF(XX(K).GT.0.0) RETURN
C     CHECK TO SEE IF THERE ARE COLD INGOTS WAITING
      IF(NNQ(6).EQ.0) RETURN
C     REMOVE FIRST HALF DRAG FORM COLD QUEUE
      CALL RMOVE(1,6,ATRIB)
C     CHANGE PIT STATUS INDICATOR TO BUSY
      XX(K)=1
C     SET ATTRIBUTE 5 OF THE ENTITY EQUAL TO
C     THE PIT NUMBER IT OCCUPIES
      ATRIB(5)=K
C     CHANGE THE PIT TEMPERATURE EQUATION
      SS(K)=.7*SSL(K)+.3*ATRIB(2)
      DRIVE(K)=2600
C     PLACE ENTITY IN PIT FILE
      CALL FILEM(K,ATRIB)
      RETURN
      END
```

D14-1.7

```
      SUBROUTINE STATE
      COMMON/SCOM1/ATRIB(100),DD(100),DDL(100),DTNOW,II,MFA,MSTOP,NCLNR
     1,NCRDR,NPRNT,NNRUN,NNSET,NTAPE,SS(100),SSL(100),TNEXT,TNOW,XX(100)
      COMMON/UCOM1/IND(4),DRIVE(4)
      DO 10 J=1,4
      DD(J)=DRIVE(J)-SS(J)
      IF(IND(J).EQ.1) DD(J)=0
10    CONTINUE
      RETURN
      END
```

D14-1.8

```
 1    GEN,ROLSTON,PROBLEM 14.1,9/10/80,1;
 2    LIMITS,6,6,200;
 3    CONT,4,0,.01,.1,.25,,.01,.01;
 4    SEVNT,4,SS(1),XP,2200,20;
 5    SEVNT,5,SS(2),XP,2200,20;
 6    SEVNT,6,SS(3),XP,2200,20;
 7    SEVNT,7,SS(4),XP,2200,20;
 8    SEVNT,8,SS(1),XN,750,10;
 9    SEVNT,9,SS(2),XN,750,10;
10    SEVNT,10,SS(3),XN,750,10;
11    SEVNT,11,SS(4),XN,750,10;
12    RECORD,TNOW,TIME,0,P,.25,0,25;
13    VAR,SS(1),F,FIRST;
14    VAR,SS(3),T,THIRD;
15    INIT,0,400;
16    ENTRY/1,0,0,0,0,1/2,0,0,0,0,0,2;
17    FIN;
```

D14-1.9

```
                    S L A M   I I   S U M M A R Y   R E P O R T

        SIMULATION PROJECT PROBLEM 14.1           BY ROLSTON

        DATE  9/10/1980                           RUN NUMBER    1 OF    1


        CURRENT TIME   0.4000E+03
        STATISTICAL ARRAYS CLEARED AT TIME  0.0000E+00


                **FILE STATISTICS**

 FILE              AVERAGE      STANDARD     MAXIMUM   CURRENT    AVERAGE
 NUMBER  LABEL/TYPE LENGTH      DEVIATION    LENGTH    LENGTH     WAITING TIME
   1               0.8788       0.3263          1         1       3.0836
   2               0.8789       0.3262          1         1       3.0839
   3               0.8493       0.3577          1         1       3.1169
   4               0.8496       0.3574          1         1       3.1179
   5               1.3835       2.0007         12         0       1.5038
   6              38.0911      20.3908         68        68     149.3770
   7    CALENDAR    5.6304       3.2265         19         5       2.1068


                **STATE AND DERIVATIVE VARIABLES**

            (I)     SS(I)         DD(I)
             1    0.2067E+04    0.5335E+03
             2    0.2067E+04    0.5335E+03
             3    0.2214E+04    0.0000E+00
             4    0.2214E+04    0.0000E+00
```

D14-1.10

14-1,Embellishment(c). Establish pit number 4 as a processor of cold ingots and let it cool down following each discharge event if ingots are waiting in the cold ingot bank. If no cold ingots are waiting, search the hot ingot file for those ingots with the lowest temperature and assign them to the cold pit processor.

14-1,Embellishment(d). A control strategy to maintain the sum of temperatures of the pits to 7000°F is an interesting problem. At 7000°F the system is very sensitive and the queues could build if a good strategy is not developed.   In the model, another state variable, SS(5), should be defined in subroutine STATE to be equal to the sum of the pit temperatures:

$$SS(5) = SS(1) + SS(2) + SS(3) + SS(4)$$

A SEVNT statement should be included in the input to determine if the value of SS(5) crosses 7000, that is,

SEVNT, 12, SS(5), XP, 7000, 20;

The logic when this state event occurs will depend upon the method to be used to maintain the sum of the temperatures below 7000°F. For example, the temperature equation for one of the pits may be changed to allow cooling. A decision rule is needed to choose which pit to cool. One alternative is to select the pit with the lowest temperature. Also, a flag should be set so that an entity in the ARRVL or LOAD routine will not change the pit temperature equation if this is not desired. Addditional logic will be required to determine when to allow the pit to heat up again. This might be added to the DSCHR routine.

14-2.   Basically, this is a single-server queueing situation in which service is characterized by a continuous process. The coding for this problem is shown in D14-2.1 through D14-2.7. In subroutine ARRL (D14-2.2) when an arrival occurs and the server is idle, SS(1) is reset. The rate at which service is given is set to a nominal value of 1.2. This means it will take less than 3 time units on the average to serve the customer if no one arrives. (The customers require 3.5 time units of service on the average.) If the service time is less than the threshold for the detection of an end-of-service (0.1) then it should be assumed that no service is required. That is, an information request is made of the attendant. If this is not included then the SEVNT statement will not detect a crossing of 0.1 (since no crossing will occur if it starts below 0.1) and the attendant will always be busy following a customer who requires such a small service time. At statement 10, additional arrivals occur and as long as the number waiting is less than 5 the actual service time will be short. The decreasing of the amount of service time remaining is programmed in subroutine STATE which is shown in D14-2.3. The computation of the value S=SRATE(N) is shown in D14-2.4. Note that once there are more than 5 customers in the system, it may take a long time to reduce the queue.

Subroutine ENDSV is listed in D14-2.5, and it follows the format for the standard single channel queueing situation. The various statistical quantities desired are collected in this subroutine. In D14-2.6, the listing of subroutine INTLC is given, and the statements for running this exercise are shown in D14-2.7. The basic SLAM II Summary Report is shown in D14-2.8. Note the large average queue length and that the maximum number of customers waiting reaches 58. The outputs indicate that further investigation of the stability of this system is warranted.

```
              SUBROUTINE EVENT(I)
              GOTO (1,2),I
       1      CALL ARRL
              RETURN
       2      CALL ENDSV
              RETURN
              END
```

D14-2.1

```
        SUBROUTINE ARRL
        COMMON/SCOM1/ATRIB(100),DD(100),DDL(100),DTNOW,II,MFA,MSTOP,NCLNR
       1,NCRDR,NPRNT,NNRUN,NNSET,NTAPE,SS(100),SSL(100),TNEXT,TNOW,XX(100)
        COMMON/UCOM1/S
        EQUIVALENCE (BUSY,XX(1))
        CALL SCHDL(1,EXPON(4.,1),ATRIB)
        ATRIB(1)=TNOW
        IF(BUSY.GT.0.) GOTO 10
        CALL FILEM(2,ATRIB)
        BUSY=1
        WAIT=0
        CALL COLCT(WAIT,1)
        SS(1)=EXPON(3.5,2)
        S=1.2
        RETURN
   10   CALL FILEM(1,ATRIB)
        XX(2)=XX(2)+1.
        N=NNQ(1)
        S=SRATE(N)
        RETURN
        END
```

D14-2.2

```
      SUBROUTINE STATE
      COMMON/SCOM1/ATRIB(100),DD(100),DDL(100),DTNOW,II,MFA,MSTOP,NCLNR
     1,NCRDR,NPRNT,NNRUN,NNSET,NTAPE,SS(100),SSL(100),TNEXT,TNOW,XX(100)
      COMMON/UCOM1/S
      SS(1)=SSL(1)-DTNOW*S
      RETURN
      END
```

D14-2.3

```
                    FUNCTION SRATE(N)
                    IF(N.GE.5) GOTO 10
                    SRATE=1.2*EXP(.173*FLOAT(N))
                    GOTO 20
               10   SRATE=1.2*(.25*EXP(-.305*FLOAT(N-5))+.5)
               20   RETURN
                    END
```

D14-2.4

```
      SUBROUTINE ENDSV
      COMMON/SCOM1/ATRIB(100),DD(100),DDL(100),DTNOW,II,MFA,MSTOP,NCLNR
     1,NCRDR,NPRNT,NNRUN,NNSET,NTAPE,SS(100),SSL(100),TNEXT,TNOW,XX(100)
      COMMON/UCOM1/S
      EQUIVALENCE (BUSY,XX(1))
      CALL RMOVE(1,2,ATRIB)
      TSYS=TNOW-ATRIB(1)
      CALL COLCT(TSYS,2)
      IF(NNQ(1).GT.0) GOTO 50
      BUSY=0
      SS(1)=0
      S=0
      RETURN
   50 CALL RMOVE(1,1,ATRIB)
      WAIT=TNOW-ATRIB(1)
      CALL COLCT(WAIT,1)
      CALL FILEM(2,ATRIB)
      XX(2)=XX(2)-1.
      N=NNQ(1)
      S=SRATE(N)
      SS(1)=EXPON(3.5,2)
      RETURN
      END
```

D14-2.5

```
SUBROUTINE INTLC
COMMON/SCOM1/ATRIB(100),DD(100),DDL(100),DTNOW,II,MFA,MSTOP,NCLNR
1,NCRDR,NPRNT,NNRUN,NNSET,NTAPE,SS(100),SSL(100),TNEXT,TNOW,XX(100)
COMMON/UCOM1/S
EQUIVALENCE (BUSY,XX(1))
BUSY=0
SS(1)=0
S=0
XX(2)=0
CALL SCHDL(1,0.0,ATRIB)
RETURN
END
```

D14-2.6

```
 1    GEN,ROLSTON,PROBLEM 14.2,8/29/80,1;
 2    LIMITS,2,1,100;
 3    STAT,1,WAIT TIME;
 4    STAT,2,TIME IN SYSTEM;
 5    TIMST,XX(1),SERVER UTIL;
 6    CONT,0,1,.01,.1,.1;
 7    RECORD,TNOW,TIME,0,P,.1,800,825;
 8    VAR,SS(1),T,REMAIN TIME;
 9    VAR,XX(1),S,SERV. STATUS;
10    VAR,XX(2),Q,NO IN QUEUE;
11    SEVNT,2,SS(1),XN,0,.01;
12    INIT,0,1000;
13    FIN;
```

D14-2.7

```
                    S L A M   I I   S U M M A R Y   R E P O R T

        SIMULATION PROJECT PROBLEM 14.2           BY ROLSTON

        DATE  8/29/1980                           RUN NUMBER    1 OF    1


        CURRENT TIME    0.1000E+04
        STATISTICAL ARRAYS CLEARED AT TIME   0.0000E+00


              **STATISTICS FOR VARIABLES BASED ON OBSERVATION**

                  MEAN        STANDARD     COEFF. OF    MINIMUM    MAXIMUM    NUMBER OF
                  VALUE       DEVIATION    VARIATION    VALUE      VALUE      OBSERVATIONS

  WAIT TIME     0.4202E+02   0.6511E+02   0.1549E+01   0.0000E+00  0.1945E+03      220
  TIME IN SYSTEM 0.4538E+02  0.6614E+02   0.1458E+01   0.6671E-01  0.1993E+03      219



              **STATISTICS FOR TIME-PERSISTENT VARIABLES**

                  MEAN        STANDARD     MINIMUM      MAXIMUM    TIME        CURRENT
                  VALUE       DEVIATION    VALUE        VALUE      INTERVAL    VALUE

  SERVER UTIL   0.8855E+00   0.3185E+00   0.0000E+00   0.1000E+01  0.1000E+04  0.1000E+01



              **FILE STATISTICS**

  FILE                AVERAGE      STANDARD     MAXIMUM    CURRENT    AVERAGE
  NUMBER  LABEL/TYPE  LENGTH       DEVIATION    LENGTH     LENGTH     WAITING TIME

    1               15.0448      18.7385        58         53        66.2764
    2                0.8855       0.3185         1          1         4.0248
    3    CALENDAR     1.0000       0.0000         1          1         3.6496



              **STATE AND DERIVATIVE VARIABLES**


          (I)        SS(I)         DD(I)
           1       0.5334E+01    0.0000E+00
```

D14-2.8

14-3.   An interpretation must be given to the time between jams, since the problem statement does not specify that the machine must be working in order that a jam occur. One interpretation is that the time between jams includes the time the machine is not working. This simplifies the modeling effort in that it is no longer necessary to monitor the number of sparkplugs produced by a machine since its last jamming but only to schedule a stop machine event, STOPM, from the start event. Thus, the change to the model under the above interpretation is to schedule a STOPM machine event (event 2) in a uniformly distributed time between 200/5.55 and 600/5.55 seconds. With this change the network model can be deleted from the input statements. The net effect of this change will be to increase the flow of sparkplugs into the barrel because machine jamming occurs more frequently.   Subroutines START and STOPM are given in D14-3.1 and D14-3.2 respectively. The SLAM II Summary Report is shown in D14-3.3.

```
SUBROUTINE START
COMMON/SCOM1/ATRIB(100),DD(100),DDL(100),DTNOW,II,MFA,MSTOP,NCLNR
1,NCRDR,NPRNT,NNRUN,NNSET,NTAPE,SS(100),SSL(100),TNEXT,TNOW,XX(100)
COMMON/UCOM1/FLOW(10),PFLOWR,NFLOW(10)
MACH=ATRIB(1)
CALL SCHDL(2,UNFRM(XX(20),XX(21),2),ATRIB)
XX(MACH+6)=1.0
IF(FLOW(MACH).LT.PFLOWR)RETURN
XX(MACH)=1.0
ATRIB(1)=ATRIB(1)+1
ATRIB(2)=-PFLOWR
CALL SCHDL(3,9.,ATRIB)
RETURN
END
```

D14-3.1

```
SUBROUTINE STOPM
COMMON/SCOM1/ATRIB(100),DD(100),DDL(100),DTNOW,II,MFA,MSTOP,NCLNR
1,NCRDR,NPRNT,NNRUN,NNSET,NTAPE,SS(100),SSL(100),TNEXT,TNOW,XX(100)
COMMON/UCOM1/FLOW(10),PFLOWR,NFLOW(10)
MACH=ATRIB(1)
XX(MACH)=0.0
XX(MACH+6)=0.0
SS(MACH)=0
CALL SCHDL(1,UNFRM(6.,24.,1),ATRIB)
IF(FLOW(MACH).LT.PFLOWR)RETURN
ATRIB(1)=ATRIB(1)+1.
ATRIB(2)=PFLOWR
CALL SCHDL(3,9.,ATRIB)
RETURN
END
```

D14-3.2

```
                    S L A M   I I   S U M M A R Y   R E P O R T

        SIMULATION PROJECT PROBLEM 14.3              BY FLOSS

        DATE  7/ 7/1986                              RUN NUMBER    1 OF    1


        CURRENT TIME   0.6000E+04
        STATISTICAL ARRAYS CLEARED AT TIME  0.0000E+00


        **STATISTICS FOR TIME-PERSISTENT VARIABLES**

                    MEAN        STANDARD    MINIMUM     MAXIMUM     TIME        CURRENT
                    VALUE       DEVIATION   VALUE       VALUE       INTERVAL    VALUE

RATE IN BAR     0.2023E+00  0.1059E+01  0.0000E+00  0.1110E+02  0.6000E+04  0.0000E+00
MTIME FOR 1     0.8337E+00  0.3724E+00  0.0000E+00  0.1000E+01  0.6000E+04  0.0000E+00
MTIME FOR 2     0.8136E+00  0.3894E+00  0.0000E+00  0.1000E+01  0.6000E+04  0.1000E+01
MTIME FOR 3     0.8229E+00  0.3817E+00  0.0000E+00  0.1000E+01  0.6000E+04  0.1000E+01
MTIME FOR 4     0.3744E+00  0.4840E+00  0.0000E+00  0.1000E+01  0.6000E+04  0.0000E+00
MTIME FOR 5     0.1053E+00  0.3069E+00  0.0000E+00  0.1000E+01  0.6000E+04  0.0000E+00


        **STATE AND DERIVATIVE VARIABLES**

                    (I)       SS(I)          DD(I)
                     1     0.0000E+00     0.0000E+00
                     2     0.3667E+03     0.0000E+00
                     3     0.3827E+03     0.0000E+00
                     4     0.0000E+00     0.0000E+00
                     5     0.1264E+03     0.0000E+00
                     6     0.1214E+04     0.0000E+00
```

D14-3.3

14-4.   By having the sparkplugs arrive in trays, a different modeling view of the system can be taken. Each tray can be considered an entity which requires a specified amount of processing time. When a machine is jammed, the tray with its remaining processing time can be routed to the next machine. Thus, a network model of the conveyor system can be developed. (A direct embellishment of Example 14-2 using discrete events can also be employed. For comparison purposes the network-continuous model approach is presented.) A network model involving the flow of trays of sparkplugs over the conveyor to one of five machines or into a barrel is shown in D14-4.1. A CREATE node generates the trays with the first tray arriving at time 18 seconds and every 20 seconds thereafter. Attribute 1 will be used to represent the machine number and attribute 2 the amount of processing time required to pack the plugs that are on the tray. The tray is then routed to ASSIGN node AS1 where the machine number is indexed. If the machine number is 6, the tray entity is routed to the barrel where the number of sparkplugs in the tray is added to SS(6). It is assumed that the sparkplugs in the tray are dumped in the barrel at once. At ASSIGN node AS1 the tray is routed to the next machine if II is less than 6 and the machine is not working on a tray. Otherwise, the tray is routed back to ASSIGN node AS1 after a delay of 9 seconds. At AWAIT node AW1, the tray is allocated the machine resource and processing begins in the following activity. The activity time is prescribed by ATRIB(2) which for a full tray is 60 seconds. Following the completion of the processing of the tray, the machine resource is freed and the tray entity exits the system.

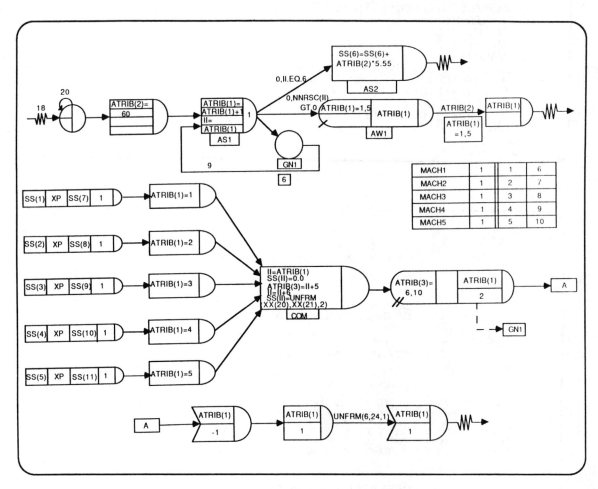

D14-4.1

The jamming of a machine is modeled with a DETECT node and ASSIGN node as in the basic model presented in Example 14-2. However, when the machine is jammed, the tray entity is preempted from the machine at a PREEMPT node. The remaining processing time is stored in attribute 2 and the preempted entity is routed to node GN1 to send the tray after a 9 second delay to the next machine. The PREEMPT entity is then routed to an ALTER node to reduce the capacity of the machine resource by 1 and then to a FREE node so that the machine resource is not considered in use while it is being repaired. The machine is then unjammed during the repair activity. An ALTER node is then used to increase the capacity of the machine by 1 to indicate that it is now available to process tray entities. The coding of subroutine STATE for this example is shown in D14-4.2 where SS(I) is used to accumulate the number of sparkplugs packed since the last jamming event. Sparkplugs packed are increased by the rate of packing plugs times a 0-1 variable which indicates whether or not the machine is packing. Subroutine INTLC is shown in D14-4.3 and the input statements are given in D14-4.4. The summary report obtained from this model is shown in D14-4.5.

```
      SUBROUTINE STATE
      COMMON/SCOM1/ATRIB(100),DD(100),DDL(100),DTNOW,II,MFA,MSTOP,NCLNR
     1,NCRDR,NPRNT,NNRUN,NNSET,NTAPE,SS(100),SSL(100),TNEXT,TNOW,XX(100)
      DO 100 I=1,5
          SS(I)=SSL(I)+FLOAT(NNACT(I))*DTNOW*5.55
 100  CONTINUE
      RETURN
      END
```

D14-4.2

```
      SUBROUTINE INTLC
      COMMON/SCOM1/ATRIB(100),DD(100),DDL(100),DTNOW,II,MFA,MSTOP,NCLNR
     1,NCRDR,NPRNT,NNRUN,NNSET,NTAPE,SS(100),SSL(100),TNEXT,TNOW,XX(100)
      XX(20)=200.0
      XX(21)=600.0
      DO 100 I=1,5
          SS(I)=0.0
          SS(I+6)=UNFRM(XX(20),XX(21),2)
 100  CONTINUE
      SS(6)=0.0
      RETURN
      END
```

D14-4.3

```
 1    GEN,PRITSKER,PROBLEM 14.4,8/6/86;
 2    LIMITS,10,3,100;
 3    RECORD,TNOW,TIME,0,B,10,2500,3000;
 4    VAR,SS(1),1,S1,0,4000;
 5    VAR,SS(2),2,S2,-1000,3000;
 6    VAR,SS(3),3,S3,-2000,2000;
 7    VAR,SS(4),4,S4,-3000,1000;
 8    VAR,SS(5),5,S5,-3000,1000;
 9    VAR,SS(6),6,BARREL,0,1000;
10    NETWORK;
11              RESOURCE/1,MACH1(1),1,6;
12              RESOURCE/2,MACH2(1),2,7;
13              RESOURCE/3,MACH3(1),3,8;
14              RESOURCE/4,MACH4(1),4,9;
15              RESOURCE/5,MACH5(1),5,10;
16    ;
17              CREATE,20,18;
18              ASSIGN,ATRIB(2)=60;
19    AS1       ASSIGN,ATRIB(1)=ATRIB(1)+1,II=ATRIB(1),1;
20              ACTIVITY,,II.EQ.6,AS2;
21              ACTIVITY,,NNRSC(II).GT.0,AW1;
22              ACTIVITY;
23    GN1       GOON;
24              ACTIVITY/6,9,,AS1;
25    AW1       AWAIT(ATRIB(1)=1,5),ATRIB(1);
26              ACTIVITY/ATRIB(1)=1,5,ATRIB(2);
27              FREE,ATRIB(1);
28              TERMINATE;
29    AS2       ASSIGN,SS(6)=SS(6)+ATRIB(2)*5.55;
30              TERMINATE;
31    ;
32              DETECT,SS(1),XP,SS(7),1;
33              ASSIGN,ATRIB(1)=1;
34              ACTIVITY,,,COM;
35              DETECT,SS(2),XP,SS(8),1;
36              ASSIGN,ATRIB(1)=2;
37              ACTIVITY,,,COM
38              DETECT,SS(3),XP,SS(9),1;
39              ASSIGN,ATRIB(1)=3;
40              ACTIVITY,,,COM;
41              DETECT,SS(4),XP,SS(10),1;
42              ASSIGN,ATRIB(1)=4;
43              ACTIVITY,,,COM;
44              DETECT,SS(5),XP,SS(11),1;
45              ASSIGN,ATRIB(1)=5;
46              ACTIVITY,,,COM;
47    COM       ASSIGN,II=ATRIB(1),SS(II)=0.0,ATRIB(3)=II+5,
48                  II=II+6,SS(II)=UNFRM(XX(20),XX(21),2);
49              PREEMPT(ATRIB(3)=6,10),ATRIB(1),GN1,2;
50              ALTER,ATRIB(1),-1;
51              FREE,ATRIB(1)/1;
52              ACTIVITY,UNFRM(6.0,24.0,1);
53              ALTER,ATRIB(1),1;
54              TERMINATE;
55              ENDNETWORK;
56    INIT,0,6000;
57    CONTINUOUS,0,11,.1;
58    FIN;
```

D14-4.4

```
                    S L A M   I I   S U M M A R Y   R E P O R T

        SIMULATION PROJECT PROBLEM 14.4          BY PRITSKER

        DATE  8/ 6/1986                          RUN NUMBER    1 OF    1


        CURRENT TIME   0.6000E+04
        STATISTICAL ARRAYS CLEARED AT TIME  0.0000E+00
```

FILE STATISTICS

FILE NUMBER	LABEL/TYPE	AVERAGE LENGTH	STANDARD DEVIATION	MAXIMUM LENGTH	CURRENT LENGTH	AVERAGE WAITING TIME
1	AW1 AWAIT	0.0000	0.0000	1	0	0.0000
2	AW1 AWAIT	0.0000	0.0000	1	0	0.0000
3	AW1 AWAIT	0.0000	0.0000	1	0	0.0000
4	AW1 AWAIT	0.0000	0.0000	1	0	0.0000
5	AW1 AWAIT	0.0000	0.0000	1	0	0.0000
6	PREEMPT	0.0000	0.0000	0	0	0.0000
7	PREEMPT	0.0000	0.0000	0	0	0.0000
8	PREEMPT	0.0000	0.0000	0	0	0.0000
9	PREEMPT	0.0000	0.0000	0	0	0.0000
10	PREEMPT	0.0000	0.0000	0	0	0.0000
11	CALENDAR	5.4851	1.0785	10	5	6.3072

REGULAR ACTIVITY STATISTICS

ACTIVITY INDEX/LABEL	AVERAGE UTILIZATION	STANDARD DEVIATION	MAXIMUM UTIL	CURRENT UTIL	ENTITY COUNT
1	0.7465	0.4350	1	1	45
2	0.6793	0.4668	1	0	36
3	0.5551	0.4970	1	1	45
4	0.4264	0.4945	1	0	67
5	0.2915	0.4544	1	0	46
6	1.2048	0.9647	5	1	803

RESOURCE STATISTICS

RESOURCE NUMBER	RESOURCE LABEL	CURRENT CAPACITY	AVERAGE UTILIZATION	STANDARD DEVIATION	MAXIMUM UTILIZATION	CURRENT UTILIZATION
1	MACH1	1	0.7465	0.4350	1	1
2	MACH2	0	0.6793	0.4668	1	0
3	MACH3	1	0.5551	0.4970	1	1
4	MACH4	1	0.4264	0.4945	1	0
5	MACH5	1	0.2915	0.4544	1	0

RESOURCE NUMBER	RESOURCE LABEL	CURRENT AVAILABLE	AVERAGE AVAILABLE	MINIMUM AVAILABLE	MAXIMUM AVAILABLE
1	MACH1	0	0.1029	-1	1
2	MACH2	0	0.1813	-1	1
3	MACH3	0	0.3137	-1	1
4	MACH4	1	0.4820	-1	1
5	MACH5	1	0.6398	-1	1

STATE AND DERIVATIVE VARIABLES

(I)	SS(I)	DD(I)
1	0.2331E+03	0.0000E+00
2	0.0000E+00	0.0000E+00
3	0.2866E+03	0.0000E+00
4	0.0000E+00	0.0000E+00
5	0.5316E+03	0.0000E+00
6	0.9288E+04	0.0000E+00
7	0.5479E+03	0.0000E+00
8	0.5821E+03	0.0000E+00
9	0.3177E+03	0.0000E+00
10	0.2645E+03	0.0000E+00
11	0.5918E+03	0.0000E+00

D14-4.5

14-5.   The redesign of the sparkplug packaging line by eliminating one of the machines is an interesting problem. The conveyor acts as a buffer storage for each of the machines. If machine 5 is eliminated, we remove the machine that has the lowest utilization but all the sparkplugs that were packed by it go into the barrel. If machine 4 is eliminated, then the conveyor portion between machine 3 and machine 5 acts as a buffer for machine 5 and the utilization of machine 5 will increase drastically. Since both machines 4 and 5 were underutilized in the basic problem, the utilization of machine 5 will tend to be the sum of the utilizations of machines 4 and 5 in the basic problem. Thus, the removal of machine 4 does not significantly increase the flow of plugs into the barrel but does delay the output of the plugs previously packed by machine 4 by the extra 9 seconds to route them to machine 5. Elimination of machine 3 does not appear to be a good alternative as the input rate is designed to keep three machines busy. From the output from the basic problem, it is seen that machines 1, 2 and 3 are utilized 80 percent of the time. Thus, the design should attempt to space machines 1, 2 and 3 so that their overflow during the time that they are jammed reaches the last machine at different times. Thus, the design problem can be restated as "the positioning of the first three machines and a fourth machine such that the repair intervals for the first three machines results in arrival to the fourth machine at different times". The difficulty of solving this design problem is that there are random variables which prescribe the time at which the repair interval begins. Thus, it is desired to minimize the probability of having concurrent arrivals to the fourth machine.

In addition to eliminating a machine, this exercise involves changing the time of scheduling arrivals to the different machines. The routine to change the input rate of sparkplugs every 1000 seconds is shown in D14-5.1. In subroutines START and STOPM, the amouunt of change, ATRIB(2), needs to be recalculated as

MIN(FLOW(M)-PFLOWR)

since the input rate to the conveyor is not a multiple of PFLOWR. This also requires changes in subroutine FLOWCHNG as shown in D14-5.2. In subroutines START and STOPM, the amount of change, ATRIB(2), needs to be recalculated as MIN(FLOW(M)-PFLOWR) since the input rate to the conveyor is not a multiple of PFLOWR. This also requires changes in subroutine FLOWCHNG as shown in D14-5.2. Results from eliminating the fourth machine and fifth machine from the model for this new input rate but keeping the locations of the machines as stated in the original problem is shown in D14-5.3 and D14-5.4 respectively. These results indicate that there is an increase in the number of arrivals so that all four machines are close to capacity and there is no statistical difference in the utilizations of the machines. Note only six samples of input rate were drawn to obtain these results.

```
      SUBROUTINE INRATE
      COMMON/SCOM1/ATRIB(100),DD(100),DDL(100),DTNOW,II,MFA,MSTOP,NCLNR
     1,NCRDR,NPRNT,NNRUN,NNSET,NTAPE,SS(100),SSL(100),TNEXT,TNOW,XX(100)
      COMMON/UCOM1/FLOW(10),PFLOWR,NFLOW(10)
      DIMENSION CPROB(3),VALUE(3)
      DATA CPROB/0.25,0.75,1.00/
      DATA VALUE/888.0,999.0,1110.0/
      CALL SCHDL(4,1000.0,ATRIB)
      ATRIB(1)=1.0
      ATRIB(2)=DPROB(CPROB,VALUE,3,7)/60.0-FLOW(1)
      CALL SCHDL(3,0.0,ATRIB)
      RETURN
      END
```

D14-5.1

```
                    S L A M   I I   S U M M A R Y   R E P O R T

          SIMULATION PROJECT PROBLEM 14.5          BY FLOSS

          DATE  7/11/1986                          RUN NUMBER   1 OF   1


          CURRENT TIME   0.6000E+04
          STATISTICAL ARRAYS CLEARED AT TIME  0.0000E+00


               **STATISTICS FOR TIME-PERSISTENT VARIABLES**

                 MEAN       STANDARD    MINIMUM     MAXIMUM     TIME        CURRENT
                 VALUE      DEVIATION   VALUE       VALUE       INTERVAL    VALUE

   RATE IN BAR   0.1702E+01  0.2829E+01  0.0000E+00  0.1850E+02  0.6000E+04  0.0000E+00
   MTIME FOR 1   0.8307E+00  0.3751E+00  0.0000E+00  0.1000E+01  0.6000E+04  0.0000E+00
   MTIME FOR 2   0.8138E+00  0.3893E+00  0.0000E+00  0.1000E+01  0.6000E+04  0.0000E+00
   MTIME FOR 3   0.8269E+00  0.3783E+00  0.0000E+00  0.1000E+01  0.6000E+04  0.1000E+01
   MTIME FOR 4   0.0000E+00  0.0000E+00  0.0000E+00  0.0000E+00  0.6000E+04  0.0000E+00
   MTIME FOR 5   0.7840E+00  0.4115E+00  0.0000E+00  0.1000E+01  0.6000E+04  0.1000E+01


               **STATE AND DERIVATIVE VARIABLES**

                 (I)      SS(I)        DD(I)
                  1     0.0000E+00   0.0000E+00
                  2     0.0000E+00   0.0000E+00
                  3     0.5250E+03   0.0000E+00
                  4     0.0000E+00   0.0000E+00
                  5     0.1091E+03   0.0000E+00
                  6     0.1021E+05   0.0000E+00
                  7     0.4693E+03   0.0000E+00
                  8     0.2292E+03   0.0000E+00
                  9     0.5956E+03   0.0000E+00
                 10     0.2606E+03   0.0000E+00
                 11     0.5315E+03   0.0000E+00
```

D14-5.2

```
                    S L A M   I I   S U M M A R Y   R E P O R T

          SIMULATION PROJECT PROBLEM 14.5          BY FLOSS

          DATE  7/11/1986                          RUN NUMBER   1 OF   1


          CURRENT TIME   0.6000E+04
          STATISTICAL ARRAYS CLEARED AT TIME  0.0000E+00


               **STATISTICS FOR TIME-PERSISTENT VARIABLES**

                 MEAN       STANDARD    MINIMUM     MAXIMUM     TIME        CURRENT
                 VALUE      DEVIATION   VALUE       VALUE       INTERVAL    VALUE

   RATE IN BAR   0.1666E+01  0.2793E+01  0.0000E+00  0.1850E+02  0.6000E+04  0.0000E+00
   MTIME FOR 1   0.8307E+00  0.3750E+00  0.0000E+00  0.1000E+01  0.6000E+04  0.0000E+00
   MTIME FOR 2   0.8168E+00  0.3868E+00  0.0000E+00  0.1000E+01  0.6000E+04  0.1000E+01
   MTIME FOR 3   0.8281E+00  0.3773E+00  0.0000E+00  0.1000E+01  0.6000E+04  0.0000E+00
   MTIME FOR 4   0.7838E+00  0.4117E+00  0.0000E+00  0.1000E+01  0.6000E+04  0.1000E+01
   MTIME FOR 5   0.0000E+00  0.0000E+00  0.0000E+00  0.0000E+00  0.6000E+04  0.0000E+00


               **STATE AND DERIVATIVE VARIABLES**

                 (I)      SS(I)        DD(I)
                  1     0.0000E+00   0.0000E+00
                  2     0.1002E+03   0.0000E+00
                  3     0.0000E+00   0.0000E+00
                  4     0.1417E+03   0.0000E+00
                  5     0.0000E+00   0.0000E+00
                  6     0.9994E+04   0.0000E+00
                  7     0.4693E+03   0.0000E+00
                  8     0.2292E+03   0.0000E+00
                  9     0.3496E+03   0.0000E+00
                 10     0.3749E+03   0.0000E+00
                 11     0.4580E+03   0.0000E+00
```

D14-5.3

```
                    S L A M   I I   S U M M A R Y   R E P O R T


        SIMULATION PROJECT PROBLEM 14.5          BY FLOSS

        DATE  7/11/1986                          RUN NUMBER    1 OF    1



        CURRENT TIME   0.6000E+04
        STATISTICAL ARRAYS CLEARED AT TIME  0.0000E+00



           **STATISTICS FOR TIME-PERSISTENT VARIABLES**

                   MEAN        STANDARD    MINIMUM     MAXIMUM     TIME        CURRENT
                   VALUE       DEVIATION   VALUE       VALUE       INTERVAL    VALUE

   RATE IN BAR     0.1702E+01  0.2829E+01  0.0000E+00  0.1850E+02  0.6000E+04  0.0000E+00
   MTIME FOR 1     0.8307E+00  0.3751E+00  0.0000E+00  0.1000E+01  0.6000E+04  0.0000E+00
   MTIME FOR 2     0.8138E+00  0.3893E+00  0.0000E+00  0.1000E+01  0.6000E+04  0.0000E+00
   MTIME FOR 3     0.8269E+00  0.3783E+00  0.0000E+00  0.1000E+01  0.6000E+04  0.1000E+01
   MTIME FOR 4     0.0000E+00  0.0000E+00  0.0000E+00  0.0000E+00  0.6000E+04  0.0000E+00
   MTIME FOR 5     0.7840E+00  0.4115E+00  0.0000E+00  0.1000E+01  0.6000E+04  0.1000E+01



           **STATE AND DERIVATIVE VARIABLES**


           (I)        SS(I)         DD(I)
            1      0.0000E+00    0.0000E+00
            2      0.0000E+00    0.0000E+00
            3      0.5250E+03    0.0000E+00
            4      0.0000E+00    0.0000E+00
            5      0.1091E+03    0.0000E+00
            6      0.1021E+05    0.0000E+00
            7      0.4693E+03    0.0000E+00
            8      0.2292E+03    0.0000E+00
            9      0.5956E+03    0.0000E+00
           10      0.2606E+03    0.0000E+00
           11      0.5315E+03    0.0000E+00
```

D14-5.4

14-6.   The SLAM II variables, files, events, and user variables used in the solution to this exercise are shown in D14-6.1. The input statements are shown in D14-6.2 which is similar to Figure 9-3. An AWAIT node follows each service activity so that car entities wait for an EXIT resource before entering the street. The capacity of five car entities is prescribed for the AWAIT node and if five cars are waiting for the resource exit and the teller completes serving a customer, the teller and the customer are blocked. Subroutine ALLOC is prescribed for the resource field at the AWAIT node in order that both the EXIT resource can be seized and jockeying between the two teller queues can be accomplished if necessary. The coding of subroutine ALLOC is shown in D14-6.3. Note the use of subroutines ULINK and LINK in subroutine ALLOC. These subroutines could have been used in Example 9-1, however they were not introduced until Chapter 12.

Subroutine EVENT is shown in D14-6.4. Event 1 is called after a car has seized the exit. The car is placed in file 5 to await entry in the street traffic. Subroutine DEPART is called to determine if the car can go into the traffic. The coding for subroutine DEPART is shown in Figure D14-6.5. First the waiting time of the first car in file 5 is determined as TNOW - ATRIB(2) for the first entry in file 5. A direct accessing of attribute 2 is made by referencing the cell of QSET that is two words past the pointer to the first entry in file 5. The gap time acceptable for the driver is computed using the table lookup function. If the interval of time until the next car arrival, TNCA-TNOW, is less than the anxiety time, the car does not enter the street, otherwise it does enter the street and event 5 is scheduled to occur after a 0.1 time delay.

```
C
C
C
C
C         ********************************************************
C
C                         SLAM VARIABLES
C                         ---------------
C         ATRIB(1) = MARK TIME
C         ATRIB(2) = TIME WAITING AT EXIT OR 1.0 FOR CAR EXITING
C                    BANK
C         XX(1)    = LIGHT STATUS  0 - RED  1 - GREEN
C         XX(2)    = CAR PASSING THROUGH EXIT
C                                  0 - NO CAR PASSING THROUGH
C                                  1 - CAR PASSING THROUGH EXIT
C
C         FILES
C         -----
C             1      - TELLER 1 QUEUE
C             2      - TELLER 2 QUEUE
C             3      - QUEUE FOR CARS WAITING TO EXIT TO STREET
C             4      - CARS WAITING FOR LIGHT
C             5      - CAR WAITING FOR STREET TRAFFIC TO CLEAR
C
C         EVENTS
C         ------
C             1      - DEPARTURE FROM TELLER
C             2      - LIGHT CHANGE TO RED
C             3      - LIGHT CHANGE TO GREEN
C             4      - STREET TRAFFIC ARRIVAL
C             5      - ARRIVAL TO LIGHT FROM BANK OR STREET
C             6      - PASSAGE THROUGH LIGHT
C
C
C                         USER VARIABLES
C                         ---------------
C
C
C         TPTL   - TIME TO PASS THROUGH LIGHT
C         TTNA   - TIME TO NEXT ARRIVAL
C         TNCA   - TIME OF NEXT CAR ARRIVAL
C
C         ********************************************************
C
C
```

D14-6.1

```
 1 GEN,PRITSKER,PROBLEM 14.6,7/10/86;
 2 LIMITS,6,5,150;
 3 TIMST,USERF(1),NO. OF CUST;
 4 SEEDS,4367651(1),6121137(2);
 5 NETWORK;
 6         RESOURCE/EXIT(1),3;
 7         CREATE,EXPON(.5,1),.1,1;
 8         SELECT,SNQ,,BALK(NBALK),LEFT,RIGHT;
 9 LEFT    QUEUE(1),2,3;
10         ACT/1,RNORM(1,.3,2),,AWA;
11 RIGHT   QUEUE(2),2,3;
12         ACT/2,RNORM(1,.3,2);
13 AWA     AWAIT(3/5),ALLOC(1),BLOCK;
14         EVENT(1);
15         TERM;
16 NBALK COLCT,BET,TIME BET. BALKS;
17         TERM;
18 C
19         ENTER,1;
20         COLCT,BET,TBD;
21         COLCT,INT(1),TIS;
22         TERM;
23         ENDNETWORK;
24 INIT,0,1000;
25 FIN;
```

D14-6.2

```
      SUBROUTINE ALLOC(I,IFLAG)
      COMMON/SCOM1/ATRIB(100),DD(100),DDL(100),DTNOW,II,MFA,MSTOP,NCLNR
     1,NCRDR,NPRNT,NNRUN,NNSET,NTAPE,SS(100),SSL(100),TNEXT,TNOW,XX(100)
      DIMENSION A(10)
      IFLAG=0
      IF(NNRSC(1).EQ.0) GO TO 10
      CALL SEIZE(1,1)
      IFLAG=1
10    NL1=NNQ(1)+NNACT(1)
      NL2=NNQ(2)+NNACT(2)
C******IF THE NUMBER IN LANE 2 EXCEEDS LANE 1 BY 2
      IF (NL2.LT.NL1+2) GO TO 20
C******THEN JOCKEY FROM 2 TO 1
      CALL ULINK(NNQ(2),2)
      CALL LINK(1)
      RETURN
C******IF THE NUMBER IN LANE 1 EXCEEDS LANE 2 BY 2
20    IF (NL1.LT.NL2+2) RETURN
C******THEN JOCKEY FROM 1 TO 2
      CALL ULINK(NNQ(1),1)
      CALL LINK(2)
      RETURN
      END
```

D14-6.3

```
        SUBROUTINE EVENT(I)
        COMMON/SCOM1/ATRIB(100),DD(100),DDL(100),DTNOW,II,MFA,MSTOP,NCLNR
       1,NCRDR,NPRNT,NNRUN,NNSET,NTAPE,SS(100),SSL(100),TNEXT,TNOW,XX(100)
        COMMON/UCOM1/TPTL,TTNA,TNCA
        GO TO (1,2,3,4,5,6),I
1       ATRIB(2)=TNOW
        CALL FILEM(5,ATRIB)
        CALL DEPART
        RETURN
C*****CHANGE LIGHT TO RED - SCHEDULE LIGHT TO GREEN EVENT
2       XX(1)=0.0
        CALL SCHDL(3,0.5,ATRIB)
        RETURN
C*****CHANGE LIGHT TO GREEN - SCHEDULE LIGHT TO RED EVENT
3       XX(1)=1.0
        CALL SCHDL(2,.65,ATRIB)
        IF(NNQ(4).EQ.0) RETURN
C*****PUT FIRST IN LIGHT QUEUE THROUGH THE LIGHT
        CALL SCHDL(6,TPTL,ATRIB)
        RETURN
C*****ARRIVAL OF STREET TRAFFIC
4       TTNA=EXPON(0.7,3)
        CALL SCHDL(4,TTNA,ATRIB)
        TNCA=TNOW+TTNA
        CALL LGHTARRV
        IF(NNQ(5).GT.0.AND.XX(2).NE.1.0) CALL DEPART
        RETURN
5       XX(2)=0.0
        CALL LGHTARRV
        RETURN
6       CALL PASSLGHT
        RETURN
        END
```

D14-6.4

```
        SUBROUTINE DEPART
        COMMON/SCOM1/ATRIB(100),DD(100),DDL(100),DTNOW,II,MFA,MSTOP,NCLNR
       1,NCRDR,NPRNT,NNRUN,NNSET,NTAPE,SS(100),SSL(100),TNEXT,TNOW,XX(100)
        COMMON/UCOM1/TPTL,TTNA,TNCA
        DIMENSION NSET(1)
        COMMON QSET(1)
        EQUIVALENCE(NSET(1),QSET(1))
        DIMENSION ANXT(4), WAIT(4)
        DATA WAIT/.25,.50,.75,1.0/
        DATA ANXT/.4,.3,.2,.1/
        WAITTIME=TNOW-QSET(MMFE(5)+2)
C*****COMPUTE GAP TIME NEEDED AS ANXIETYT
        ANXIETYT=GGTBL(ANXT,WAIT,WAITTIME,4)
        IF(TNCA-TNOW.LT.ANXIETYT) RETURN
        XX(2)=1.0
        ATRIB(2)=1.0
        CALL SCHDL(5,.1,ATRIB)
        RETURN
        END
```

D14-6.5

Events 2 and 3 change the traffic light at the intersection from green to red and red to green, respectively. When the light turns to green, the traffic waiting for the light in file 4 is scheduled to pass through the light. Cars passing through the light are processed in event 6, PASSLGHT. Subroutine PASSLGHT is shown in D14-6.6. Cars waiting in file 4 are removed and pass through the light by rescheduling event 6. If a car is waiting to enter the street traffic and there is no car passing through the exit, subroutine DEPART is called.

Event 4 models the arrival process of cars on the street. As each car arrives, the time of the next car arrival is established. The current arrival is scheduled to arrive at the light. A decision is made as to whether or not a car waiting to depart the bank can enter the street by a call to subroutine DEPART.

Event 5 represents a car departing the bank and arriving to the light. It is coded in subroutine LGHTARRV shown in D14-6.7. If the light is red or other cars are waiting to pass through the light, then the arriving car is placed in file 4. Otherwise, event 6 is scheduled for the car to pass through the light. If attribute 2 is less than 1 then the car is an arrival from street traffic. Otherwise, the car is a departure from the bank which requires that the EXIT resource be freed, the car removed from file 5, and an entity placed back into the network at ENTER node 1 where statistics on the time in the bank system are collected for the car entity.

Subroutine INTLC and function USERF are shown in Figure D14-6.8 and D14-6.9 and the summary report in D14-6.10.

```
      SUBROUTINE PASSLGHT
      COMMON/SCOM1/ATRIB(100),DD(100),DDL(100),DTNOW,II,MFA,MSTOP,NCLNR
     1,NCRDR,NPRNT,NNRUN,NNSET,NTAPE,SS(100),SSL(100),TNEXT,TNOW,XX(100)
      COMMON/UCOM1/TPTL,TTNA,TNCA
C*****IF LIGHT IS RED OR NO ONE WAITING TERM ARRIVAL BY NOT PROCESSING
C*****ENTITY AND RETURN
      IF (XX(1).EQ.0.0.OR.NNQ(4).EQ.0) RETURN
      CALL RMOVE(1,4,ATRIB)
      DT=TPTL*0.8
      CALL SCHDL(6,DT,ATRIB)
      IF(NNQ(5).GT.0.AND.XX(2).NE.1.0) CALL DEPART
      RETURN
      END
```

D14-6.6

```
      SUBROUTINE LGHTARRV
      COMMON/SCOM1/ATRIB(100),DD(100),DDL(100),DTNOW,II,MFA,MSTOP,NCLNR
     1,NCRDR,NPRNT,NNRUN,NNSET,NTAPE,SS(100),SSL(100),TNEXT,TNOW,XX(100)
      COMMON/UCOM1/TPTL,TTNA,TNCA
      IF(XX(1).EQ.0.0.OR.NNQ(4).GT.0) THEN
         CALL FILEM(4,ATRIB)
      ELSE
         CALL SCHDL(6,TPTL,ATRIB)
      ENDIF
      IF (ATRIB(2).LT.1.0) RETURN
       CALL FREE(1,1)
       CALL RMOVE(1,5,ATRIB)
       CALL ENTER(1,ATRIB)
      RETURN
      END
```

D14-6.7

```
SUBROUTINE INTLC
COMMON/SCOM1/ATRIB(100),DD(100),DDL(100),DTNOW,II,MFA,MSTOP,NCLNR
1,NCRDR,NPRNT,NNRUN,NNSET,NTAPE,SS(100),SSL(100),TNEXT,TNOW,XX(100)
COMMON/UCOM1/TPTL,TTNA,TNCA
CALL SCHDL(4,0.,ATRIB)
TPTL=0.125
XX(1)=1
CALL SCHDL(2,.65,ATRIB)
RETURN
END
```

D14-6.8

```
      FUNCTION USERF(I)
C*****
C  CALCULATES THE TOTAL NUMBER OF CUSTOMERS IN THE SYSTEM
C*****
      USERF=NNQ(1)+NNACT(1)+NNQ(2)+NNACT(2)
      RETURN
      END
```

D14-6.9

```
                    S L A M   I I   S U M M A R Y   R E P O R T


        SIMULATION PROJECT PROBLEM 14.6          BY PRITSKER

        DATE  7/10/1986                          RUN NUMBER   1 OF   1


        CURRENT TIME   0.1000E+04
        STATISTICAL ARRAYS CLEARED AT TIME  0.0000E+00
```

STATISTICS FOR VARIABLES BASED ON OBSERVATION

	MEAN VALUE	STANDARD DEVIATION	COEFF. OF VARIATION	MINIMUM VALUE	MAXIMUM VALUE	NUMBER OF OBSERVATIONS
TIME BET. BALKS	0.6741E+01	0.1310E+02	0.1943E+01	0.4517E-02	0.7082E+02	147
TBD	0.5401E+00	0.4512E+00	0.8353E+00	0.9998E-01	0.7146E+01	1849
TIS	0.2571E+01	0.1146E+01	0.4455E+00	0.3293E+00	0.8543E+01	1850

STATISTICS FOR TIME-PERSISTENT VARIABLES

	MEAN VALUE	STANDARD DEVIATION	MINIMUM VALUE	MAXIMUM VALUE	TIME INTERVAL	CURRENT VALUE
NO. OF CUST	0.4327E+01	0.2158E+01	0.0000E+00	0.8000E+01	0.1000E+04	0.6000E+01

FILE STATISTICS

FILE NUMBER	LABEL/TYPE	AVERAGE LENGTH	STANDARD DEVIATION	MAXIMUM LENGTH	CURRENT LENGTH	AVERAGE WAITING TIME
1	LEFT QUEUE	1.3314	1.0488	3	2	1.1264
2	RIGH QUEUE	1.1397	0.9841	3	2	1.2163
3	AWA AWAIT	0.1376	0.6234	5	0	0.0744
4		0.8455	1.0917	7	0	0.3893
5		0.3030	0.4596	1	0	0.1638
6		0.0000	0.0000	0	0	0.0000
7	CALENDAR	5.4821	0.8103	9	5	0.2709

SERVICE ACTIVITY STATISTICS

ACTIVITY INDEX	START NODE OR ACTIVITY LABEL	SERVER CAPACITY	AVERAGE UTILIZATION	STANDARD DEVIATION	CURRENT UTILIZATION	AVERAGE BLOCKAGE	MAXIMUM IDLE TIME/SERVERS	MAXIMUM BUSY TIME/SERVERS	ENTITY COUNT
1	LEFT QUEUE	1	0.9377	0.2417	1	0.0034	1.9385	72.4390	931
2	RIGH QUEUE	1	0.9110	0.2847	1	0.0042	4.1143	69.1348	919

RESOURCE STATISTICS

RESOURCE NUMBER	RESOURCE LABEL	CURRENT CAPACITY	AVERAGE UTILIZATION	STANDARD DEVIATION	MAXIMUM UTILIZATION	CURRENT UTILIZATION
1	EXIT	1	0.3030	0.4596	1	0

RESOURCE NUMBER	RESOURCE LABEL	CURRENT AVAILABLE	AVERAGE AVAILABLE	MINIMUM AVAILABLE	MAXIMUM AVAILABLE
1	EXIT	1	0.6970	0	1

D14-6.10

14-7.   This exercise is an embellishment of Illustration 6-3. The definitions of the SLAM variables, files, events and user-variables are shown in D14-7.1. The input statements for this model are shown in D14-7.2. The network variable ARRAY is used to store the travel time between machine locations, the travel time from a machine location to the wash station and from the wash station to a machine location. The latter two rows of array are used to avoid recomputing the travel times for each move of the cart when the cart is loaded. Casting entities waiting for operation I wait in file I. There is a correspondence between resource number, mill number and operation number. For the flexible mill, MILLF, operations are given the following priority: operation 20: operation 30, and operation 10. The resource statements are set up to model five mills for performing operation 10, one mill for performing operation 20, two mills for performing operation 30 and two flexible mills. The allocation of mills to castings is performed in subrouting ALLOC(1) which is associated with the AWAIT node AMILL. This allocation process will be described later. After a casting entity has been assigned a machine, it is placed in file 4 at a QUEUE node waiting for the cart to move it to the allocated mill. The movement of a cart will be modeled using discrete events and a casting will be removed from file 4 using subroutine RMOVE. The casting is placed back into the network at ENTER node 1. The casting is then sent to the wash station where it exits the system or is inspected and washed prior to its next operation. Before the casting entity is reentered into the network, the acquiring and scheduling of the cart for the move is modeled in a discrete event.

Subroutine INTLC is shown in D14-7.3. The cart entity is created and scheduled to arrive to the wash station. The type of machine at each location is then defined in variable TYPE. The processing time for each operation type is then set in PROCTIME. Subroutine ALLOC is shown in D14-7.4. IR is defined as the operation number to be performed. If a dedicated machine is available, it is seized and the resource type is allocated to the casting entity requesting the resource at the AWAIT node. Otherwise, a check is made to determine if a flexible machine is available. If it is, it is assigned and IFLAG is set to 1. Otherwise, IFLAG is set to 0.

There are two events in this model as shown in D14-7.5 Event 1 is to move the cart from location to location. Event 2 is an end-of- processing by a mill on a casting. The end-of-processing event is shown in D14-7.6, and it will be discussed first. When a mill completes processing of a casting, the location of the mill is identified, the status indicator is set to the pointer to where the entity that completed processing is to be stored, and the casting entity is placed in file 5 awaiting a cart to move it to its next destination. This completes the end-of-processing event.

The movement of the cart is coded in subroutine MOVE shown in D14-7.7. The first move event is scheduled in subroutine INTLC with ATRIB(4) being the location to which the cart is arriving. It is assumed that the cart continues to move around the track looking for machines that have completed processing. If the cart has a load, it goes directly to the machine that has been assigned to perform the operation on the load. Subroutine MOVE is heavily commented and therefore a direct explanation of each statement will not be given. Basically the subroutine accomplishes the following: 1) if the cart is unloaded and no item is at the station, the cart moves to the next station; 2) if the cart is unloaded and there is a load at the station, the cart takes the load and returns to the wash station; 3) if the cart is loaded and the station is without a casting, the cart unloads a casting and moves to the next station; 4) if the cart is loaded and the mill has a casting, then the cart is unloaded, the casting is loaded onto the cart and the cart is scheduled to go to the wash station; 5) if the cart arrives to the wash station and a casting has been assigned a mill, the cart takes the casting to the mill.

The summary report for this exercise is shown in D14-7.8.

```
                          SLAM II VARIABLES

        XX(2)              Time to move from machine to wash station
        ATRIB(1)           Operation number:  1
        ATRIB(2)           Resource type:  1,2,3, or 4
        ATRIB(3)           Cart load indicator:  0--> not loaded,
                                                 1--> loaded
        ATRIB(4)           Machine location: 1,2,..., or 11
        ATRIB(5)           Mark time

        ARRAY(1,J)         Travel times for cart to move from location J to J+1
        ARRAY(2,J)         Travel times to wash station from location J
        ARRAY(3,J)         Travel times to location J from wash station


        FILES

        File  1            Castings waiting for Operation 1
        File  2            Castings waiting for Operation 2
        File  3            Castings waiting for Operation 3
        File  4            Castings waiting for cart move to machine
        File  5            Castings waiting to be moved from machine
        File  6            Castings waiting for wash/inspection


        EVENTS

              1            Cart move (MOVE)
              2            End-of-machine processing (ENDPROC)


        USER VARIABLES

        ISTATUS(J)         Status of machine at location J
                           0--> idle;
                           1--> busy
                           -pointer-->finished processing


        PROCTIME(K)        Processing time for machine type K
        TYPE(J)            Machine type at location J
```

D14-7.1

```
    1   GEN,PRITSKER,PROBLEM 14.7,7/10/86,1;
    2   LIMITS,6,5,200;
    3   INIT,0,4800;
    4   ARRAY(1,11)/0.3,0.2,0.2,0.2,0.2,0.6,0.2,0.2,0.2,0.2,0.3;
    5   ARRAY(2,10)/2.5,2.3,2.1,1.9,1.7,1.1,0.9,0.7,0.5,0.3;
    6   ARRAY(3,10)/0.3,0.5,0.7,0.9,1.1,1.7,1.9,2.1,2.3,2.5;
    7   EQUIVALENCE/ATRIB(1),OPERATION;
    8   EQUIVALENCE/ATRIB(2),MACHINE;
    9   NETWORK;
   10          RESOURCE/MILL1(5),1/MILL2(1),2;
   11          RESOURCE/MILL3(2),3/MILLF(2),2,3,1;
   12          CREATE,22.,,5;
   13   SETA   ASSIGN,ATRIB(4)=0.0,OPERATION=OPERATION+1;
   14   AMILL  AWAIT(OPERATION=1,3),ALLOC(1);
   15          QUEUE(4);
   16   ;
   17          ENTER,1,1;
   18          ACT,XX(2),OPERATION.LT.3,TOLL;
   19          ACT,XX(2);
   20          COLCT,INT(5),TIME IN SYSTEM;
   21          TERM;
   22   TOLL   QUEUE(6);
   23          ACT(2)/1,10.0,,SETA;
   24          END;
   25   FIN;
```

D14-7.2

```
   SUBROUTINE INTLC
   COMMON/SCOM1/ATRIB(100),DD(100),DDL(100),DTNOW,II,MFA,MSTOP,NCLNR
  1,NCRDR,NPRNT,NNRUN,NNSET,NTAPE,SS(100),SSL(100),TNEXT,TNOW,XX(100)
   COMMON/UCOM1/ISTATUS(10),TYPE(10),PROCTIME(3)
   ATRIB(3)=0.0
   ATRIB(4)=11.0
   CALL SCHDL(1,0.0,ATRIB)
   DO 10 J=1,10
      ISTATUS(J)=0
10 CONTINUE
   DO 20 J=1,5
      TYPE(J)=1.0
20 CONTINUE
   TYPE(6)=2.0
   DO 30 J=7,8
      TYPE(J)=3.0
30 CONTINUE
   DO 40 J=9,10
      TYPE(J)=4.0
40 CONTINUE
   PROCTIME(1)=120.0
   PROCTIME(2)=40.0
   PROCTIME(3)=56.0
   RETURN
   END
```

D14-7.3

```
      SUBROUTINE ALLOC(ICODE,IFLAG)
      COMMON/SCOM1/ATRIB(100),DD(100),DDL(100),DTNOW,II,MFA,MSTOP,NCLNR
     1,NCRDR,NPRNT,NNRUN,NNSET,NTAPE,SS(100),SSL(100),TNEXT,TNOW,XX(100)
      COMMON/UCOM1/ISTATUS(10),TYPE(10),PROCTIME(3)
      IR=ATRIB(1)
C
C     SEIZE DEDICATED MACHINE IF AVAILABLE
C
      IF (NNRSC(IR).GT.0) THEN
        CALL SEIZE (IR,1)
        ATRIB(2)=IR
        IFLAG=1
      ELSE IF (NNRSC(4).GT.0) THEN
C
C     OTHERWISE, SEIZE FLEXIBLE MACHINE IF AVAILABLE
C
        CALL SEIZE (4,1)
        ATRIB(2)=4
        IFLAG=1
      ELSE
        IFLAG=0
      ENDIF
      RETURN
      END
```

D14-7.4

```
             SUBROUTINE EVENT(I)
             GO TO (10,20),I
          10 CALL MOVE
             RETURN
          20 CALL ENDPROC
             RETURN
             END
```

D14-7.5

```
      SUBROUTINE ENDPROC
      COMMON/SCOM1/ATRIB(100),DD(100),DDL(100),DTNOW,II,MFA,MSTOP,NCLNR
     1,NCRDR,NPRNT,NNRUN,NNSET,NTAPE,SS(100),SSL(100),TNEXT,TNOW,XX(100)
      COMMON/UCOM1/ISTATUS(10),TYPE(10),PROCTIME(3)
      J=ATRIB(4)
      ISTATUS(J)=-MFA
      CALL FILEM(5,ATRIB)
      RETURN
      END
```

D14-7.6

```
      SUBROUTINE MOVE
      COMMON/SCOM1/ATRIB(100),DD(100),DDL(100),DTNOW,II,MFA,MSTOP,NCLNR
     1,NCRDR,NPRNT,NNRUN,NNSET,NTAPE,SS(100),SSL(100),TNEXT,TNOW,XX(100)
      COMMON/UCOM1/ISTATUS(10),TYPE(10),PROCTIME(3)
C**** SET J = MACHINE NUMBER OF CART ARRIVAL
      J=ATRIB(4)
      IF (J.EQ.11) GO TO 100
C**** IF CART IS LOADED, GO TO 400
      IF (ATRIB(3).GT.0.0) GO TO 400
C**** CART NOT LOADED - DOES MACHINE HAVE LOAD?
      IF (ISTATUS(J).GE.0) GO TO 200
C**** MACHINE HAS LOAD AND CART EMPTY
      NPTR=ISTATUS(J)
      ISTATUS(J)=0
C**** FREE MACHINE AS LOAD IS REMOVED
      IR=TYPE(J)
      CALL FREE(IR,1)
   10 CALL RMOVE(NPTR,5,ATRIB)
C**** SEND CART TO WASH
      ATRIB(4)=11.0
      XX(2)=GETARY(2,J)
      CALL ENTER(1,ATRIB)
      CALL SCHDL(1,XX(2),ATRIB)
      RETURN
C**** ARRIVAL TO LOAD/WASH STATION - LOAD SENT OVER NETWORK
  100 ATRIB(3)=0.0
      ATRIB(4)=0.0
      IF (NNQ(4).EQ.0) GO TO 200
      CALL RMOVE(1,4,ATRIB)
      ATRIB(3)=1.0
  200 ATRIB(4)=ATRIB(4)+1.0
      TNM=GETARY(1,INT(ATRIB(4)))
      CALL SCHDL(1,TNM,ATRIB)
      RETURN
C**** IF MACHINE IS BUSY OR NOT OF RIGHT TYPE, MOVE TO NEXT
C**** IF MACHINE IS IDLE, UNLOAD
C**** IF MACHINE HAS LOAD, UNLOAD AND LOAD
  400 IF (ISTATUS(J).EQ.1 .OR. TYPE(J).NE.ATRIB(2)) GO TO 200
      CALL SCHDL(2,PROCTIME(INT(ATRIB(1))),ATRIB)
      IF (ISTATUS(J).LT.0) GO TO 500
      ISTATUS(J)=1
      ATRIB(3)=0.0
      GO TO 200
  500 NPTR=ISTATUS(J)
      ISTATUS(J)=1
      GO TO 10
      END
```

D14-7.7

```
                        S L A M   I I   S U M M A R Y   R E P O R T

           SIMULATION PROJECT PROBLEM 14.7           BY PRITSKER

           DATE  7/10/1986                           RUN NUMBER   1 OF    1


           CURRENT TIME   0.4800E+04
           STATISTICAL ARRAYS CLEARED AT TIME  0.0000E+00


              **STATISTICS FOR VARIABLES BASED ON OBSERVATION**

                    MEAN        STANDARD      COEFF. OF     MINIMUM      MAXIMUM      NUMBER OF
                    VALUE       DEVIATION     VARIATION     VALUE        VALUE        OBSERVATIONS

TIME IN SYSTEM    0.7443E+03   0.3116E+03    0.4186E+00   0.2524E+03   0.1268E+04        161


                          **FILE STATISTICS**

FILE                      AVERAGE      STANDARD    MAXIMUM    CURRENT    AVERAGE
NUMBER  LABEL/TYPE        LENGTH       DEVIATION   LENGTH     LENGTH     WAITING TIME

  1     AMIL AWAIT        11.2825       6.9773       24         24       247.2880
  2     AMIL AWAIT         0.6563       0.6476        3          1        16.5797
  3     AMIL AWAIT        10.8988       7.6180       25         23       281.2606
  4        QUEUE           0.1462       0.3564        2          0         1.2829
  5                        0.2265       0.4438        3          1         2.0168
  6     TOLL QUEUE         0.0049       0.0699        1          0         0.0626
  7        CALENDAR       11.1982       1.3341       14         11         2.7644


                        **SERVICE ACTIVITY STATISTICS**

ACTIVITY  START NODE OR   SERVER    AVERAGE      STANDARD   CURRENT     AVERAGE    MAXIMUM IDLE   MAXIMUM BUSY   ENTITY
INDEX     ACTIVITY LABEL  CAPACITY  UTILIZATION  DEVIATION  UTILIZATION BLOCKAGE   TIME/SERVERS   TIME/SERVERS   COUNT

  1       TOLL QUEUE         2       0.7852       0.6846       1        0.0000       2.0000         2.0000        376


                          **RESOURCE STATISTICS**

RESOURCE  RESOURCE  CURRENT   AVERAGE      STANDARD    MAXIMUM      CURRENT
NUMBER    LABEL     CAPACITY  UTILIZATION  DEVIATION   UTILIZATION  UTILIZATION

  1       MILL1       5       4.9436       0.3806         5            5
  2       MILL2       1       0.9594       0.1975         1            1
  3       MILL3       2       1.9115       0.4009         2            2
  4       MILLF       2       1.9442       0.3138         2            2


RESOURCE  RESOURCE  CURRENT    AVERAGE     MINIMUM     MAXIMUM
NUMBER    LABEL     AVAILABLE  AVAILABLE   AVAILABLE   AVAILABLE

  1       MILL1       0        0.0564         0           5
  2       MILL2       0        0.0406         0           1
  3       MILL3       0        0.0885         0           2
  4       MILLF       0        0.0558         0           2
```

CHAPTER 15

Simulation Languages

15-1. A GPSS model of the inspection and adjustment stations is included in Schriber, T., Simulation Using GPSS, John Wiley, 1974, pages 134-139. A SIMSCRIPT model would require events representing arrival to inspection; completion of inspection which would also contain the possibility of an arrival to the adjustment station; and completion of adjustment which would also contain the recycling of the TV as an arrival to the inspection event.

15-2. The TERMINATE block of GPSS is similar to the TERMINATE node of SLAM II. However, the TERMINATE block of GPSS reduces a counter by a specified number in order to determine if a desired number is reached. All TERMINATE blocks reduce a common counter. In SLAM II, each TERMINATE node has its own counter.

The CREATE statement in SIMSCRIPT involves finding storage space for storing the attributes of an entity. The CREATE node in SLAM II introduces an entity to the system.

The SEIZE block of GPSS is used to allocate a FACILITY to a transaction. The AWAIT node in SLAM II is used to allocate a number of resources to an entity. One difference is that the SEIZE is restricted to a single resource allocation. Also, in the AWAIT node statement, a file can be referenced and the ranking of entities in the file can be established. By using a file, statistics are automatically collected on the items waiting for the resource. The ACCUMULATE statement in SIMSCRIPT is used for statistical purposes only whereas the ACCUMULATE node in SLAM II provides a logic operation for combining entities and for routing entities from the node.

15-3. In SLAM II and SIMSCRIPT, events are placed on an event calendar and simulation time is advanced when a new event is to be processed. Events are placed on the event calendar in SLAM II by starting an activity or through a call to subroutine SCHDL. The SCHEDULE statement is used in SIMSCRIPT to place an event on the event calendar. For GPSS, there are different mechanisms for causing simulated time to change. In later versions, an event calendar was employed. When a transaction enters an ADVANCE block, a future event is scheduled. Time advances in GPSS when there are no further current events to be processed.

In continuous simulation languages, time advancement is done implicitly. Thus, in a continuous model programmed in SLAM II, a step is taken to update the values of the state variables from time TTLAS to time TNOW. The step size taken is dependent on accuracy requirements as well as the time at which time and state events occur. In a CSSL, time advance is dependent on the numerical integration algorithm employed to update the values of the state variables. In DYNAMO, a fixed step size (DT) is specified by the modeler. The level, rates and auxiliary statements are evaluated at the end of each DT time units.

15-4. One procedure for evaluating simulation languages is presented by R. Shannon in Systems Simulation: The Art and The Science, Prentice-Hall, 1975. If at all possible, measures of each of the features included in Table 15-6 of the Simulation and SLAM II text should be hypothesized and a grading scale developed. There have been many corporate evaluations of simulation languages but these normally are confidential and difficult to obtain.

15-5. The answer to this exercise follows along the lines of the solutions to Exercises 1-3, 1-5, and 1-9. In Chapter 3, it was stated that a world view provides a way of describing a system. A language's world view or world views could restrict how a model is constructed.

15-6. To modify the SIMSCRIPT II queueing example to include two parallel servers, there are several possibilities. Given the same preamble, we can accomplish the programming by letting the variable STATUS also take on the value of 2. In event ARRIVAL at statement 5, the test condition would then be:

    IF STATUS = BUSY+BUSY

At statement 9, the following change should be made

    LET STATUS = STATUS+BUSY

and, at statement 6 in event DEPARTURE, the change would be

    LET STATUS = STATUS - BUSY.

The new parameter values for the uniformly distributed service time would be inserted in statement 10 in EVENT ARRIVAL and in statement 10 of EVENT DEPARTURE.

A more appealing approach to this problem is to define the server number as an attribute of the customer entity and to maintain the status of each server by having two variables which refer to the status of each server. Then in EVENT ARRIVAL, a random number would be generated when both servers are idle to select the server to make busy.

CHAPTER 16

Material Handling Extension To Slam II

16.1. The QBATCH node combines the BATCH node and AWAIT node functions. It adds on to the BATCH node the capability of an intermediate threshold which can be used to satisfy a batch requirement if the resource requested is available. In the model of the transfer car, discrete events will be used to determine the number of entities to include in a batch to take care of the threshold and maxsize requirements prescribed by a QBATCH node.

The statement model is shown in D16-1.1 and subroutine EVENT is shown in D16-1.2. Entities are created and routed to EVENT node 1. In EVENT node 1, ATRIB(1) is assigned a value which is the weight associated with the entity. The entity is then stored in file 1. XX(1) is used to accumulate the weight of all entities in file 1. If XX(1) is less than 30 then a batch will only be formed if XX(1) is greater than or equal to 20 and the resource is free. If XX(1) is equal to 30, a BATCH should be formed and all entities in file 1 included in the batch. A transfer is made to statement 3 to set N equal to the number of entries in file 1. If XX(1) is greater than 30 then all but the last arriving entity should be included in the batch and a transfer is made to statement 4 where N is set equal to 1 less than the number of entities in file 1. Following statement 25, an entity is removed from file 1. The value of XX(1) is reduced by the weight of the entity removed from file 1 and the entity is reentered into the network at ENTER node 1 with attribute 2 equal to the number of entities to include in the batch.

```
 1    GEN,PRITSKER,PROBLEM 16.1,8/9/1986,1,Y,Y,Y,N;
 2    ;
 3    ;    TRANSFER CAR EXAMPLE
 4    ;
 5    LIMITS,2,3,50;
 6    INIT,0.,300.;
 7    MONTR,CLEAR,100;
 8    NETWORK;
 9    ;
10    ;    RESOURCE DEFINITIONS
11    ;
12         RESOURCE,TRANCAR(1),2;
13    ;
14         CREATE,EXPON(2.7),1;
15         EVENT,1;
16         TERM;
17    ;
18         ENTER,1;
19         BATCH,,ATRIB(2),,FIRST/1,ALL(3);
20         AWAIT(2),TRANCAR;
21           ACT/2,UNFRM(2.8,3.9);               TRANSPORT
22         GOON,2;
23           ACT/3,TRIAG(1.0,1.9,2.3),,FTC;      CAR RETURN
24           ACT/4,,,UNB;                        TO UNBATCH
25    ;
26 FTC  FREE,TRANCAR;                            RELEASE CAR
27         EVENT,2;
28         TERMINATE;
29    ;
30 UNB  UNBATCH,3;                               UNBATCHING
31           ACT/5;
32         TERMINATE;
33         ENDNETWORK;
34    FIN;
```

D16-1.1

```
      SUBROUTINE EVENT(I)
      COMMON/SCOM1/ATRIB(100),DD(100),DDL(100),DTNOW,II,MFA,MSTOP,NCLNR
     1,NCRDR,NPRNT,NNRUN,NNSET,NTAPE,SS(100),SSL(100),TNEXT,TNOW,XX(100)
      DIMENSION A(100)
      GO TO(1,2),I
    1 ATRIB(1)=UNFRM(5.0,8.0,9)
      CALL FILEM(1,ATRIB)
      XX(1)=XX(1)+ATRIB(1)
      IF (XX(1)-30.) 2,3,4
    4 N=NNQ(1)-1
      GO TO 25
    2 IF (XX(1).LT.20.0 .OR. NNRSC(1).LE.0) RETURN
    3 N=NNQ(1)
   25 DO 10 J=1,N
         CALL RMOVE(1,1,A)
         XX(1)=XX(1)-A(1)
         A(2)=N
         CALL ENTER(1,A)
   10 CONTINUE
      RETURN
      END
```

D16-1.2

Following the ENTER node in D16-1.1, the entities are sent to a BATCH node where they are formed into a batch and routed to an AWAIT node where 1 unit of TRANCAR is requested. When TRANCAR is allocated to the batched entity, the batched entity moves over activity 2 to a GOON node. At the GOON node, the batched entity is routed to a FREE node after a delay to model the return of the transfer car. The batched entity is also sent to the UNBATCH node where each of the entities in the batch is released by indicating that attribute 3 contains the pointer to the individual entities in the batch. Following the freeing of the transfer car, an entity is sent to EVENT node 2. In D16-1.2, EVENT 2 is coded starting at statement 2 where a check is made to see if entities are waiting in file 1 which have a sum of weights greater than or equal to 20. The code that was described above is used again.

16-2.   The input statements are given in D16-2.1. The solution is basically the same as the model given in Illustration 16-1 for the transfer car. The SLAM II Summary Report is given in D16-2.2.

16-2,Embellishment(a).   The resource UNLOAD is added to the model which is required to perform the second operation. A disjoint network is added to preempt the unload operation when it breaks down. The statement model for the situation is shown in D16-2.3.

16-2,Embellishment(b).   The statement model for this embellishment is shown in D16-2.4 and involves the addition of a new CREATE node for the oak trees. Attribute 3 is assigned the type of tree and the two batches segregated on attribute 3 are maintained at the QBATCH node QBAT.

16-2,Embellishment(c).   The throughput obtained from running the model with minimum truckloads of 15, 20, and 25 tons are:

Minimum Load Size	Throughput
15	560
20	580
25	563

```
 1    GEN,FLOSS,PROBLEM 16.2,7/17/86,1;
 2    LIMITS,1,5,50;
 3    NETWORK;
 4         RESOURCE,TRUCK(5),1;
 5
 6         CREATE,RNORM(25,10);                      TREE CUT
 7         ASSIGN,ATRIB(1)=UNFRM(1000,1500,1);       ASSIGN WEIGHT
 8         QBATCH(1),TRUCK,,30000,60000,1,,ALL(2);   LOAD TRUCK
 9         ACTIVITY/1,150;                           DELIVER
10         GOON;
11         ACTIVITY/2,20;                            UNLOAD
12         GOON,2;
13         ACTIVITY,,,TREE;
14         ACTIVITY/3,120;                           RETURN
15         FREE,TRUCK/1;                             TRUCK AVAIL
16         TERMINATE;
17    ;
18    TREE  UNBATCH,2;
19         ACTIVITY/4;                               THROUGHPUT
20         TERMINATE;
21         END;
22    INIT,0,14400;
23    FIN;
```

D16-2.1

```
                    S L A M   I I   S U M M A R Y   R E P O R T

        SIMULATION PROJECT PROBLEM 16.2          BY FLOSS

        DATE  7/17/1986                          RUN NUMBER    1 OF    1


        CURRENT TIME   0.1440E+05
        STATISTICAL ARRAYS CLEARED AT TIME  0.0000E+00


                          **FILE STATISTICS**
```

FILE NUMBER	LABEL/TYPE	AVERAGE LENGTH	STANDARD DEVIATION	MAXIMUM LENGTH	CURRENT LENGTH	AVERAGE WAITING TIME
1	QBATCH	11.7141	7.1321	26	1	287.8557
2	CALENDAR	1.4662	0.4989	28	2	11.4309

```
                    **REGULAR ACTIVITY STATISTICS**
```

ACTIVITY INDEX/LABEL	AVERAGE UTILIZATION	STANDARD DEVIATION	MAXIMUM UTIL	CURRENT UTIL	ENTITY COUNT
1 DELIVER	0.2426	0.4286	1	1	23
2 UNLOAD	0.0319	0.1759	1	0	23
3 RETURN	0.1917	0.3936	1	0	23
4 THROUGHPUT	0.0000	0.0000	26	0	560

```
                    **RESOURCE STATISTICS**
```

RESOURCE NUMBER	RESOURCE LABEL	CURRENT CAPACITY	AVERAGE UTILIZATION	STANDARD DEVIATION	MAXIMUM UTILIZATION	CURRENT UTILIZATION
1	TRUCK	5	0.4662	0.4989	1	1

RESOURCE NUMBER	RESOURCE LABEL	CURRENT AVAILABLE	AVERAGE AVAILABLE	MINIMUM AVAILABLE	MAXIMUM AVAILABLE
1	TRUCK	4	4.5338	4	5

D16-2.2

```
 1   GEN,FLOSS,PROBLEM 16.2A,7/17/86,1;
 2   LIMITS,3,5,50;
 3   NETWORK;
 4        RESOURCE,TRUCK(5),1;
 5        RESOURCE,UNLOAD(1),2;
 6
 7        CREATE,RNORM(25,10);                  TREE CUT
 8        ASSIGN,ATRIB(1)=UNFRM(1000,1500,1);   ASSIGN WEIGHT
 9        QBATCH(1),TRUCK,,30000,60000,1,,ALL(2);  LOAD TRUCK
10        ACTIVITY/1,150;                       DELIVER
11        AWAIT(2),UNLOAD;
12        ACTIVITY/2,20;                        UNLOAD
13        FREE,UNLOAD/1,2;
14        ACTIVITY,,,TREE;
15        ACTIVITY/3,120;                       RETURN
16        FREE,TRUCK/1;                         TRUCK AVAIL
17        TERMINATE;
18
19   TREE UNBATCH,2;
20        ACTIVITY/4;                           THROUGHPUT
21        TERMINATE;
22
23        CREATE,UNFRM(1800,3000,2);
24        PREEMPT(3),UNLOAD;                    BREAKDOWN
25        ACTIVITY/5,120;                       REPAIR
26        FREE,UNLOAD/1;                        COMPLETE REPAIR
27        TERMINATE;
28        END;
29   INIT,0,14400;
30   FIN;
```

D16-2.3

```
 1   GEN,FLOSS,PROBLEM 16.2B,7/17/86,1;
 2   LIMITS,1,6,100;
 3   NETWORK;
 4        RESOURCE,TRUCK(5),1;
 5
 6        CREATE,RNORM(25,10);
 7        ASSIGN,ATRIB(1)=UNFRM(1500,2000,1),
 8             ATRIB(3)=1;                      ASSIGN WEIGHT
 9        ACTIVITY,,,QBAT;
10        CREATE,RNORM(25,10);                  TREE CUT
11        ASSIGN,ATRIB(1)=UNFRM(1000,1500),
12             ATRIB(3)=2;                      ASSIGN WEIGHT
13   QBAT QBATCH(1),TRUCK,2/3,30000,60000,1,,ALL(2);  LOAD TRUCK
14        ACTIVITY/1,150;                       DELIVER
15        GOON;
16        ACTIVITY/2,20;                        UNLOAD
17        GOON,2;
18        ACTIVITY,,,TREE;
19        ACTIVITY/3,120;                       RETURN
20        FREE,TRUCK/1;                         TRUCK AVAIL
21        TERMINATE;
22
23   TREE UNBATCH,2;
24        ACTIVITY/4;                           THROUGHPUT
25        TERMINATE;
26        END;
27   INIT,0,14400;
28   FIN;
```

D16-2.4

16.3   The input statements for this exercise are shown in D16-3.1. The statement model is heavily commented and explains the use of each statement directly. The resource definitions were given in the text. The summary reports at time 480 and 2400 are given in D16-3.2 and D16-3.3. From these summary reports, it is seen that it takes an initial period for the system to become loaded as the throughputs and number of observations in 2400 time units is greater than 5 times the values obtained in 400 time units.

```
 1   GEN,STEIN,PROBLEM 16.3,4/22/86,1,Y,Y,Y,N;
 2   INIT,0,2400;
 3   LIMITS,6,10,100;
 4   EQUIVALENCE/1.5,PICK/2.5,DROP/UNFRM(12.,18.),FINISH;
 5   NETWORK;
 6   ;
 7   ; *** ALUMINUM COIL PROCESSING FACILITY ***
 8   ;
 9   ;
10   ; *** AREA DEFINITIONS ***
11   ;
12        AREA,HVYSLIT/1,1,6,4;
13        AREA,LGTSLIT/2,1,6,4;
14        AREA,SHEAR/3,1,6,4;
15        AREA,FIN_STOR/4,2,2,6;
16        AREA,COOLRACK/5,3,2,3;
17   ;
18   ; *** PILE DEFINITIONS ***
19   ;
20        PILE,1,HVYSLIT,1.0,75.0,20.0;
21        PILE,2,LGTSLIT,1.0,110.0,20.0;
22        PILE,3,SHEAR,1.0,225.0,20.0;
23        PILE,4,FIN_STOR,1000.0,350.0,50.0;
24        PILE,5,FIN_STOR,1000.0,350.0,20.0;
25        PILE,6,COOLRACK,800.0,150.0,55.0;
26        PILE,7,COOLRACK,600.0,200.0,55.0;
27        PILE,8,COOLRACK,600.0,275.0,55.0;
28   ;
29   ; *** RESOURCE DEFINITIONS ***
30   ;
31        RESOURCE,OPER(3),5;
32        RESOURCE,HELPER(3),5;
33   ;
34   ; *** CRANE DEFINITIONS ***
35   ;
36        CRANE,CRNW,1,25.,35.,300.,0.,0.,300.,1;
37        CRANE,CRNE,1,300.,50.,300.,2000.,2000.,300.,2;
38   ;
39   ; *** NETWORK FLOW ***
40   ;
41   CR1  CREATE,20,,5;
42        ACT,,0.3,HVY;                         HEAVY GAUGE COIL
43        ACT,,0.3,LGT;                         LIGHT GAUGE COIL
44        ACT,,0.4,SHR;                         SHEAR COIL
45   ;
46   ; *** ASSIGN ATTRIBUTES TO INCOMING COILS ***
47   ;
48   ;    ATTRIBUTE           DESCRIPTION
49   ;    ---------           -----------
50   ;        1               COIL TYPE
51   ;                          1=HEAVY GAUGE
52   ;                          2=LIGHT GAUGE
53   ;                          3=SHEAR
54   ;        2               COIL SIZE (VOLUME,TCI)
55   ;        3               PROCESSING TIME
56   ;        4               PILE # AT FINISHED PROCESSING
57   ;        5               MARKING ATTRIBUTE
58   ;        6               QUANTITY TO REQUEST AT PROCESSING (=1)
59   ;
60   HVY  ASSIGN,ATRIB(1)=1.,ATRIB(2)=UNFRM(90.,110.),
61             ATRIB(3)=TRIAG(38.,40.,42.),
62             ATRIB(4)=4.,ATRIB(6)=1.;
63        ACT,,,COOL;
64   LGT  ASSIGN,ATRIB(1)=2.,ATRIB(2)=UNFRM(90.,110.),
65             ATRIB(3)=TRIAG(24.,26.,28.),
66             ATRIB(4)=4.,ATRIB(6)=1.;
67        ACT,,,COOL;
68   SHR  ASSIGN,ATRIB(1)=3.,ATRIB(2)=UNFRM(90.,110.),
69             ATRIB(3)=TRIAG(22.,25.,28.),
70             ATRIB(4)=5.,ATRIB(6)=1.;
71        ACT,,,COOL;
72   COOL GWAIT(3),COOLRACK/FMA(1);              INITIAL LOCATION
73   MACH GWAIT(4),ATRIB(1),CRNW/CRNE;           SEIZE CRANE AND MACHINE
74        ACT,PICK;                              PICK COIL FROM RACKS
75   GFR1 GFREE/MOVE,COOLRACK;                   MOVE COIL OVERHEAD MACH
76        ACT,DROP;                              DROP COIL ON MACHINE
77   GFR2 GFREE,,CRNW/CRNE;                      FREE CRANE
78   PERS GWAIT(5),,,OPER,HELPER,AND;            SEIZE OPER AND HELPER
79        ACT,ATRIB(3);                          CUTTING OPERATION
80   GFR3 GFREE,,,OPER,HELPER,AND;               FREE OPER AND HELPER
81   STOR GWAIT(6),FIN_STOR/MSW(4),CRNW/CRNE;    SEIZE RACK AND CRANE
82        ACT,PICK;                              PICK COIL FROM MACHINE
83   GFR4 GFREE/MOVE,ATRIB(1);                   MOVE OVERHEAD RACK
84        ACT,DROP;                              DROP COIL IN RACK
85   GFR5 GFREE,,CRNW/CRNE;                      FREE CRANE
86        ACT,FINISH;                            FINISH/PACK PROCESSING
87   GFR6 GFREE,FIN_STOR;                        FREE STORAGE AREA
88   STAT COLCT,INT(5),TIM_IN_SYS;               COLLECT TIME IN SYSTEM
89        TERM;
90        ENDNETWORK;
91   MONTR,SUMRY,480,480;
92   FIN;
```

```
                    S L A M   I I   S U M M A R Y   R E P O R T

            SIMULATION PROJECT PROBLEM 16.3              BY STEIN

            DATE  4/22/1986                              RUN NUMBER   1 OF   1


            CURRENT TIME   0.4800E+03
            STATISTICAL ARRAYS CLEARED AT TIME  0.0000E+00


                 **STATISTICS FOR VARIABLES BASED ON OBSERVATION**

                MEAN        STANDARD     COEFF. OF    MINIMUM      MAXIMUM      NUMBER OF
                VALUE       DEVIATION    VARIATION    VALUE        VALUE        OBSERVATIONS

  TIM_IN_SYS    0.9583E+02  0.3395E+02   0.3543E+00   0.5316E+02   0.1570E+03        19


                              **FILE STATISTICS**

  FILE                      AVERAGE      STANDARD     MAXIMUM   CURRENT    AVERAGE
  NUMBER  LABEL/TYPE        LENGTH       DEVIATION    LENGTH    LENGTH     WAITING TIME

    1                       0.1217       0.5028          4        0        1.0428
    2                       0.1217       0.5028          4        0        1.0428
    3     COOL GWAIT        0.0000       0.0000          1        0        0.0000
    4     MACH GWAIT        2.1333       1.0854          4        3       42.6659
    5     PERS GWAIT        0.0000       0.0000          1        0        0.0000
    6     STOR GWAIT        0.0000       0.0000          1        0        0.0000
    7         CALENDAR      4.3610       0.8717          7        4        2.8211


                            **RESOURCE STATISTICS**

  RESOURCE  RESOURCE  CURRENT   AVERAGE      STANDARD    MAXIMUM      CURRENT
  NUMBER    LABEL     CAPACITY  UTILIZATION  DEVIATION   UTILIZATION  UTILIZATION

    1       OPER         3      1.2932       0.6851         3            1
    2       HELPER       3      1.2932       0.6851         3            1


  RESOURCE  RESOURCE  CURRENT    AVERAGE     MINIMUM     MAXIMUM
  NUMBER    LABEL     AVAILABLE  AVAILABLE   AVAILABLE   AVAILABLE

    1       OPER         2       1.7068         0           3
    2       HELPER       2       1.7068         0           3


                           **AREA/PILE STATISTICS**

  AREA    AREA     PILE     PILE      CURRENT      AVERAGE      MINIMUM      MAXIMUM      NUMBER OF TRANSACTIONS
  NUMBER  LABEL    NUMBER   CAPACITY  UTILIZATION  UTILIZATION  UTILIZATION  UTILIZATION  ARRIVING    LEAVING

    1     HVYSLIT    1        1.00       1.00         0.67         0.00         1.00          7          6
    2     LGTSLIT    2        1.00       0.00         0.08         0.00         1.00          1          1
    3     SHEAR      3        1.00       0.00         0.92         0.00         1.00         13         13
    4     FIN_STOR   4     1000.00       0.00        31.41         0.00       193.80          7          7
                     5     1000.00     100.10        50.08         0.00       109.08         13         12
    5     COOLRACK   6      800.00     101.15       149.45         0.00       309.61         14         13
                     7      600.00     100.87        75.83         0.00       308.21          7          6
                     8      600.00     108.37         5.26         0.00       108.37          3          2


                              **CRANE STATISTICS**

  CRANE                 NUMBER OF       PICK   TO PICK  TO PICK   DROP    TO DROP  TO DROP   TOTAL
  NUMBER  CRANE LABEL  PICKS   DROPS    UTIL   UTIL     INTERF    UTIL    UTIL     INTERF    UTIL

    1     CRNW          26   .   26     0.08    0.02     0.01     0.14     0.03     0.00      0.27
    2     CRNE          15       15     0.05    0.02     0.07     0.08     0.01     0.01      0.23
```

```
                          S L A M   I I   S U M M A R Y   R E P O R T

                  SIMULATION PROJECT PROBLEM 16.3          BY STEIN

                  DATE  4/22/1986                          RUN NUMBER   1 OF   1


                  CURRENT TIME   0.2400E+04
                  STATISTICAL ARRAYS CLEARED AT TIME  0.0000E+00



                     **STATISTICS FOR VARIABLES BASED ON OBSERVATION**

                  MEAN        STANDARD      COEFF. OF     MINIMUM       MAXIMUM      NUMBER OF
                  VALUE       DEVIATION     VARIATION     VALUE         VALUE        OBSERVATIONS

  TIM_IN_SYS   0.9353E+02   0.4696E+02   0.5021E+00   0.4770E+02   0.2461E+03      118



                             **FILE STATISTICS**

  FILE                    AVERAGE      STANDARD     MAXIMUM    CURRENT    AVERAGE
  NUMBER  LABEL/TYPE      LENGTH       DEVIATION    LENGTH     LENGTH     WAITING TIME

    1                     0.0935       0.4069          4          0       0.8163
    2                     0.0935       0.4069          4          0       0.8163
    3     COOL GWAIT      0.0000       0.0000          1          0       0.0000
    4     MACH GWAIT      1.7890       1.3193          6          1      35.7801
    5     PERS GWAIT      0.0000       0.0000          1          0       0.0000
    6     STOR GWAIT      0.0069       0.0828          1          0       0.1406
    7         CALENDAR    4.7467       0.9067          8          3       2.6938



                             **RESOURCE STATISTICS**

  RESOURCE  RESOURCE  CURRENT   AVERAGE       STANDARD     MAXIMUM       CURRENT
  NUMBER    LABEL     CAPACITY  UTILIZATION   DEVIATION    UTILIZATION   UTILIZATION

    1       OPER         3       1.4775        0.7521          3             1
    2       HELPER       3       1.4775        0.7521          3             1



  RESOURCE  RESOURCE  CURRENT    AVERAGE     MINIMUM     MAXIMUM
  NUMBER    LABEL     AVAILABLE  AVAILABLE   AVAILABLE   AVAILABLE

    1       OPER         2        1.5225         0           3
    2       HELPER       2        1.5225         0           3



                             **AREA/PILE STATISTICS**

  AREA     AREA      PILE     PILE       CURRENT       AVERAGE       MINIMUM       MAXIMUM      NUMBER OF TRANSACTIONS
  NUMBER   LABEL     NUMBER   CAPACITY   UTILIZATION   UTILIZATION   UTILIZATION   UTILIZATION  ARRIVING    LEAVING

    1      HVYSLIT     1        1.00        0.00          0.74          0.00          1.00          36        36
    2      LGTSLIT     2        1.00        0.00          0.45          0.00          1.00          31        31
    3      SHEAR       3        1.00        1.00          0.74          0.00          1.00          52        51
    4      FIN_STOR    4     1000.00        0.00         57.22          0.00        216.29          67        67
                       5     1000.00        0.00         43.93          0.00        109.84          51        51
    5      COOLRACK    6      800.00      108.21        125.72          0.00        393.46          66        65
                       7      600.00        0.00         56.15          0.00        308.21          40        40
                       8      600.00        0.00         16.93          0.00        213.76          14        14



                             **CRANE STATISTICS**

  CRANE                   NUMBER OF       PICK     TO PICK    TO PICK    DROP     TO DROP    TO DROP    TOTAL
  NUMBER   CRANE LABEL   PICKS   DROPS    UTIL     UTIL       INTERF     UTIL     UTIL       INTERF     UTIL

    1      CRNW          158     158      0.10     0.03       0.03       0.16     0.03       0.00       0.36
    2      CRNE           79      79      0.05     0.02       0.08       0.08     0.01       0.00       0.25
```

D16-3.3

16-3,Embellishment. The key to this embellishment is to recognize that the operator and helper work as a team and can be modeled as a CRANE to provide the movement between areas. The statement model for this embellishment is shown in D16-3.4. The changes to the model involve defining three additional areas in lines 17, 18, and 19, three additional piles in lines 31, 32, and 33 and two cranes to model the operator and helper combinations in lines 42 and 43. The change in the entity flow model is embodied in the statements in lines 85 through 88 where the cranes OP&HLP1 or OP&HLP2 are requested to perform the cutting operation. In the original problem GWAIT and GFREE nodes were used to request the OPER and HELPER resources. (See lines 78, 79, and 80 of D16-3.1.) Note that attribute 7 has been added for this embellishment to define the area number associated with positions for the operator and helper crane. Separate runways were used for OP&HLP1 and OP&HLP2 since interference between personnel combinations or the crane is not included in this model. The summary report at time 2400 is shown in D16-3.5

16-4.   The solution to this exercise is to add 10 seconds to each pick activity and 15 seconds to each drop activity. A more interesting problem is to define a y-location as the height of a rack and to include a vertical movement time for the crane to move from one rack height to another.

```
  1  GEN,STEIN,PROBLEM 16.3A,4/22/86,1,Y,Y,Y,N;
  2  INIT,0,2400;
  3  LIMITS,8,11,100;
  4  EQUIVALENCE/1.5,PICK/2.5,DROP/UNFRM(12.,18.),FINISH;
  5  NETWORK;
  6  ;
  7  ; *** ALUMINUM COIL PROCESSING FACILITY ***
  8  ;
  9  ;
 10  ; *** AREA DEFINITIONS ***
 11  ;
 12        AREA,HVYSLIT/1,1,6,4;
 13        AREA,LGTSLIT/2,1,6,4;
 14        AREA,SHEAR/3,1,6,4;
 15        AREA,FIN_STOR/4,2,2,6;
 16        AREA,COOLRACK/5,3,2,3;
 17        AREA,OPHVYSLIT/6,1,6,5;
 18        AREA,OPLGTSLIT/7,1,6,5;
 19        AREA,OPSHEAR/8,1,6,5;
 20  ;
 21  ; *** PILE DEFINITIONS ***
 22  ;
 23        PILE,1,HVYSLIT,1.0,75.0,20.0;
 24        PILE,2,LGTSLIT,1.0,110.0,20.0;
 25        PILE,3,SHEAR,1.0,225.0,20.0;
 26        PILE,4,FIN_STOR,1000.0,350.0,50.0;
 27        PILE,5,FIN_STOR,1000.0,350.0,20.0;
 28        PILE,6,COOLRACK,800.0,150.0,55.0;
 29        PILE,7,COOLRACK,600.0,200.0,55.0;
 30        PILE,8,COOLRACK,600.0,275.0,55.0;
 31        PILE,9,OPHVYSLIT,1.0,75.0,20.0;
 32        PILE,10,OPLGTSLIT,1.0,110.0,20.0;
 33        PILE,11,OPSHEAR,1.0,225.0,20.0;
 34  ;
 35  ; *** CRANE DEFINITIONS ***
 36  ;
 37        CRANE,CRNW,1,25.,35.,300.,0.,0.,300.,1;
 38        CRANE,CRNE,1,300.,50.,300.,2000.,2000.,300.,2;
 39  ;
 40  ; *** OPERATOR AND HELPER DEFINITIONS ***
 41  ;
 42        CRANE,OP&HLP1,2,75.0,20.0,200.,0.,0.,200.,7;
 43        CRANE,OP&HLP2,3,225.0,20.0,200.,0.,0.,200.,8;
 44  ;
 45  ; *** NETWORK FLOW ***
 46  ;
 47  CR1  CREATE,20,,5;
 48       ACT,,0.3,HVY;                              HEAVY GAUGE COIL
 49       ACT,,0.3,LGT;                              LIGHT GAUGE COIL
 50       ACT,0.4,SHR;                               SHEAR COIL
 51  ;
 52  ; *** ASSIGN ATTRIBUTES TO INCOMING COILS ***
 53  ;
 54  ;    ATTRIBUTE           DESCRIPTION
 55  ;    ---------           -----------
 56  ;       1                COIL TYPE
 57  ;                          1=HEAVY GAUGE
 58  ;                          2=LIGHT GAUGE
 59  ;                          3=SHEAR
 60  ;       2                COIL SIZE (VOLUME,TCI)
 61  ;       3                PROCESSING TIME
 62  ;       4                PILE # AT FINISHED PROCESSING
 63  ;       5                MARKING ATTRIBUTE
 64  ;       6                QUANTITY TO REQUEST AT PROCESSING (=1)
 65  ;       7                ATRIB(1)+5.0
 66  ;
 67  HVY  ASSIGN,ATRIB(1)=1.,ATRIB(2)=UNFRM(90.,110.),
 68            ATRIB(3)=TRIAG(38.,40.,42.),
 69            ATRIB(4)=4.,ATRIB(6)=1.,ATRIB(7)=6.;
 70       ACT,,,COOL;
 71  LGT  ASSIGN,ATRIB(1)=2.,ATRIB(2)=UNFRM(90.,110.),
 72            ATRIB(3)=TRIAG(24.,26.,28.),
 73            ATRIB(4)=4.,ATRIB(6)=1.,ATRIB(7)=7.;
 74       ACT,,,COOL;
 75  SHR  ASSIGN,ATRIB(1)=3.,ATRIB(2)=UNFRM(90.,110.),
 76            ATRIB(3)=TRIAG(22.,25.,28.),
 77            ATRIB(4)=5.,ATRIB(6)=1.,ATRIB(7)=8.;
 78       ACT,,,COOL;
 79  COOL GWAIT(3),COOLRACK/FMA(1);                  INITIAL LOCATION
 80  MACH GWAIT(4),ATRIB(1),CRNW/CRNE;               SEIZE CRANE AND MACHINE
 81       ACT,PICK;                                  PICK COIL FROM RACKS
 82  GFR1 GFREE/MOVE,COOLRACK;                        MOVE COIL OVERHEAD MACH
 83       ACT,DROP;                                  DROP COIL ON MACHINE
 84  GFR2 GFREE,,CRNW/CRNE;                           FREE CRANE
 85  PERS GWAIT(5),ATRIB(7),OP&HLP1/OP&HLP2;         SEIZE OPER AND HELPER
 86       GFREE/MOVE;                                OPER AND HELPER MOVE
 87       ACT,ATRIB(3);                              CUTTING OPERATION
 88  GFR3 GFREE,ATRIB(7),OP&HLP1/OP&HLP2;            FREE OPER AND HELPER
 89  STOR GWAIT(6),FIN_STOR/MSW(4),CRNW/CRNE;        SEIZE RACK AND CRANE
 90       ACT,PICK;                                  PICK COIL FROM MACHINE
 91  GFR4 GFREE/MOVE,ATRIB(1);                        MOVE OVERHEAD RACK
 92       ACT,DROP;                                  DROP COIL IN RACK
 93  GFR5 GFREE,,CRNW/CRNE;                           FREE CRANE
 94       ACT,FINISH;                                FINISH/PACK PROCESSING
 95  GFR6 GFREE,FIN_STOR;                             FREE STORAGE AREA
 96  STAT COLCT,INT(5),TIM_IN_SYS;                   COLLECT TIME IN SYSTEM
 97       TERM;
 98       ENDNETWORK;
 99  MONTR,SUMRY,480,480;
100  FIN;
```

S L A M   I I   S U M M A R Y   R E P O R T

SIMULATION PROJECT PROBLEM 16.3A   BY STEIN

DATE 4/22/1986   RUN NUMBER   1 OF   1

CURRENT TIME   0.2400E+04
STATISTICAL ARRAYS CLEARED AT TIME 0.0000E+00

STATISTICS FOR VARIABLES BASED ON OBSERVATION

	MEAN VALUE	STANDARD DEVIATION	COEFF. OF VARIATION	MINIMUM VALUE	MAXIMUM VALUE	NUMBER OF OBSERVATIONS
TIM_IN_SYS	0.8619E+02	0.4271E+02	0.4955E+00	0.4634E+02	0.2470E+03	118

FILE STATISTICS

FILE NUMBER	LABEL/TYPE	AVERAGE LENGTH	STANDARD DEVIATION	MAXIMUM LENGTH	CURRENT LENGTH	AVERAGE WAITING TIME
1		0.0359	0.2508	5	0	0.3264
2		0.0359	0.2508	5	0	0.3264
3	COOL GWAIT	0.0000	0.0000	1	0	0.0000
4	MACH GWAIT	1.3111	1.2582	6	0	26.2212
5	PERS GWAIT	0.1203	0.3253	2	0	2.4060
6	STOR GWAIT	0.0199	0.1398	1	0	0.4020
7		0.0966	0.2955	1	0	1.9327
8		0.0966	0.2955	1	0	1.9327
9	CALENDAR	4.7406	0.8686	8	4	2.2800

AREA/PILE STATISTICS

AREA NUMBER	AREA LABEL	PILE NUMBER	PILE CAPACITY	CURRENT UTILIZATION	AVERAGE UTILIZATION	MINIMUM UTILIZATION	MAXIMUM UTILIZATION	NUMBER OF ARRIVING	TRANSACTIONS LEAVING
1	HVYSLIT	1	1.00	0.00	0.71	0.00	1.00	36	36
2	LGTSLIT	2	1.00	0.00	0.44	0.00	1.00	31	31
3	SHEAR	3	1.00	1.00	0.71	0.00	1.00	53	52
4	FIN_STOR	4	1000.00	0.00	62.78	0.00	216.29	67	67
		5	1000.00	99.42	43.85	0.00	109.84	52	51
5	COOLRACK	6	800.00	0.00	104.87	0.00	393.46	80	80
		7	600.00	0.00	32.66	0.00	299.94	33	33
		8	600.00	0.00	5.19	0.00	183.15	7	7
6	OPHVYSLI	9	1.00	0.00	0.67	0.00	1.00	36	36
7	OPLGTSLI	10	1.00	0.00	0.39	0.00	1.00	31	31
8	OPSHEAR	11	1.00	1.00	0.60	0.00	1.00	53	52

CRANE STATISTICS

CRANE NUMBER	CRANE LABEL	NUMBER OF PICKS	NUMBER OF DROPS	PICK UTIL	TO PICK UTIL	TO PICK INTERF	DROP UTIL	TO DROP UTIL	TO DROP INTERF	TOTAL UTIL
1	CRNW	152	152	0.09	0.02	0.01	0.16	0.02	0.01	0.32
2	CRNE	87	87	0.05	0.02	0.06	0.09	0.02	0.00	0.25
3	OP&HLP1	70	69	0.00	0.01	0.00	0.82	0.00	0.00	0.83
4	OP&HLP2	50	50	0.00	0.01	0.00	0.66	0.00	0.00	0.66

D16-3.5

16-5.   To model the FMS using AGVs, it is necessary to define each machine as a resource and to provide a description of the guidepath with control points before each machine and segments connecting the guidepath. A description of the AGV fleet is then required. These definitions and the statement model are shown in D16-5.1. Two rows of ARRAY are used. The first row provides the processing times and the second row defines the type of machine at each machine location. From the ARRAY(2,10) statement, it is seen that machines 1 through 5 are dedicated to operation 1, machine 6 is dedicated to operation 2, machine 7 and 8 are dedicated to operation 3, and machines 9 and 10 are flexible.

Entities are created every 22 time units and sent to an ASSIGN node where OPERATION is set to 1 to indicate that this entity needs to have operation 1 performed on it. OPERATION is equivalenced to ATRIB(2). The next set of statements is a direct search to find an available machine. If a dedicated machine is not available, a flexible machine is investigated. If neither are available, the entity is placed in activity 1 waiting for a machine to be freed. The machine is then assigned at AWAIT node AWA1 and a request for an AGV is made at VWAIT node for an AGV to come to control point 11. When the AGV arrives, the entity will be moved to the assigned machine by the VMOVE activity. Upon arrival to the machine, the AGV is freed at the VFREE node. The entity continues on the network and the AGV is reassigned in accordance with the VFLEET resource statement.

The part is processed in activity 2 after which a request for an AGV is made at the VWAIT node. The location of the part is at control point MACHINE. When the AGV arrives, the MACHINE is made available and will be assigned to any entity waiting in activity 1 that requires this machine type. The AGV moves the part to the wash station (control point 11) and upon arrival the AGV is freed. The part awaits inspection, is inspected and washed, and then either cycles back to have its next operation performed or leaves the system because all the operations have been performed. The SLAM II Summary Report for this model is shown in D16-5.2.

An alternative model is shown in D16-5.3 where a user function is employed to allow additional logic to be placed in the model for assigning machines to parts. The statements in D16-5.3 replace line number 63 through 76 of D16-5.1. The function USERF is given in D16-5.4.

Another way to achieve more control on the allocation of flexible machines is to use the STOPA capability of SLAM II. Activity 1 the indeterminant time of activity 1 can be replaced by the specification STOPA(MACHINE). Following the statement, FREE,MACHINE; in line 82 of D16-5.1, the statements shown in D16-5.5 would be used. The order in which the values are assigned to STOPA prescribes the order in which the machines will be allocated to castings waiting for the operation. In D16-5.5, the flexible machine will first attempt to process any casting for which operation 2 is required to be performed, then operation 3 and lastly operation 1.

16-5,Embellishment(a).   This embellishment only requires that the number of vehicles as prescribed on the VFLEET statement be changed to 2.

16-5,Embellishment(b).   The modification required to accomplish this embellishment is to change the direction on the VSGMENT resource statement to BID from UNI. This modification allows the vehicle to move to the machines in either direction and cuts down on the travel time. An approximate three percent increase in throughput was observed when running this embellishment.

16-5,Embellishment(c).   The alternative solution shown for the basic problem involving the processing of castings that have had machining operations performed on them is one form of solution to this exercise. If it is assumed that value has been added by processing, the flexible machine should be assigned to perform operation 3, then operation 2, then operation 1 and the carts should be allocated to the movement of those entities which have had more processing performed on them. This can be accomplished by ranking job requests for vehicles by high value of attribute 2 which signifies that the higher the operation number the greater the job request priority for the AGV.

```
 1    GEN,PRITSKER,PROBLEM 16.5,7/23/86,1;
 2    LIMITS,5,7,300;
 3    INIT,0,4800;
 4    VCONTROL,0.01,0.1;
 5    CONTINUOUS,0,1,,,N;
 6    ARRAY(1,3)/120.0,40.0,56.0;
 7    ARRAY(2,10)/1,1,1,1,1,2,3,3,4,4;
 8    EQUIVALENCE/ATRIB(2),OPERATION/
 9                 ATRIB(3),MACHINE/
10                 ATRIB(4),TYPE/
11                 ARRAY(1,OPERATION),ACTIVITY_TIME/
12                 ARRAY(2,MACHINE),MACHINE_TYPE;
13    NETWORK;
14    ;
15    ;   DEFINE INSPECT RESOURCE
16    ;
17          RESOURCE/1,MACH1(1),1;
18          RESOURCE/2,MACH2(1),1;
19          RESOURCE/3,MACH3(1),1;
20          RESOURCE/4,MACH4(1),1;
21          RESOURCE/5,MACH5(1),1;
22          RESOURCE/6,MACH6(1),1;
23          RESOURCE/7,MACH7(1),1;
24          RESOURCE/8,MACH8(1),1;
25          RESOURCE/9,MACH9(1),1;
26          RESOURCE/10,MACH10(1),1;
27          RESOURCE/11,INSP(2),4;
28    ;
29    ;   DEFINE VEHICLE CONTROL POINTS
30    ;
31          VCPOINT,1/MACH#1;
32          VCPOINT,2/MACH#2;
33          VCPOINT,3/MACH#3;
34          VCPOINT,4/MACH#4;
35          VCPOINT,5/MACH#5;
36          VCPOINT,6/MACH#6;
37          VCPOINT,7/MACH#7;
38          VCPOINT,8/MACH#8;
39          VCPOINT,9/MACH#9;
40          VCPOINT,10/MACH#10;
41          VCPOINT,11/INSPECT;
42    ;
43    ;   DEFINE VEHICLE GUIDEPATH SEGMENTS
44    ;
45          VSGMENT,1,1,2,20,UNI;
46          VSGMENT,2,2,3,20,UNI;
47          VSGMENT,3,3,4,20,UNI;
48          VSGMENT,4,4,5,20,UNI;
49          VSGMENT,5,5,6,60,UNI;
50          VSGMENT,6,6,7,20,UNI;
51          VSGMENT,7,7,8,20,UNI;
52          VSGMENT,8,8,9,20,UNI;
53          VSGMENT,9,9,10,20,UNI;
54          VSGMENT,10,10,11,30,UNI;
55          VSGMENT,11,11,1,30,UNI;
56    ;
57    ;   DEFINE AUTOMATED GUIDED VEHICLE
58    ;
59          VFLEET,AGV,1,100,100,0,0,4,,,5/CLOSEST,CRUISE;
60    ;
61    ;   DEFINE NETWORK
62    ;
63          CREATE,22,,1;
64          ASSIGN,OPERATION=1,MACHINE=1,TYPE=1;
65    GON1  ASSIGN,II=MACHINE,1;
66          ACTIVITY,,TYPE.EQ.MACHINE_TYPE.AND.NNRSC(II).GT.0,AWA1;
67          ACTIVITY,,MACHINE.EQ.10.AND.TYPE.EQ.4,WAT1;
68          ACTIVITY,,MACHINE.EQ.10,ASN1;
69          ACTIVITY,,NXT1;
70    NXT1  ASSIGN,MACHINE=MACHINE+1;
71          ACTIVITY,,,GON1;
72    ASN1  ASSIGN,MACHINE=1,TYPE=4;
73          ACTIVITY,,,GON1;
74    WAT1  ASSIGN,MACHINE=1,TYPE=OPERATION;
75          ACTIVITY/1,REL(FRE1),,GON1;
76    AWA1  AWAIT(1),MACHINE;
77          VWAIT(2),AGV,11,CLOSEST;
78          VMOVE,MACHINE;
79          VFREE,AGV;
80          ACTIVITY/2,ACTIVITY_TIME;
81          VWAIT(3),AGV,MACHINE,CLOSEST;
82    FRE1  FREE,MACHINE;
83          VMOVE,11;
84          VFREE,AGV;
85          AWAIT(4),INSP;
86          ACTIVITY/3,10;
87          FREE,INSP,1;
88          ACTIVITY,,OPERATION.EQ.3,TRM1;
89          ACTIVITY,,,ASN2;
90    ASN2  ASSIGN,OPERATION=OPERATION+1,MACHINE=1,TYPE=OPERATION;
91          ACTIVITY,,,GON1;
92    TRM1  COLCT,INT(1),TIME IN SYSTEM;
93          TERMINATE;
94          ENDNETWORK;
95    FIN;
```

```
                        S L A M  I I  S U M M A R Y  R E P O R T

                SIMULATION PROJECT PROBLEM 16.5          BY PRITSKER

                DATE  7/23/1986                          RUN NUMBER   1 OF   1


                CURRENT TIME   0.4800E+04
                STATISTICAL ARRAYS CLEARED AT TIME  0.0000E+00


            **STATISTICS FOR VARIABLES BASED ON OBSERVATION**

                MEAN        STANDARD     COEFF. OF    MINIMUM     MAXIMUM      NUMBER OF
                VALUE       DEVIATION    VARIATION    VALUE       VALUE        OBSERVATIONS

TIME IN SYSTEM  0.4937E+03  0.1213E+03   0.2457E+00   0.2604E+03  0.7168E+03      187
```

FILE STATISTICS

FILE NUMBER	LABEL/TYPE	AVERAGE LENGTH	STANDARD DEVIATION	MAXIMUM LENGTH	CURRENT LENGTH	AVERAGE WAITING TIME
1	AWA1 AWAIT	0.0000	0.0000	1	0	0.0000
2	VWAIT	0.1439	0.3517	2	0	1.1508
3	VWAIT	0.3121	0.5787	4	0	2.5391
4	AWAIT	0.1234	0.3612	3	0	1.0041
5	VEHICLE	0.3054	0.6305	4	0	1.2318
6	CALENDAR	11.3143	1.5819	34	12	0.1949

REGULAR ACTIVITY STATISTICS

ACTIVITY INDEX/LABEL	AVERAGE UTILIZATION	STANDARD DEVIATION	MAXIMUM UTIL	CURRENT UTIL	ENTITY COUNT
1	10.2876	5.8682	23	21	6498
2	9.0856	1.3169	10	10	590
3	1.2287	0.7267	2	1	589

RESOURCE STATISTICS

RESOURCE NUMBER	RESOURCE LABEL	CURRENT CAPACITY	AVERAGE UTILIZATION	STANDARD DEVIATION	MAXIMUM UTILIZATION	CURRENT UTILIZATION
1	MACH1	1	0.9944	0.0746	1	1
2	MACH2	1	0.9897	0.1008	1	1
3	MACH3	1	0.9848	0.1224	1	1
4	MACH4	1	0.9801	0.1396	1	1
5	MACH5	1	0.9755	0.1547	1	1
6	MACH6	1	0.9715	0.1664	1	1
7	MACH7	1	0.9581	0.2003	1	1
8	MACH8	1	0.9514	0.2151	1	1
9	MACH9	1	0.9771	0.1496	1	1
10	MACH10	1	0.9669	0.1789	1	1
11	INSP	2	1.2287	0.7267	2	1

RESOURCE NUMBER	RESOURCE LABEL	CURRENT AVAILABLE	AVERAGE AVAILABLE	MINIMUM AVAILABLE	MAXIMUM AVAILABLE
1	MACH1	0	0.0056	0	1
2	MACH2	0	0.0103	0	1
3	MACH3	0	0.0152	0	1
4	MACH4	0	0.0199	0	1
5	MACH5	0	0.0245	0	1
6	MACH6	0	0.0285	0	1
7	MACH7	0	0.0419	0	1
8	MACH8	0	0.0486	0	1
9	MACH9	0	0.0229	0	1
10	MACH10	0	0.0331	0	1
11	INSP	1	0.7713	0	2

VEHICLE UTILIZATION REPORT

————— AVERAGE NUMBER OF VEHICLES —————

VEHICLE FLEET LABEL	NUMBER AVAILABLE	TRAVELING TO LOAD (EMPTY)	LOADING	TRAVELING TO UNLOAD (FULL)	UNLOADING	TOTAL PRODUCTIVE
AGV	1	0.148	0.000	0.347	0.000	0.495

VEHICLE PERFORMANCE REPORT

————— AVERAGE NUMBER OF VEHICLES —————

VEHICLE FLEET LABEL	NUMBER OF LOADS	NUMBER OF UNLOADS	TRAVELING EMPTY BLOCKED	TRAVELING FULL BLOCKED	TRAVELING IDLE	STOPPED IDLE	TOTAL NON-PRODUCTIVE
AGV	1190	1190	0.003	0.006	0.000	0.497	0.505

SEGMENT STATISTICS

SEGMENT NUMBER	SEGMENT LABEL	CONTROL END POINTS	NUMBER OF ENTRIES	AVERAGE UTILIZATION	MAXIMUM UTIL.	CURRENT UTIL.
1		1 / 2	849	0.035	1	0
2		2 / 3	849	0.035	1	0
3		3 / 4	849	0.035	1	0
4		4 / 5	849	0.035	1	0
5		5 / 6	849	0.106	1	0
6		6 / 7	849	0.034	1	0
7		7 / 8	849	0.035	1	0
8		8 / 9	848	0.035	1	0
9		9 / 10	848	0.035	1	0
10		10 / 11	848	0.052	1	0
11		11 / 1	849	0.049	1	0

CONTROL POINT STATISTICS

CONTROL POINT NUMBER	CONTROL POINT LABEL	NUMBER OF ENTRIES	AVERAGE UTILIZATION	MAXIMUM NUMBER WAITING	CURRENT UTILIZATION
1	MACH#1	849	0.076	0	0
2	MACH#2	849	0.043	0	0
3	MACH#3	849	0.066	0	0
4	MACH#4	849	0.025	0	0
5	MACH#5	849	0.025	0	0
6	MACH#6	849	0.090	0	0
7	MACH#7	849	0.059	0	0
8	MACH#8	849	0.079	0	1
9	MACH#9	848	0.073	0	0
10	MACH#10	848	0.095	0	0
11	INSPECT	849	0.029	0	0

*** VEHICLE TRIP REPORT MATRIX ***

TABLE 1

FROM CP \ TO CP	1	2	3	4	5	6	7	8	9	10	11	TOTAL
MACH#1 1.	0	2	0	0	0	13	0	10	4	8	40	77
MACH#2 2.	0	0	1	0	0	8	13	0	10	4	39	75
MACH#3 3.	0	0	0	0	0	15	1	3	9	9	38	75
MACH#4 4.	0	0	0	0	31	1	0	2	1	2	38	75
MACH#5 5.	1	0	0	0	0	11	22	0	2	1	38	75
MACH#6 6.	11	13	9	9	0	0	20	11	17	13	102	205
MACH#7 7.	4	0	3	0	0	10	0	35	8	12	77	149
MACH#8 8.	3	6	4	8	1	18	0	0	15	15	78	148
MACH#9 9.	12	8	11	7	3	11	10	4	0	12	77	155
MACH#10 10.	6	7	9	12	2	15	8	7	10	0	79	155
INSPECT 11.	40	39	38	39	38	103	75	77	79	79	583	1190
TOTAL.	77	75	75	75	75	205	149	149	155	155	1189	2379

STATE AND DERIVATIVE VARIABLES

(I)	SS(I)	DD(I)
1	0.0000E+00	0.0000E+00

```
61          CREATE,22,,1;
62          ASSIGN,OPERATION=1,TYPE=1;
63   ASG1   ASSIGN,MACHINE=USERF(1),1;
64          ACTIVITY/1,STOPA(OPERATION),MACHINE.EQ.0,ASG1;
65          ACTIVITY;
66   AWA1   AWAIT(1),MACHINE;
```

D16-5.3

```
     FUNCTION USERF(I)
     COMMON/SCOM1/ATRIB(100),DD(100),DDL(100),DTNOW,II,MFA,MSTOP,NCLNR
    1,NCRDR,NPRNT,NNRUN,NNSET,NTAPE,SS(100),SSL(100),TNEXT,TNOW,XX(100)
     DIMENSION NBRKP(5)
     EQUIVALENCE(ATRIB(2),OPERATION),(ATRIB(4),TYPE)
     DATA NBRKP/1,6,7,9,11/
C**** SET USERF TO AVAILABLE MACHINE
C**** AND TYPE TO 4 IF FLEXIBLE MACHINE IS ASSIGNED
C**** SET USERF TO 0 IF NO MACHINE AVAILABLE
     INTOP=INT(OPERATION)
   5 DO 10 J=NBRKP(INTOP),NBRKP(INTOP+1)-1
     USERF=J
     IF (NNRSC(J).GT.0) RETURN
  10 CONTINUE
     IF (TYPE.EQ.4) THEN
         TYPE=OPERATION
         USERF=0.0
         RETURN
     ELSE
         TYPE=4.0
         INTOP=4
         GO TO 5
     ENDIF
     END
```

D16-5.4

```
73          ACTIVITY,,TYPE.EQ.4,ASG2;
74          ACTIVITY;
75          ASSIGN,STOPA=OPERATION;
76          ACTIVITY,,,VMV1;
77   ASG2   ASSIGN,STOPA=2,STOPA=3,STOPA=1;
78   VMV1   VMOVE,11;
```

D16-5.5

16-6. The integrated model of Example 6-1 and 6-2 is shown in D16-6.1.

```
 1  GEN,PRITSKER & ASSOC.,PROBLEM 16.6,7/29/86,1;
 2  ;
 3  ;   ASRS AND AGV INTEGRATION
 4  ;
 5  ;   (ALL TIMES ARE IN SECONDS, DISTANCES IN FEET.)
 6  ;
 7  LIMITS,20,8,200;
 8  PRIORITY/10,LVF(3)/11,HVF(3)/12,LVF(4);
 9  INIT,0.,89400.;
10  VCONT,.1,2.;
11  CONTINUOUS,0,3,,,,N;
12  ;
13  ;   EQUIVALENCE AND ARRAY DECLARATIONS
14  ;
15  ARRAY(1,6)/7,8,9,11,12,13;
16  EQUIVALENCE /6, NMACH/
17             ATRIB(5), MACH/
18             ARRAY(1,MACH), CP;
19  NETWORK;
20  ;
21  ;   MATERIAL HANDLING EXTENSION RESOURCE DEFINITIONS
22  ;
23  ;  AREAS
24  ;                 AREA        NUMBER       QUANTITY    ASSOCIATED
25  ;                 LABEL       OF PILES     ATTRIBUTE    FILES
26  ;                 ======      ========     ==========  ==========
27        AREA,       ENTRY,      1,           2,           1;
28        AREA,       EXIT,       1,           2,           9;
29        AREA,       LATHA,      1,           2,           3;
30        AREA,       MILLA,      1,           2,           6;
31        AREA,       RACK,       9,           2,           8,5,2;
32  ;
33  ; PILES
34  ;
35  ;                 PILE        AREA         PILE        X-COOR      Y-COOR
36  ;.                NUMBER      LABEL        CAPACITY    POSITION    POSITION
37  ;                 ======      =====        ========    ========    ========
38        PILE,       1,          RACK,        5,           90,         5;
39        PILE,       2,          RACK,        3,          120,         5;
40        PILE,       3,          RACK,        5,           75,         5;
41        PILE,       4,          RACK,        5,          105,         5;
42        PILE,       5,          RACK,        5,          135,         5;
43        PILE,       6,          RACK,        3,          165,         5;
44        PILE,       7,          RACK,        5,          150,         5;
45        PILE,       8,          RACK,        5,          180,         5;
46        PILE,       9,          RACK,        5,          195,         5;
47        PILE,       10,         ENTRY,       1,           45,         5;
48        PILE,       11,         EXIT,        1,           45,         5;
49        PILE,       12,         LATHA,       2,          120,         5;
50        PILE,       13,         MILLA,       2,          165,         5;
51  ;
52  ; CRANES
53  ;
54  ;       CRANE RUNWAY INITIAL INITIAL  MAXIMUM   NOMINAL NOMINAL TROLLEY JOB F
55  ;       LABEL NUMBER X POS.  Y POS.  BRDG. SPD.  ACCEL.  DECELL.  SPEED  NUMB
56  ;       ===== ====== ======= ======= ========== ======= ======= ======= =====
57        CRA, CR1,    1,     60,     5,   .8333,      0,      0,    .1667,   10;
58        CRA, CR2,    1,    140,     5,   .8333,      0,      0,    .1667,   11;
59  ;
60  ; DEFINE REGULAR SLAM II RESOURCES
61  ;
62        RESOURCE,LATHE(1),4;
63        RESOURCE,MILL(1),7;
64  ;
65  ; DEFINE MACHINE RESOURCES
66  ;
67        RESOURCE/3,MACH1(1),14;
68        RESOURCE/4,MACH2(1),14;
69        RESOURCE/5,MACH3(1),14;
70        RESOURCE/6,MACH4(1),14;
71        RESOURCE/7,MACH5(1),14;
72        RESOURCE/8,MACH6(1),14;
73        RESOURCE/9,SFIX(1),12;
74        RESOURCE/10,SFIXSP(5),13;
75  ;
```

```
76   ;  DEFINE THE VEHICLE CONTROL POINTS
77   ;
78   ;              NUMBER     LABEL
79   ;
80      VCPOINT,    1    /FIXTURE;
81      VCPOINT,    2/;
82      VCPOINT,    3/;
83      VCPOINT,    4    /STAGE;
84      VCPOINT,    5/;
85      VCPOINT,    6/;
86      VCPOINT,    7    /MACH1;
87      VCPOINT,    8    /MACH2;
88      VCPOINT,    9    /MACH3;
89      VCPOINT,   10/;
90      VCPOINT,   11    /MACH4;
91      VCPOINT,   12    /MACH5;
92      VCPOINT,   13    /MACH6;
93      VCPOINT,   14    /INGRT;
94   ;
95   ;  DEFINE THE VEHICLE GUIDEPATH SEGMENTS
96   ;
97   ;;                       BEGINNING        ENDING
98   ;          NUMBER    CONTROL POINT   CONTROL POINT   LENGTH    TYPE
99      VSGMENT,    1,         1,              2,          20,     UNI;
100     VSGMENT,    2,         2,              3,          82,     UNI;
101     VSGMENT,    3,        10,              2,         115,     UNI;
102     VSGMENT,    4,         3,              4,          27,     UNI;
103     VSGMENT,    5,         3,              5,          16,     UNI;
104     VSGMENT,    6,         4,              5,          27,     UNI;
105     VSGMENT,    7,         5,              6,          13,     UNI;
106     VSGMENT,    8,         6,              7,          35,     UNI;
107     VSGMENT,    9,         7,              8,          35,     UNI;
108     VSGMENT,   10,         8,              9,          35,     UNI;
109     VSGMENT,   11,         9,             10,          35,     UNI;
110     VSGMENT,   12,         6,             11,          75,     UNI;
111     VSGMENT,   13,        11,             12,          35,     UNI;
112     VSGMENT,   14,        12,             13,          35,     UNI;
113     VSGMENT,   15,        13,             10,          75,     UNI;
114     VSGMENT,   16,         2,              1,          20,     UNI;
115     VSGMENT,   17,        14,              1,         250,     BI;
116  ;
117  ;  DEFINE AGV
118  ;
119     VFLEET,AGV1,2,4.5,4.0,,,4.0,4.5,,18/CLOSEST,STOP(4),4(2,4);
120     VFLEET,AGV2,1,4.5,4.0,,,4.0,4.5,,19/CLOSEST,STOP(14),14;
121  ;
122  ;  NETWORK STATEMENTS
123  ;
124     CREATE,900.,0,1;
125  ;
126  ;  PART ARRIVAL PROCESSING
127  ;
128     ASSIGN,ATRIB(2) = 1,
129            ATRIB(3) = 0;          SET PALLET AREA = 1
130      ACT/6,,NNQ(3).LE.10,ARR2;    ENTERING
131      ACT/5,,NNQ(3).GT.10;         BYPASS CELL
132     TERM;
133  ;
134  ;  MOVE PALLETS TO ENTRY AREA
135  ;
136  ARR2  GWAIT(1),ENTRY/FAW;
137        ACT,15.;                    INDEX
138  ARR3  GWAIT(2),RACK/FAW,CR1/CR2;  GET RACK & CRANE
139        ACT,30.;                    PICK
140  ARR4  GFREE/MOVE,ENTRY;           FREE ENTRY
141        ASSIGN,ATRIB(3)=2.0;        PALLET: R -> L
142        ACT,30.;                    PLACE
143        GFREE,,CR1/CR2;             FREE CRANE
144  ;
145  ;  MOVE PALLETS FROM RACK TO LATHE
146  ;
147  LATH  GWAIT(3),LATHA/FAW,CR1/CR2; GET LATHE AREA & CRANE
148        ACT,30.;
149        GFREE/MOVE,RACK;            FREE RACK
150        ASSIGN,ATRIB(3) = 4;        PALLET: L -> R
```

```
151              ACT,30.;                    PLACE
152              GFREE,,CR1/CR2;             FREE CRANE
153     ;
154     ;   PROCESS PARTS AT LATHE
155     ;
156  LAT4  AWAIT(4),LATHE/1;                 REQUEST LATHE
157              ACT/1, RNORM(600.,30.,1);   LATHING
158              GOON,1;
159              ACT/3,UNFRM(900.,1500.,2),.2; MAINTAIN LATHE
160              ACT,,.8;
161  LAT5  FREE,LATHE/1;                     FREE LATHE
162     ;
163     ;   MOVE PALLETS TO RACK
164     ;
165  LST0  GWAIT(5),RACK/LAW,CR1/CR2;        GET RACK AND CRANE
166              ACT,30.;                    PICK
167  LST1  GFREE/MOVE,LATHA;                 FREE LATHE AREA
168  LST2  ASSIGN, ATRIB(3) = 5;            PALLET : R -> M
169              ACT,30.;                    PICK
170  LST3  GFREE,,CR1/CR2;                   FREE CRANE
171     ;
172     ;   MOVE PALLETS FROM RACK
173     ;
174  LST4  GWAIT(6),MILLA/FAW,CR1/CR2;       GET MILL AREA AND CRANE
175              ACT,30.;                    PICK
176  LST5  GFREE/MOVE,RACK;                  FREE RACK
177  LST6  ASSIGN, ATRIB(3) = 3;            PALLET: M -> R
178              ACT,30.;                    PLACE
179  LST7  GFREE,,CR1/CR2;                   FREE CRANE
180     ;
181     ;   PROCESS PARTS AT MILL
182     ;
183  MILL  AWAIT(7),MILL/1;                  REQUEST MILL MACHINE
184              ACT/2,RNORM(780.,30.,1);    MILLING
185              GOON,1;
186              ACT/4,UNFRM(600.,1200.,2),.1; MAINTAIN MILL
187              ACT,,.9;
188              FREE,MILL/1;                FREE MILL
189     ;
190     ;   MOVE PALLETS FROM MILL TO RACK
191     ;
192  MIL2  GWAIT(8),RACK/FAW,CR1/CR2;        GET RACK AND CRANE
193              ACT,30.;                    PICK
194  MIL3  GFREE/MOVE,MILLA;                 FREE MILL AREA
195  MIL4  ASSIGN,ATRIB(3) = 1;             PALLET: R -> EXIT
196              ACT,30.;                    PLACE
197  MIL5  GFREE,,CR1/CR2;                   FREE CRANE
198     ;
199     ;   GROUP PALLETS AND MOVE VIA AGV
200     ;
201  BTCH  BATCH,,5,,,ALL(4);                BATCH 5 PALLETS
202              ACT,UNFRM(0.,1200.,3);       AGV TIME TO ARRIVE
203              UNBATCH,4;                  REESTABLISH PALLETS
204     ;
205     ;   MOVE PALLETS TO EXIT AREA
206     ;
207  UNB1  GWAIT(9),EXIT/FAW,CR1/CR2;        GET EXIT AREA & CRANE
208              ACT,30.;                    PICK
209              GFREE/MOVE,RACK;            FREE RACK
210              GOON,1;
211              ACT,30.;                    PLACE
212              GFREE,,CR1/CR2;             FREE CRANE
213              ACT,15.0;                   INDEX
214              GFREE,EXIT;                 FREE EXIT AREA
215              VWAIT(20),AGV2,14,CLOSEST,MATCH;
216              VMOVE,1;
217              VFREE,AGV2;
218  AW02  AWAIT(12),SFIX/1;                 WAIT FOR FIXTURING STATION
219     ;
220              ACT/7,220.;                 LOAD PART    INTO FIXTURE
221
222  AW03  AWAIT(13),SFIXSP/1;               WAIT FOR OUTPUT BUFFER SPACE
223  FR04  FREE,SFIX/1;                      RELEASE THE FIXTURING STATION
224     ;
225     ;   MACHINE SELECTION LOGIC
```

```
226  ;
227  ASO    ASSIGN,MACH = 0;                   INITIALIZE MACH
228  AS1    ASSIGN,MACH = MACH+1,
229            II = MACH,1;                    INCREMENT MACHINE NUMBER
230         ACT,,MACH.GT.NMACH,Q6;             NO MACHINES WAIT IN QUEUE #6
231         ACT,,NNRSC(II).GT.O,AW05;          MACHINE OPEN
232         ACT,,NNRSC(II).EQ.0,AS1;           MACHINE BUSY TRY NEXT MACHINE
233  ;
234  ;   HOLD PARTS FOR NEXT AVAILABLE MACHINE
235  ;
236  Q6     QUEUE(17);
237         ACT(1)/10,REL(FR11),,ASO;  HOLD PART    FOR MACHINE RELEASE
238  ;
239  ;   SEIZE MACHINE AND TRANSPORT PART
240  ;
241  AW05   AWAIT(14),MACH/1;                  SEIZE MACHINE NUMBER MACH
242  VW06   VWAIT(15),AGV1,1,FIFO,TOP;         REQUEST A VEHICLE AT CONTROL POINT 1
243         ACT,45.;                           LOAD PART    ONTO AGV
244  ;
245  FR07   FREE,SFIXSP/1;                     RELEASE FIXTURE OUTPUT BUFFER SPACE
246  ;
247  VM08   VMOVE,CP;                          MOVE PART TO MACHINE CONTROL POINT
248         ACT,45.;                           UNLOAD PART FROM AGV
249  VF09   VFREE,AGV1;                        RELEASE AGV
250  ;
251         ACT/8,TRIAG(920.,1860.,2720.);  PROCESS PART
252  ;
253  VW10   VWAIT(16),AGV1,CP,FIFO,MATCH;REQUEST A VEHICLE FOR TRANSPORT BACK
254  ;                                         TO FIXTURING STATION
255         ACT,45.;                           LOAD COMPLETED PART ONTO AGV
256  FR11   FREE,MACH/1;                       RELEASE THE MACHINE
257  ;
258  VM12   VMOVE,1;                           MOVE PART TO FIXTURE STATION CONTROL POINT
259  ;
260         ACT,45.;                           UNLOAD PART AT FIXTURE STATION
261  ;
262  VF13   VFREE,AGV1;                        RELEASE AGV
263  ;
264  AW14   AWAIT(12),SFIX/1;                  WAIT FOR FIXTURE STATION
265         ACT/9,180.;                        UNLD FIXTURE FROM AGV
266  FR15   FREE,SFIX/1;                       RELEASE FIXTURE STATION
267  ;
268         COLCT,INT(1),TIME IN SYSTEM;  COLLECT TIME SYSTEM STATISTICS
269  ;
270         ENDNETWORK;
271  ;
272  FIN;
```

D16-6.1(4)

16-7.   To change Example 16-2 so that vehicles cruise the guidepath, the rule for idle vehicle routing, RIDL, on the VFLEET statement is changed to CRUISE.

When the AGVs are cruising, a time event can be used to specify when they should be charged. A charging entity will be created at time 0 and at time 7200. The entity making the request is given a unique ATRIB(3) value so that a match can be made at a VWAIT node where the AGV is requested for charging. The statements for processing entities that represent the request for charging are shown in D16-7.1. Following the ASSIGN node, a VWAIT node is used to request one unit from fleet AGV1 at control point 4. User rule 1 for vehicle requests is specified and will be described after the network model is presented. MATCH is specified to ensure the job requesting the AGV is loaded on to it when it arrives. Activity 5 represents the five minute charging time after which the AGV is freed and the next charging time for the vehicle is scheduled to occur in 7200 time units. The code for URVREQ is shown in D16-7.2. A vehicle unit number is established as the one that was not charged last which is maintained as XX(1). When the charging entity arrives to the VWAIT node, subroutine URVREQ is called. The next vehicle to charge is established as NVUNIT and XX(1) is set to this value. The current status of this vehicle unit is then determined by calling subroutine GETASG which returns the variable ISTATVU as the status of the AGV. If this status is greater than 2, then the vehicle is assigned a job. If it is not, then the number of the vehicle unit, NVUNIT, is returned to the calling subroutine to request that the vehicle be sent to control point 4 as requested at VWAIT node VW16.

```
        133            CREATE,0,7200,,2;
        134            ASSIGN,XX(2)=XX(2)+1.0,ATRIB(3)=XX(2);
        135    VW16    VWAIT(8),AGV1,4,URVREQ(1),MATCH;
        136            ACT/5,300;                    CHARGE AGV1
        137            VFREE,AGV1;
        138            ACT,7200,,VW16;
```

D16-7.1

```
     SUBROUTINE URVREQ(NR,NVFNUM,NCPNUM,BTRIB,NVUNIT)
     COMMON/SCOM1/ATRIB(100),DD(100),DDL(100),DTNOW,II,MFA,MSTOP,NCLNR
    1,NCRDR,NPRNT,NNRUN,NNSET,NTAPE,SS(100),SSL(100),TNEXT,TNOW,XX(100)
     IF (XX(1).EQ.2.0) THEN
         XX(1)=1.0
         NVUNIT=1
     ELSE
         XX(1)=2.0
         NVUNIT=2
     ENDIF
     CALL GETASG(NVUNIT,NCPDES,IPLOADE,ISTATVU)
     IF (ISTATVU.GT.2) THEN
         NVUNIT=0
     ENDIF
     RETURN
     END
```

D16-7.2

If a vehicle cannot be obtained when the charging entity is routed to VWAIT node VW16, then when a vehicle becomes idle, the rule for job requests for assignment is used to send the vehicle to the charging control point. This is established through the specification of URJREQ on the VFLEET statement. The code for subroutine URJREQ is shown in D16-7.3 where a search of the jobs waiting for vehicle unit number,NVUNIT, is examined to see if any of the waiting entities have an attribute 3 value equal to the unit number. If the vehicle is not to be charged, then no assignment is made and the AGV cruises the guidepath looking for a job.

16-7,Embellishment. To establish a priority of AGV moves requires that entities waiting for the AGV be given attributes to rank requests. Code similar to that provided for URJREQ in D16-7.3 would be used to test the attributes of waiting jobs and NRANKR would be returned as the pointer to the job which has the highest priority. These priorities could be based on distance, the value added to the product, and other characteristics of the job entity and AGV resource.

```
      SUBROUTINE URJREQ(NR,NVFNUM,NVUNIT,IFLREQ,NRANKR)
      COMMON/SCOM1/ATRIB(100),DD(100),DDL(100),DTNOW,II,MFA,MSTOP,NCLNR
     1,NCRDR,NPRNT,NNRUN,NNSET,NTAPE,SS(100),SSL(100),TNEXT,TNOW,XX(100)
      COMMON QSET(5000)
      NRANKR=-MMFE(IFLREQ)
      NEXT=-NRANKR
   10 IF (NEXT.EQ.0) RETURN
      NVUTEST=QSET(NEXT+3)
      IF (NVUTEST.EQ.NVUNIT) THEN
          NRANKR=-NEXT
          RETURN
      ELSE
          NEXT=NSUCR(NEXT)
          GO TO 10
      ENDIF
      END
```

D16-7.3

16-8.   To eliminate most of the searching for the machine number, the value of XX(1) is set to the machine number whenever a machine is freed. When a request is made for a machine, the last freed machine is assigned to MACH and the value of XX(1) is set to 0. A search is then necessary only when there are two job arrivals prior to a machine being freed. To eliminate all searching, a list of available machine numbers would have to be maintained.

16-9.   The call to subroutine ALLOC is accomplished by specifying the resource field in AWAIT node AWO5 as ALLOC, that is,

     AWO5 AWAIT(3),ALLOC(1);

Subroutine ALLOC to search for an available machine is shown in D16-9.1. Subroutine ALLOC is called when an entity arrives to the AWAIT node or when a machine is freed. IFLAG is set to 0 to indicate that no assignment can be made. Each of the machine resources is then tested to see if one is available. If NNRSC(I) is greater than 0 then a machine is available. The machine is seized by a call to subroutine SEIZE. ATRIB(2) is set equal to the machine number and IFLAG is set to 1 to indicate that the first entity waiting in file 3 of the AWAIT node is to be removed and that its attribute vector is to be changed. Recall that the attribute vector of the first entity in the file is passed to subroutine ALLOC when it is called.

```
      SUBROUTINE ALLOC(I,IFLAG)
      COMMON/SCOM1/ATRIB(100),DD(100),DDL(100),DTNOW,II,MFA,MSTOP,NCLNR
     1,NCRDR,NPRNT,NNRUN,NNSET,NTAPE,SS(100),SSL(100),TNEXT,TNOW,XX(100)
      IFLAG=0
      DO 10 I=1,6
         IF (NNRSC(I).GT.0) THEN
             CALL SEIZE(I,1)
             ATRIB(2)=I
             IFLAG=1
             RETURN
         ENDIF
   10 CONTINUE
      RETURN
      END
```

D16-9.1

16-9,Embellishment. A machine allocation rule could use the information on the number of parts waiting and the number of idle machines to improve the machine allocation logic. If few parts are waiting then the machine farthest from the load/unload station should be used. Similarly, if the AGVs are not in use then longer travel times for the AGVs are not as critical at the machine allocation time. If there is a large number of parts waiting at the load/unload station then the shortest possible time by the AGVs is desirable. When all machines are busy, it may be desirable to place one AGV at control point 7 and one at control point 11 to wait for a machine to finish. Thus, it may be efficient to load machine 1 and machine 4 first. Alternatively, machine 3 is the closest to the unload station and it may be desirable to load it first. The above discussion presents thoughts on how to employ information about the status of the machines, AGVs, and number of parts waiting in the rules used to effectively allocate parts to machines.

16-10.   No solution given.

16-11.   No solution given.

CHAPTER 17

Simulation Support Systems

17-1.   The structural framework of a decision support system is shown in D17-1.1. A functional model can be developed to delineate the functions of a simulation support system. D17-1.2 and D17-1.3 are two high-level functional models taken from the report "Computer Program Development Specifications for ICAM Decision Support System (IDSS) Version 2.0". The function model shows a function in a rectangle with the inputs on the left, the outputs on the right, the resources required for the performance of the function on the bottom, and the controls on the top. The name of the function is in the rectangle. Functions are aggregated to show overall capability without time dependencies. Each function can then be decomposed to a greater level of detail using the same modeling definitions.

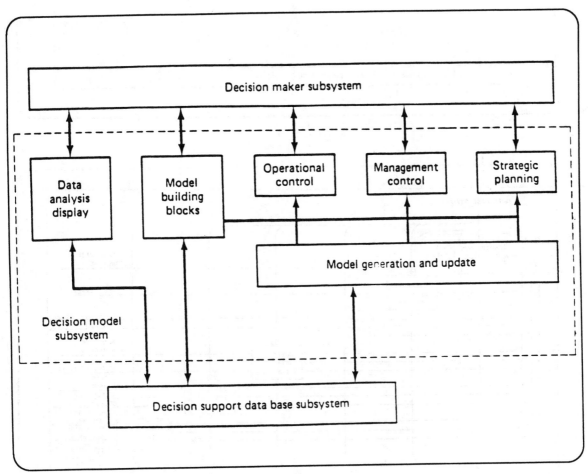

D17–1.1

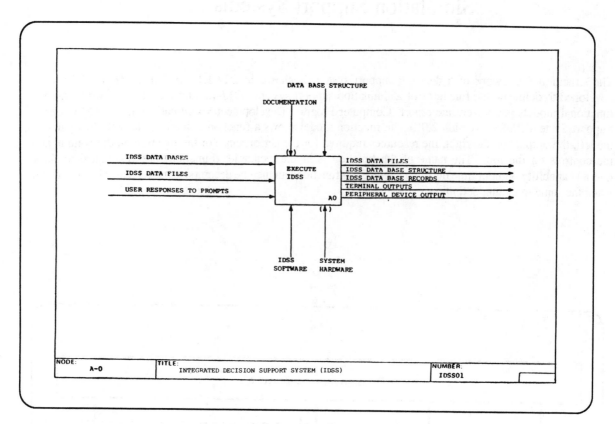

| NODE: A-0 | TITLE: INTEGRATED DECISION SUPPORT SYSTEM (IDSS) | NUMBER: IDSS01 |

D17-1.2

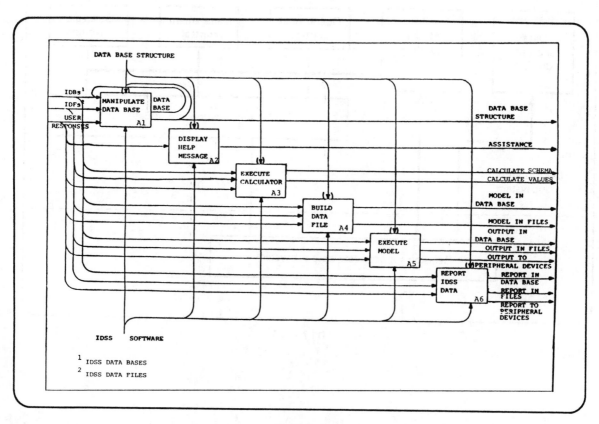

D17-1.3

17-2.   An information model describes how data elements (called entity classes in information model terminology) relate to one another. The data elements of TESS are: NETWORK, FACILITY, DEFINITION, CONTROL, SCENARIO, DATA, SUMMARY, FORMAT, MACRO, ICON, and RULE. Since TESS is well defined, it is easier to develop an information model for the TESS data elements than for a general simulation support system. A simplified information model for TESS is given in D17-2.1. The attributes and key attributes of the data elements should also be listed. The relation between the data elements shown in D17-2.1 is defined to be 1-1, 1-n, n-1 or n-n. For example, a network data element could have one or more scenarios associated with it. A scenario can have only one network model associated with it. Thus, there is an n-1 relation between network models and scenarios. This establishes that in a simulation support system designed to accommodate the above relation that different network models should be capable of being related to a single scenario and that a scenario can have only one network model included in its definition.

17-3.   Analysis and reporting tasks are well defined. To support these tasks, it is necessary to implement algorithms to perform the analysis and to provide graphics and report writing programs that support the presentation of data. Many programs are available to perform such tasks. The supporting of analysis and reporting tasks can be made complicated if it is necessary to support the selection of which analysis technique should be used and the selection of the presentation form. It is in these areas that artificial intelligence (AI)/expert systems will find their greatest use.

The modeling tasks in a simulation project involve a selection of a world view, a level of detail to be included in the model, performance measures, and definition of alternatives. These subtasks are subjective and system dependent. If algorithms exist for performing these tasks, they are specified in a fuzzy way. The support system for modeling should also include capabilities to document the historical aspects of model development and provide capabilities for documenting the model. Such documentation tasks usually are difficult to design and implement. The support interface for modeling tasks also tends to be more complex as decisions in modeling are at many different levels whereas decisions for analysis and reporting tend to be at one level.

17-4.   The displaying of the movement of entities on a SLAM II network is similar to the displaying of transactions on a Q-GERT network or entities in an IDSS 2.0 network. References 3, 4, and 5 of Chapter 17 describe the concepts used in Q-GRAF and IDSS 2.0 for such displays.

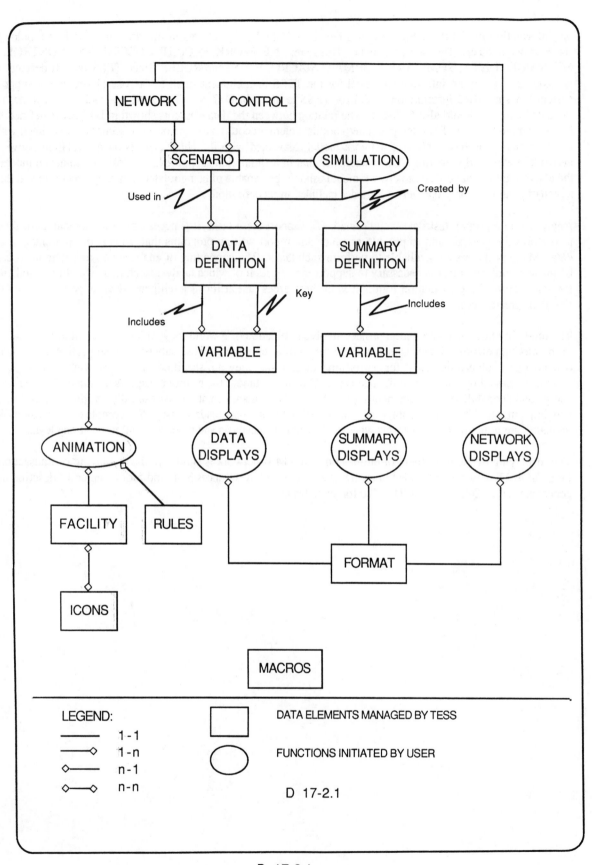

D 17-2.1

17-5.   Two good starting places for obtaining information in order to prepare a detailed specification for a simulation support system is the TESS User's Manual and general decision support systems such as IFPS, Knowledgeman II, and IDSS Build 1. The paradigm described in Exercise 17-6 could also be used in explaining the functions of simulation support systems. A paradigm for decision support for industrial engineers is shown in D17-5.1.

DECISION SUPPORT FOR INDUSTRIAL ENGINEERS				
		Function		
	Building	Satisfying	Optimizing	Planning
Definitional				
Design				
Modeling				
Analysis				

(ACTIVITY)

D17-5.1

17-6.   Definitions of the terms used in this exercise are given below (See Reference 3 and Bennett, John L., Building Decision Support Systems, Addison-Wesley, 1983, p.161.)

Strategic planning is the process of deciding on the objectives of an organization, on changes in these objectives, on the resources used to obtain these objectives, and on the policies that are to govern the acquisition, use, and disposition of resources. Management control is the process by which managers assure that the required resources are obtained and used effectively and efficiently in the accomplishment of the organization's objectives. Operational control is the process of assuring that specific tasks are carried out effectively and efficiently. Structured decisions are those which are understood well enough to be automated. Semistructured decisions are those which involve some judgment and subjective analysis and are sufficiently well defined to enable the use of models. Unstructured decisions are those for which the alternatives, objectives, and consequences are ambiguous.

A list of areas of decisions for procedural systems categorized by planning and control level is shown in D17-6.1. A discussion of how simulation can support such decisions should take into account the type of decision as defined above and the role of a model as discussed in Chapter 1, that is, explanatory devices, capability analyzers, design assesors, or future predictors. The principal focus of a simulation support system is depicted in D17-6.2.

AREAS OF DECISION MAKING FOR PROCEDURAL SYSTEMS

STRATEGIC PLANNING

     1.Design of new processes
     2.Design of new policies
     3.Determination of effect of different priorities
     4.Design of new systems
     5.Forecast of production levels
     6.Determination of required resources
     7.Estimation of cost alternatives

MANAGEMENT CONTROL

     8.Determination of how to improve throughput
     9.Determination of effect of changes in resource capacities
     10.Determination of effect of delays in raw materials
     11.Determination of how to relieve bottlenecks
     12.Determination of effect of change in demand
     13.Determination of effect of equipment failures
     14.Determination of system efficiency

OPERATIONAL CONTROL

     15.Determination of capacity
     16.Determination of bottlenecks
     17.Determination of operational requirements
     18.Assessment of in-process inventories
     19.Determination of utilizations
     20.Determination of critical operation rate
     21.Determine best staffing configurations

D17-6.1

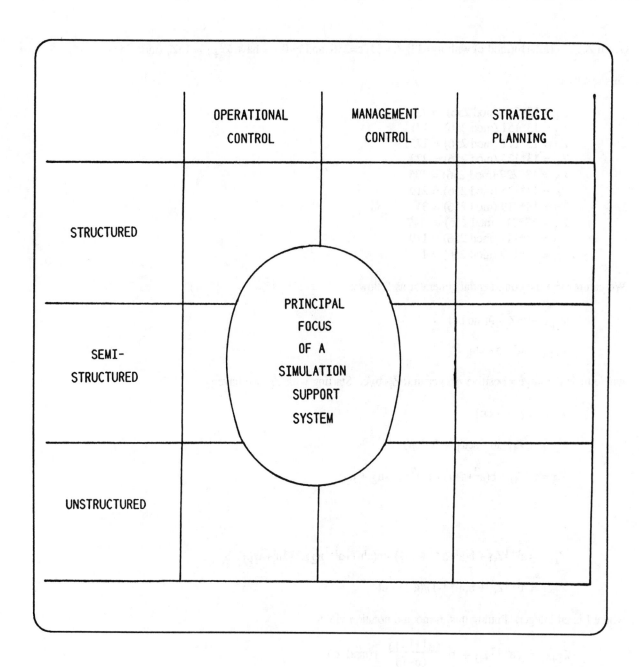

CHAPTER 18

Random Sampling From Distributions

18-1.   Given $Z_{i+1} = (aZ_i+b)(\text{mod } c)$ with $a = 13$, $Z_0=51$, $c=256$, and $b=0$ we have $Z_{i+1} = 13Z_i \ (\text{mod } 256)$.

This results in

$$Z_1 = 13*51 \ \ (\text{mod } 256) \ = 151$$
$$Z_2 = 13*151 \ (\text{mod } 256) = 171$$
$$Z_3 = 13*171 \ (\text{mod } 256) = 175$$
$$Z_4 = 13*175 \ (\text{mod } 256) = 227$$
$$Z_5 = 13*227 \ (\text{mod } 256) = 135$$
$$Z_6 = 13*135 \ (\text{mod } 256) = 219$$
$$Z_7 = 13*219 \ (\text{mod } 256) = 31$$
$$Z_8 = 13*31 \ \ (\text{mod } 256) \ = 147$$
$$Z_9 = 13*147 \ (\text{mod } 256) = 119$$
$$Z_{10} = 13*119 \ (\text{mod } 256) = 11$$

18-2.   We can rewrite the congruential generator as follows:

$$Z_{i+1} = (aZ_i+b)(\text{mod } c)$$

$$Z_{i+1} = aZ_i+b - cn_i$$

where n_i is the largest positive integer in $(aZ_i+b)/c$. Starting with Z_0, we have

$$Z_1 = aZ_0+b - cn_1$$

$$Z_2 = a^2Z_0+ab - acn_1 + b - cn_2$$

$$Z_3 = a^3Z_0 + b(a^2+a+1) - c(a^2n_1+an_2 + n_3)$$

.

.

.

$$Z_{i+1} = a^{i+1}Z_0 + b(a^i+a^{i-1}+...+1) - c(a^in_1+a^{i-1}n_2+...+an_i+n_{i+1})$$

$$Z_{i+1} = a^{i+1}Z_0 + b(a^{i+1}-1)/(a-1) - cI$$

where I is an integer. Putting this in modulo notation yields

$$Z_{I+1} \equiv (a^{i+1}Z_0 + b \ \frac{(a^{i+1}-1)}{(a-1)} \ )(\text{mod } c)$$

Since $cI(\text{mod } c) \equiv 0$.

18-3.   Given the probability density function

$$f(x) = \begin{cases} \dfrac{3x^2}{8} & 0 \le x \le 2 \\[2mm] 0 & \text{otherwise,} \end{cases}$$

the cumulative distribution function is

$$F(x) = \int_0^x \frac{3y^2}{8}\, dy = \begin{cases} 0 \; ; & x < 0 \\[2mm] \dfrac{x^3}{8} \; ; & 0 \le x \le 2 \\[2mm] 1 \; ; & x > 2 \end{cases}$$

Using the inverse transformation method, we equivalence a random number, r, to F(x) and solve for x. Thus,

$$r = \frac{x^3}{8}$$

and $x = (8r)^{1/3}$. Each value in Exercise 18-1 is divided by 256 to obtain a string of 10 random numbers from which x_i can be computed. The following vector gives the values of x_i obtained: {1.677, 1.748, 1.762, 1.922, 1.615, 1.898, 0.989, 1.662, 1.549, 0.701}

18-4.   Given the probability mass function P(0)=.2; P(1)=.2; P(2)=.4; and P(3)=.2, the cumulative distribution function is {.2, .4, .8, 1.0} and the sequence of samples using the random numbers of Exercise 18-1 is: {2,2,2,3,2,3,0,2,2,0}.

18-5.   The function EXPONT below provides a routine for sampling from a truncated exponential distribution.

```
        FUNCTION EXPONT(XMN,XLO,XHI,ISTRM)
C       LIMIT RANDOM NUMBER TO PROVIDE
C       SAMPLE IN DESIRED RANGE
        A = 1. - EXP(-XLO/XMN)
        B = 1. - EXP(-XHI/XMN)
        R = UNFRM(A,B,ISTRM)
        EXPONT = -ALOG(1.-R)/XMN
        RETURN
        END
```

A second subroutine using rejection is shown below.

```
        FUNCTION EXPONR(XMN,XLO,XHI,ISTRM)
C       REJECT EXPONENTIAL SAMPLE IF
C       OUTSIDE DESIRE LIMITS
    10  EXPONR = -ALOG(DRAND(ISTRM))/XMN
        IF(EXPONR.LT.XLO) GO TO 10
        IF(EXPONR.GT.XHI) GO TO 10
        RETURN
        END
```

18-6.   Let $Y = \max(U_1, U_2)$
where

$$f_U(u) = \begin{cases} 1, & 0 \leq u \leq 1 \\ 0, & \text{otherwise} \end{cases}$$

$$P[Y \leq y] = P[U_1 \leq y; U_2 \leq y]$$

$$= P[U_1 \leq y]\, P[U_2 \leq y] \quad \text{by independence of } U_i$$

$$= y^2, \quad 0 \leq y \leq 1$$

Taking the derivative yields

$$f_y(y) = \begin{cases} 2y, & 0, \leq y \leq 1 \\ 0, & \text{otherwise} \end{cases}$$

which is a form of the triangular distribution with a=0, m=1, and b=1, and $\mu = 2/3$ and $\sigma^2 = 1/18$.

18-7.   A process which has exponential interarrival times and a Poisson number of entities per arrival is referred to as a compound Poisson process. The procedure for modeling batch arrivals from a compound Poisson process is to generate arrival events using the exponential sampling routine, EXPON, and at each arrival event generate a sample from the Poisson sampling routine, NPSSN. If λ is the mean arrival rate and μ is the mean batch size then the expected number of arrivals in T time units is $\mu(\lambda T)$.

18-8.   The program to generate a histogram of the values using SLAM II routines is given below.

```
              .
              .
              .
         TEND = 1000.
         TCUR = 0.
         DO 10  I=1,100
         T = UNFRM(TCUR,TEND,1)
         TBA = T-TCUR
         CALL COLCT(TBA,1)
         TCUR = T
     10 CONTINUE
              .
              .
              .
```

Testing involves the use of goodness-of-fit procedures or programs. In the above code, each value of TBA could be written out to a file for direct input into a goodness-of-fit program.

18-9.   The procedure for generating bivariate normal samples X_i with means μ_i and variances σ_i^2 and a correlation coefficient of ρ is as follows.

1. Generate Z_1 and Z_2 as standard normal samples.
2. Set $X_1 = \mu_1 + \sigma_1 Z_1$
3. Set $X_2 = \mu_2 + \sigma_2 [\rho Z_1 + (1 - \rho^2)^{\frac{1}{2}} Z_2]$

For multinormal samples, see Scheuer, E. and D. Stoller,, "On the Generation of Normal Random Vectors", TECHNOMETRICS, 4, 1962, pp. 278- 281 and the work of Fishman. For references on sampling from multivariate distributions, see the survey by Schmeiser.

18-10. An ARMA process has an autoregressive part with p terms, a moving average part with q terms and a random component. After initializing the process, samples are obtained by generating a sample from the random component and substituting the values previously computed into the process equation.

18-11. Perform standard tests on samples obtained from the substitution of pseudorandom numbers into the equation given for RNORMS.

CHAPTER 19

Statistical Aspects Of Simulation

19-1.

Term	Definition
Reliable	The amount of credence placed in a result; reliability indicates a measure of trustworthiness and dependability.
Batch	A group of observations considered as a single unit for the purpose of analysis.
Stochastic Process	A set of ordered random variables.
Ergodic	A property of a stochastic process that allows estimation for the process variables to be made from a single time series observed for the process.
Stationarity	A property of a stochastic process implying that the underlying behavior of the process is time invariant.
Steady State	A time period in which the probability distributions associated with a stochastic process do not change.
Time Series	A finite realization of a stochastic process; the observations over a single run.
Sample Mean	The average of a set of random variables.
Average	The arithmetic mean.
Expectation	The mean value.
Mean Square Error	The expected value of the square of the difference between the sample mean and the theoretical mean.
Kurtosis	The fourth central moment divided by the square of the second central moment minus three.
Spectrum	$g(\lambda) = \dfrac{1}{2\pi} \displaystyle\sum_{h=-\infty}^{\infty} R_h \, e^{-i\lambda h} \quad -\pi \leq \lambda \leq \pi$ where R_h is the covariance of lag h.
Regeneration Point	A time at which the system state is such that future behavior of the system is independent of past behavior.
Parametric Modeling	The development of a model from observations. In the context of simulation output analysis, it refers to the constuction of a model that is an abstraction of the simulation model based on observations obtained from fromexperiments using the simulation model.

Spectral Density Function	$f(\lambda) = g(\lambda)/R_0$ where $g(\lambda)$ is the spectrum and R_0 is the variance (covariance of lag 0).
ARMA Model	Combined autoregressive and moving average model.
White Noise	A stochastic process in which the random variables are independent over time with a mean of zero and the same variance at each time instant.
VRT	Variance Reduction Technique.
Stratified Sampling	A technique that classifies observations into classes (strata) in accordance with prior knowledge about the number of observations expected in each strata.
CATCH-22	A book by Joseph Heller; the concept that to get X you need Y but to get Y you need X. For example, to estimate the variance of the sample mean with a prescribed confidence, you need to determine the number of observations required, but to determine the number of observations, you need the variance of the sample mean.
Bias	The difference between the expected value of an estimator and its theoretical value.

19-2. Let (i,j) define the states of the system where i is the number in the first teller subsystem, and j is the number in the second teller subsystem. Since jockeying is assumed, i and j can only differ by one and the differential-difference equations for the problem are shown below.

$$\frac{dP_{00}(t)}{dt} = -\lambda P_{00}(t) + \mu P_{10}(t) + \mu P_{01}(t)$$

$$\frac{dP_{ij}(t)}{dt} = +\lambda P_{jj}(t) - (\lambda + 2\mu)P_{ij}(t) + \mu P_{ii}(t), \quad i=j+1$$

Alternatively, the system can be viewed as an M/M/2/20 Markov Process since the jockeying does not affect the number in the system. For this variable, a single queue model provides equivalent results.

The steady-state probabilities, mean, variance and m values presented in the exercise statement were obtained from a program written by Hazen and modified by Wilson to evaluate this situation. (See Reference 40.) A generalization of the results is given in Reference 32.

To establish run lengths, we specify the number of batches, I, to meet the condition that the estimates be within 10 percent of the theoretical values using the relation $\frac{\sigma^2_{\bar{X}}}{} = \frac{m}{I}$. For the average, we have

$\qquad$ I = 3805 for $\alpha = 0.05$

To set I for the condition on $\hat{m}$, the estimate of m, we use

$\qquad$ $P[m-.1m \leq \hat{m} \leq m + .1m] \geq 1-\alpha$

$\qquad$ $P[.9 \leq \frac{\hat{m}}{m} \leq 1.1] \geq 1-\alpha$

Since $\dfrac{\hat{m}}{m}$ (I-1) is χ^2 distributed with mean I-1 and variance 2(I-1), we obtain the following:

$$P[z \leq -.1 \sqrt{\dfrac{(I-1)}{2(I-1)}}\;] \leq \dfrac{\alpha}{2}$$

Assuming I is sufficiently large so that the normal approximation to the χ^2 holds, the smallest I that satisfies this inequality for $\alpha=0.05$ is approximately 800.

19-3.   This exercise involves collecting statistics from experiments on a model and applying the various procedures for estimating the variance of the sample mean. The amount of simulation time is related to the number of observations and this question is addressed in Exercise 19-2. For replication analysis, the time for each replication depends on the operation of the system. If the bank is operated over an 8-hour period then statistics over the 8-hour period should be collected, that is, the system has a natural termination point. If the analysis is to obtain estimates of the steady state values then the time until steady state is achieved must be considered as well as the possible truncation of initial observations.

When determining the batch size, the procedure discussed in Section 19.3.2 should be considered. The value of m for batches of size b is b times the value of m associated with the individual observations.

19-3,Embellishment(a). A spectral analysis performed on individual observations has not provided good estimates of the variance of the sample mean. However, a spectral analysis on batched observations has yielded reliable results.

19-3,Embellishment(b). Since exponential times are used in this example, any state-of-the-system can be used as a regeneration point. Estimates of m using different states tends to provide insight into the estimation process. Also the use of jackknife techniques or the direct calculation of the bias associated with the ratio estimator have been found to be enlightening. For long simulation runs, the bias has tended to be very small.

19-3,Embellishment(c). The use of the Box-Jenkins package to perform parametric modeling on the outputs of the simulation data has been found to be convenient. Again, fitting models to the individual observations does not produce good results. However, when the observations are batched, the parametric models yield satisfactory estimates of the quantities of interest.

19-3,Embellishment(d). The procedure for obtaining 95% confidence limits on m is similar to the procedure presented in the solution to Exercise 19-2.

19-4.   The discussion of the solution for Exercise 19-3 provides information about the experimental results when analyzing individual observations and batched observations.

19-5.   The modal state is empty and idle and there is no problem in using this condition as the intial state-of-the-system. The expected state is 4.232 customers and assumptions are required to set the initial state to the expected state. The choices are 2 customers in each tellers' system or 3 customers in the first teller system and 2 customers in the second teller system. The effects of initial conditions on final outputs for this problem are not severe.

19-6.   A procedure for evaluating truncation rules is contained in Wilson, J.R. and A.A.B. Pritsker, "Evaluation of Startup Policies in Simulation Experiments", Simulation, Vol. 31, No. 3, September, 1978, pages 79-89. As can be seen from that paper, a formal evaluation of truncation rules can be a complicated process.

19-7.   This exercise is useful for understanding variance reduction techniques and for demonstrating that the variance of the sample mean is a function of the experimental procedure, that is, the variance of the sample mean depends on the run length and the way in which the run is performed.

19-8.   In solving this problem, the experiment is run and estimates of the variance, s^2, are obtained. It is easier to apply the rules when using batching which approximates the independence case.

19-9.   Performing Exercise 19-2 through 19-8 for a problem in which the solution is not known is a much more difficult exercise. As with other problem situations, when a solution is not available, there is increased uncertainty and greater difficulty in making comparisons between runs.

19-10. Switching the random number streams for arrivals and service times was studied by Page, E.S., "On Monte Carlo Methods in Congestion Problems: Simulation of Queueing Systems", Operations Research, Vol. 13, 1965, pages 300-305. The variance reduction obtained by Page was on the order of 50 to 70 percent for traffic intensities of 0.75 to 0.90. This variance reduction is due in effect to short service times resulting in low congestions which, on the second run, result in small times between arrivals leading to high congestion. By averaging a low congestion run with a high congestion run, lower variance estimates should be obtained.

19-11. To perform this simulation, the service time for the customer should be established as an attribute. When the customer leaves the system, the time in the system should be stratified according to the service time for the customer. It is suggested that six stratification levels be used when performing this exercise.

19-12. When employing an optimization procedure in conjunction with SLAM II, it is necessary to set the decision variables between runs. There have been several methods for implementing such procedures in subroutines INTLC or OTPUT. The book by Biles and Swain for performing optimizations in conjunction with simulations is a good source for background information for this exercise. (See Biles, W.E. and J.J. Swain Optimization and Industrial Experimentation, New York, John Wiley, 1980.)

19-13. The joint use of common streams and antithetic samples has been studied by numerous authors. Schruben and Margolin indicate that experimental designs exist in which the joint use will result in variance reductions. (Schruben, L.W., "Designing Correlation Induction Strategies for Simulation Experiments", Current Issues in Computer Simulation, Chapter 16 edited by Adams and Dogmenci, Academic Press, 1979, and Schruben, L.W. and B.H. Margolin, "Pseudo-random Number Assignment in Statistically Designed Simulation and Distribution Sampling Experiments," JASA, Vol. 73, 1978, pages 504-522.)